AF522071

COMPUTATIONAL BIOTECHNOLOGY

COMPUTATIONAL BIOTECHNOLOGY

By

Dr. Sharad Srivastava

Dept. of Zoology
Vikram University
Ujjain

DISCOVERY PUBLISHING HOUSE PVT. LTD.
NEW DELHI-110 002

Published by:
Tilak Wasan
DISCOVERY PUBLISHING HOUSE PVT. LTD.
4383/4B, Ansari Road, Darya Ganj
New Delhi-110 002 (India)
Phone : +91-11-23279245; 23253475; 43596065
E-mail: discoverybooksindia@gmail.com
discoverypublishinghouse@gmail.com
namitwasan9@gmail.com
web : www.discoverypublishinggroup.com

First Edition: **2009**

Reprinted: **2020**

ISBN: 978-81-8356-471-7

Computational Biotechnology

Printed at:
Infinity Imaging Systems
Delhi

Preface

The present title *"Computational Biotechnology"* is an exciting new field bringing together life scientists, mathematicians, computer scientists and engineers to explore a new and deeper understanding of biological systems. The present text has been compiled for computer science and mathematics graduate and level undergraduate students. It will also be of interest to molecular biologists interested in bioinformatics.

The rationable of the book is to present algorithmic ideas in computational biology and to show how they are connected to molecular biology and to biotechnology. To achieve this goal, the book has a substantial "computating biology without formulas" component that presents biological motivation and computational ideas in a simple way. This simplified presentation of biology and computing aims to make the book accessible to computer scientists entering this new areas and to biologists who do not have sufficient background for more involved computational techniques. Every chapter has an introductory section that describes both computational and biological ideas without any formulas. Thus book concentrate on computational ideas rather than details of the algorithms and makes special efforts to prevent there ideas in a simple way.

To make the work more comprehensive and informative, the author has consulted many authoritative books, research journals, abstracts, monographs etc. He is grateful to all those great scholars whose work are cited or substantially reproduced.

There can be no claim to originality except in the manner of treatment and much of the information has been obtained from the books and scientific journals available in the different libraries.

The author expresses his thanks to his friends and colleagues whose continue inspirations have initiated him to bring out this book.

The author expresses his gratitude to Mr. Wasan and staff of M/s Discovery Publishing House Pvt. Ltd. for their whole hearted cooperation in the publication of this book.

In the mean time, the author will remain sincerely responsible for any shortcomings of the book and be grateful to the readers for their suggestions and constructive criticism for the continuous betterment of the book. He takes this opportunity to appeal to the readers to send their suggestions straightaway to his Publisher.

Author

Contents

Chapter 1

Introduction

Both living and non-living things are present in nature. Living things can move, reproduce, grow, eat, and so on—they have an *active* participation in their environment, as opposed to non-living things. Yet research in the past centuries reveals that both kinds of matter are composed by the same atoms and conform to the same physical and chemical rules. What is the difference then? For a long time in human history, people thought that some sort of extra matter bestowed upon living beings their active characteristics—that they were "animated" by such a thing. But nothing of the kind has ever been found. Instead, our current understanding is that living beings act the way they do due to a complex array of chemical reactions that occur inside them. These reactions never cease. It is often the case that the products of one reaction are being constantly consumed by another reaction, keeping the system going. A living organism is also constantly exchanging matter and energy with its surroundings. In contrast, anything that is in equilibrium with its surrounding can generally be considered dead. (Some notable exceptions are vegetative forms, like seeds, and viruses, which may be completely inactive for long periods of time, and are not dead.

Modern science has shown that life started some 3.5 billions of years ago, shortly (in geological terms) after the Earth itself was formed. The first life forms were very simple, but over billions of years a continuously acting process called evolution made them evolve and diversify, so that today we find very complex organisms as well as very simple ones. Both complex and simple organisms have a similar molecular chemistry, or *bio-chemistry*. The main actors in the chemistry of life are molecules called **proteins** and **nucleic adds**. Roughly speaking, proteins are responsible for what a living being is and does in a physical sense. (The distinguished scientist Russell Doolittle once wrote that "we *are* our proteins.") Nucleic acids, on the other hand, encode the information necessary to produce proteins and are responsible for passing along this "recipe" to subsequent generations. Molecular biology research is basically devoted to the understanding of the structure and function of proteins and nucleic acids. These molecules are therefore the fundamental objects of this book, and we now proceed to give a basic and brief description of the current state of knowledge regarding them.

STRUCTURAL PROTEINS

Most substances in our bodies are proteins, of which there are many different kinds. *Structural proteins* act as tissue building blocks, whereas other proteins known as *enzymes* act as catalysts of chemical reactions. A catalyst is a substance that speeds up a chemical reaction. Many biochemical reactions, if left unattended, would take too long to complete or not happen at all and would therefore not be useful to life. An enzyme can speed up this process by orders of magnitude, thereby making life possible. Enzymes are very specific—usually a given enzyme can help only one kind of biochemical reaction. Considering the large number of reactions that must occur to sustain life, we need a lot of enzymes. Other examples of protein function are oxygen transport and antibody defence. But what exactly are proteins? How are they made? And how do they perform their functions? This section tries briefly to answer these questions.

A protein is a chain of simpler molecules called **amino adds**. Every amino acid has one central carbon atom, which is known as the alpha carbon, or C_α. To the C_α atom are attached a hydrogen atom, an amino group (NH_2), a carboxyl group (COOH), and a *side chain*. It is the side chain that distinguishes one amino acid from another. Side chains can be as simple as one hydrogen atom (the case of amino acid glycine) or as complicated as two carbon rings (the case of tryptophan). In nature we find 20 different amino acids, which are listed in Table 1.1. These 20 are the most common in proteins; exceptionally a few non-standard amino acids might be present.

Table 1.1: The twenty amino acids commonly found in proteins.

	One-letter code	Three-letter code	Name
1	A	Ala	Alanine
2	C	Cys	Cysteine
3	D	Asp	Aspartic Acid
4	E	Glu	Glutamic Acid
5	F	Phe	Phenylalanine
6	G	Gly	Glycine
7	H	His	Histidine
8	I	Ile	Isoleucine
9	K	Lys	Lysine
10	L	Leu	Leucine
11	M	Met	Methionine
12	N	Asn	Asparagine
13	P	Pro	Proline
14	Q	Gln	Glutamine
15	R	Arg	Arginine
16	S	Ser	Serine

(contd.)

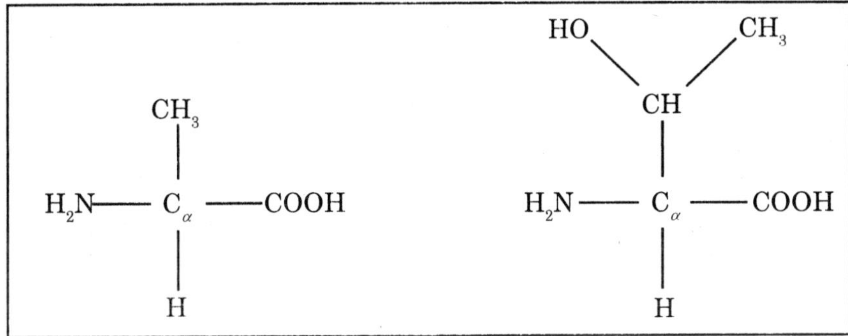

Fig. 1.1: Examples of amino acids: alanine (left) and threonine.

	One-letter code	Three-letter code	Name
17	T	Thr	Threonine
18	V	Val	Valine
19	W	Trp	Tryptophan
20	Y	Tyr	Tyrosine

In a protein, amino acids are joined by *peptide bonds.* For this reason, proteins are polypeptidic chains. In a peptide bond, the carbon atom belonging to the carboxy group of amino acid A_i bonds to the nitrogen atom of amino acid A_{i+1}'s amino group. In such a bond, a water molecule is liberated, because the oxygen and hydrogen of the carboxy group joins the one hydrogen from the amino group. Hence, what we really find inside a polypeptide chain is a **residue** of the original amino acid. Thus, we generally speak of a protein having 100 residues, rather than 100 amino acids. Typical proteins contain about 500 residues, but there are proteins with as few as 100 or with as many as 5,000 residues.

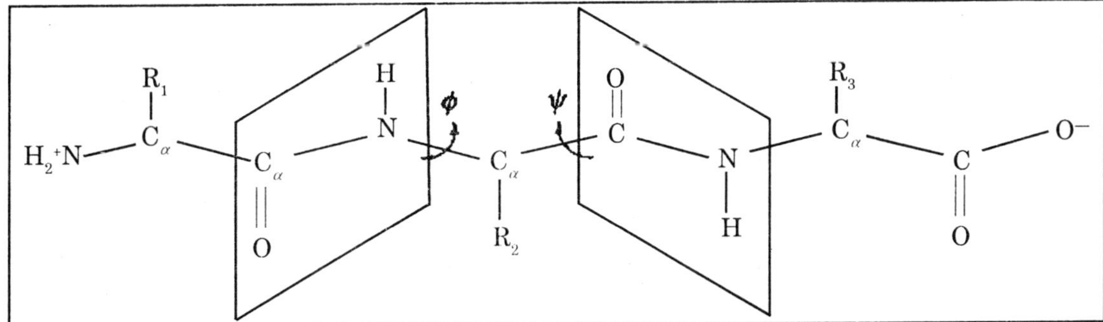

Fig. 1.2: **A polypeptide chain. The R_i side chains identify the component amino acids. Atoms inside each quadrilateral are on the same plane, which can rotate according to angles f and y**

The peptide bond makes every protein have a *backbone,* given by repetitions of the basic block —N—C_α (CO)—. To every C_α there corresponds a side chain. See Fig. 1.2 for a schematic view of a polypeptide chain. Because we have an amino group at one end of the backbone and a carboxy group at the other end, we can distinguish both ends of a polypeptide chain and thus give it a direction. The convention is that polypeptides begin at the amino group *(N-terminal)* and end at the carboxy group *(C-terminal).*

A protein is not just a linear sequences of residues. This sequence is known as its **primary structure**. Proteins actually fold in three dimensions, presenting **secondary**, **tertiary**, and **quaternary** structures. A protein's secondary structure is formed through interactions between backbone atoms only and results in "local" structures such as helices. Tertiary structures are the result of secondary structure packing on a more global level. Yet another level of packing, or a group of different proteins packed together, receives the name of quaternary structure. Proteins can fold in three dimensions because the plane of the bond between the C_α atom and the nitrogen atom may rotate, as can the plane between the C_α atom and the other C atom. These rotation angles are known as ϕ and ψ, respectively. Side chains can also move, but it is a secondary movement with respect to the backbone rotation. Thus, if we specify the values of all ϕ — ψ pairs in a protein, we know its exact folding.

Determining the folding, or three-dimensional structure, of a protein is one of the main research areas in molecular biology, for three rea-sons. First, the three-dimensional shape of a protein is related to its function. Second, the fact that a protein can be made out of 20 different kinds of amino acids makes the resulting three-dimensional structure in many cases very complex and without symmetry. Third, no simple and accurate method for determining the three-dimensional structure is known. These methods try to predict a molecule's structure from its primary sequence. The three-dimensional shape of a protein determines its function in the following way.

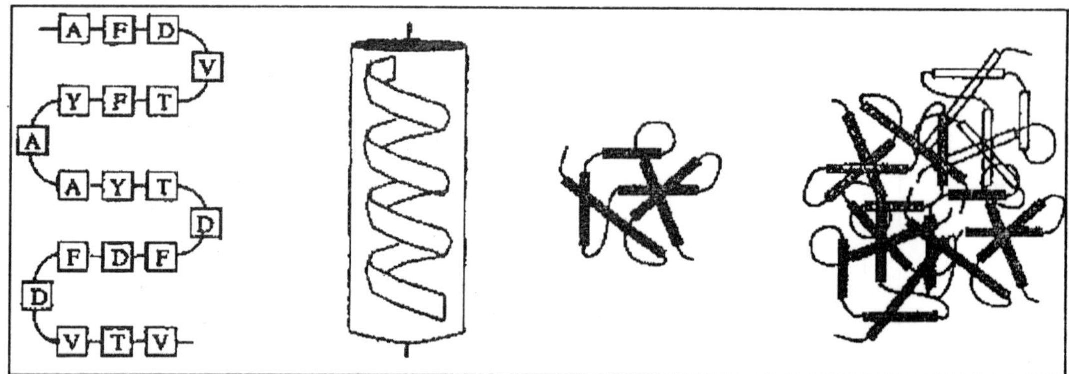

Fig. 1.3: Primary, secondary, tertiary, and quaternary structures of proteins.

A folded protein has an irregular shape. This means that it has varied nooks and bulges, and such shapes enable the protein to come in closer contact with, or **bind to** some other specific molecules. The kinds of molecules a protein can bind to depend on its shape. For example, the shape of a protein can be such that it is able to bind with several identical copies of itself, building, say, a thread of hair. Or the shape can be such that molecules *A* and *B* bind to the protein and thereby start exchanging atoms. In other words, a reaction takes place between A and *B*, and the protein is fulfilling its role as a catalysts. But how do we get out proteins? Proteins are produced in a cell structure called ri-*bosome.* In a ribosome the component amino acids of a protein are assembled one by one thanks to information contained in an important molecule called messenger ribonucleic acid. To explain how this happens, we need to explain what nucleic acids are.

NUCLEIC ACIDS

Living organisms contain two kinds of nucleic acids: **ribonucleic acid**, abbreviated by RNA, and **deoxyribonucleic acid**, or DNA. We describe DNA first.

DNA

Like a protein, a molecule of DNA is a chain of simpler molecules. Actually it is a *double* chain, but let us first understand the structure of one simple chain, called **strand**. It has a backbone consisting of repetitions of the same basic unit. This unit is formed by a sugar molecule called 2'-deoxyribose attached to a phosphate residue. The sugar molecule contains five carbon atoms, and they are labelled 1' through 5' (see Fig. 1.4 The bond that creates the backbone is between the 3' carbon of one unit, the phosphate residue, and the 5' carbon of the next unit. For this reason, DNA molecules also have an orientation, which by convention, starts at the 5' end and finishes at the 3' end. When we see a single stranded DNA sequence in a technical paper, book, or a sequence database file, it is always written in this canonical.

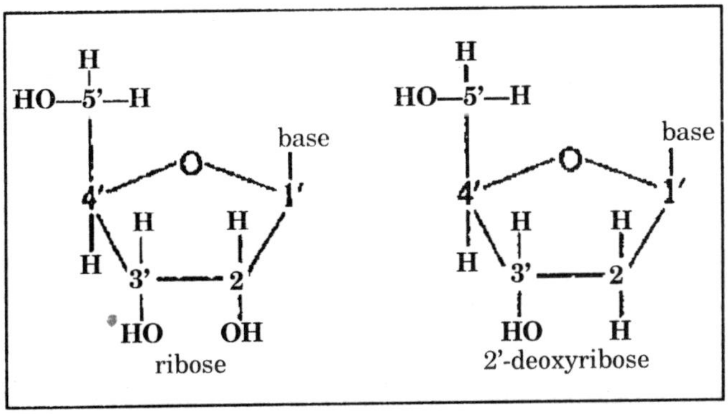

ribose

2'-deoxyribose

Fig. 1.4: Sugars present in nucleic acids. Symbols 1' through 5' represent carbon atoms. The only difference between the two sugars is the oxygen in carbon 2'. Ribose is present in RNA and 2'-deoxyribose is found in DNA.

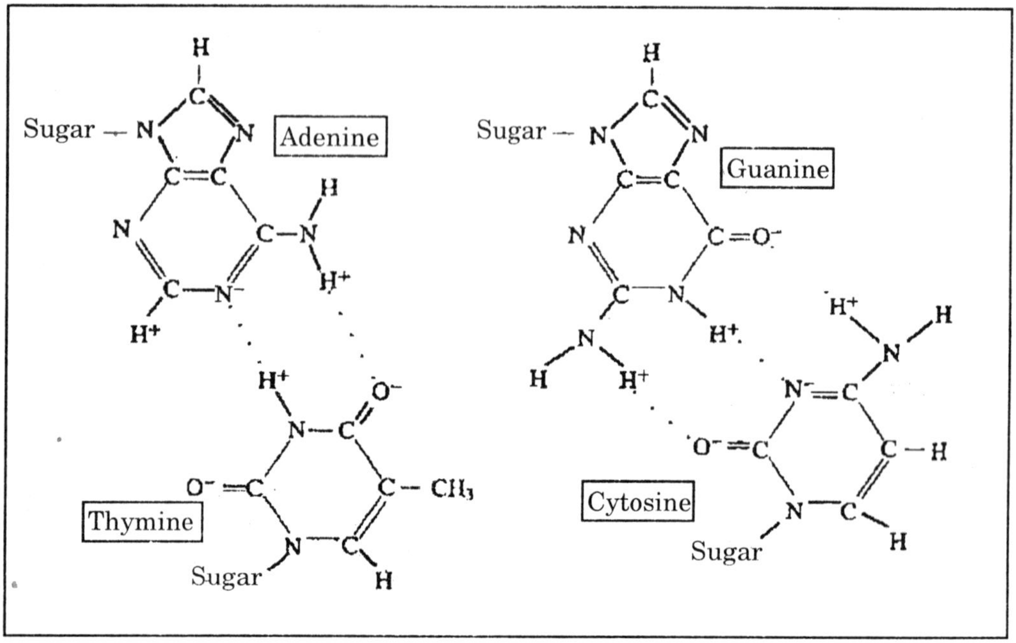

Fig. 1.5: Nitrogenated bases present in DNA. Notice the bonds that con form between adenine and thymine and between guanine and cytosine, indicated by the dotted lines.

5' $\rightarrow$ *3'* direction, unless otherwise stated. Attached to each 1' carbon in the backbone are other molecules called **bases**. There are four kinds of bases: adenine (A), guanine (G), cytosine (C), and thymine (T). In Fig 1.5 we show the schematic molecular structure of each base, and in Fig. 1.6 show a schematic view of the single DNA strand described so far. Bases A and G belong to a larger group of substances called *purines*, whereas C and T belong to the *pyrimidines*. When we see the basic unit of a DNA molecule as consisting of the sugar, the phosphate, and its base, we call it a **nucleotide**. Thus, although bases and nucleotides are not the same thing, we can speak of a DNA molecule having 200 bases or 200 nucleotides.

A DNA molecule having a few (tens of) nucleotides is referred to as an *oligonucleotide*. DNA molecules in nature are very long, much longer than proteins. In a human cell, DNA molecules have hundreds of millions of nucleotides. As already mentioned, DNA molecules are double strands. The two strands are tied together in a helical structure, the famous double helix discovered by James Watson and Francis Crick in 1953. How can the two strands hold together? Because each base in one strand is paired with (or *bonds to*) a base in the other strand. Base A is always paired with base T, and C is always paired with G. Bases A and T are said to be the complement of each other, or a pair of complementary bases. Similarly, C and G are **complementary bases**. These pairs are known as *Watson-Crick base pairs*.

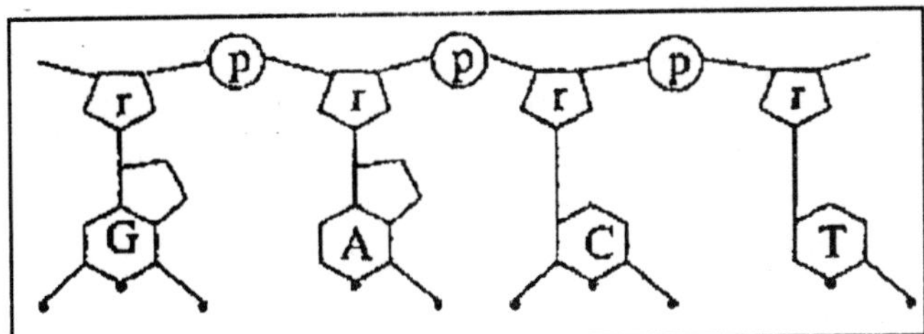

Fig. 1.6: A schematic molecular structure view of one DNA strand.

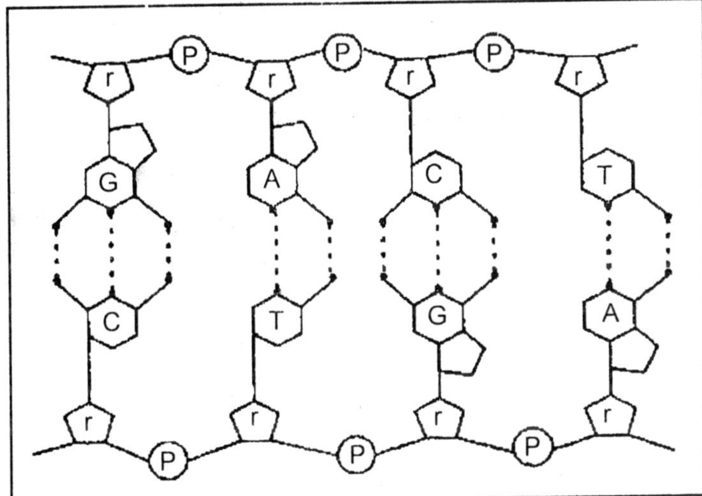

Fig. 1.7: A schematic molecular structure view of a double strand of DNA.

Base pairs provide the unit of length most used when referring to DNA molecules, abbreviated to **bp.** So we say that a certain piece of DNA is 100,000 bp long, or 100 kbp. In this book we will generally consider DNA as string of letters, each letter representing a base. Figure 1.8 presents this "string-view" of DNA, showing that we represent the double strand by placing one of the strings on top of the other. Notice the base-pairing. Even though the strands are linked, each one preserves its own orientation, and the two orientations are opposite. Notice that the 3' end of one strand corresponds to the 5' end of the other strand. This property is sometimes ex-pressed by saying that the two strands are *antiparallel.* The fundamental consequence of this structure is that it is possible to infer the sequence of one strand given the other. The operation that enables us to do that is called **reverse complementation.** For example, given strand s = AGACGT in the canonical direction, we do the following to obtain its reverse complement: First we reverse s, obtaining s' = TGCAGA, and then we replace each base by its complement, obtaining $\overline{s}$ = ACGTCT. (Note that we use the bar over the *s* to denote the reverse complement of strand *s*.) It is precisely this mechanism that allows DNA in a cell to *replicate,* therefore allowing an organism that starts its life as one cell to grow into billions of other cells, each one carrying copies of the DNA molecules from the original cells. In organisms whose cells do not have a nucleus, DNA is found free-floating inside each cell. In higher organisms, DNA is found inside the nucleus and in cell organelles called *mitochondria* (animals and plants) and *chloroplasts* (plants only).

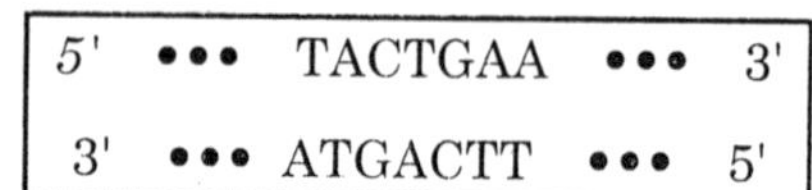

Fig. 1.8: A double-stranded DNA sequence represented by strings of letters.

RNA

RNA molecules are much like DNA molecules, with the following basic compositional and structural differences:

- In RNA the sugar is ribose instead of 2'-deoxyribose.
- In RNA we do not find thymine (T); instead, uracil (U) is present. Uracil also binds with adenine like thymine does.
- RNA does not form a double helix. Sometimes we see RNA-DNA hybrid helices; also, parts of an RNA molecule may bind to other parts of the same molecule by complementarity. The three-dimensional structure of RNA is far more varied than that of DNA.

Another difference between DNA and RNA is that while DNA performs essentially one function (that of encoding information), we will see shortly that there are different kinds of RNAs in the cell, performing different functions.

THE MECHANISMS OF MOLECULAR GENETICS

The importance of DNA molecules is that the information necessary to build each protein or RNA found in an organism is encoded in DNA molecules. For this reason, DNA is sometimes referred to as "the blueprint of life". In this section we will describe this encoding and how a protein is built out of DNA (the process *of protein synthesis).* We will see also how the information in DNA, or genetic information, is passed along from a parent to its offspring.

Genes and the Genetic Code

Each cell of an organism has a few very long DNA molecules. Each such molecule is called a **chromosome**. We will have more to say about chromosomes later, so for the moment let us examine the encoding of genetic information from the point of view of only one very long DNA molecule, which we will simply call "the DNA." The first important thing to know about this DNA is that certain contiguous stretches along it encode information for building proteins, but others do not. The second important thing is that to each different kind of protein in an organism there usually corresponds one and only one contiguous stretch along the DNA. This stretch is known as a **gene**.

Because some genes originate RNA products, it is more correct to say that a gene is a contiguous stretch of DNA that contains the information necessary to build a protein or an RNA molecule. Gene lengths vary, but in the case of humans a gene may have something like 10,000 bp. Certain cell mechanisms are capable of recognizing in the DNA the precise points at which a gene starts and at which it ends. A protein, as we have seen, is a chain of amino acids. Therefore, to "specify" a protein all you have to do is to specify each amino acid it contains. And that is precisely what the DNA in a gene does, using triplets of nucleotides to specify each amino acid. Each nucleotide triplet is called a **codon**.

The table that gives the correspondence between each possible triplet and each amino acid is the so-called **genetic code**. In Table 1.2 you will notice that nucleotide triplets are given using RNA bases rather than DNA bases. The reason is that it is RNA molecules that provide the link between the DNA and actual protein synthesis, in a process to be detailed shortly. Before that let us study the genetic code in more detail. There are 64 possible nucleotide triplets, but there are only 20 amino acids to specify.

Table 1.2: The genetic code mapping codons to amino acids.

First position	Second position				Third position
	G	A	C	U	
G	Gly	Glu	Ala	Val	G
	Gly	Glu	Ala	Val	A
	Gly	Asp	Ala	Val	C
	Gly	Glu	Ala	Val	U
A	Arg	Lys	Thr	Met	G
	Arg	Lys	Thr	Ile	A
	Ser	Asn	Thr	Ile	C
	Ser	Asn	Thr	Ile	U
C	Arg	Gln	Pro	Leu	G
	Arg	Gln	Pro	Leu	A
	Arg	His	Pro	Leu	C
	Arg	His	Pro	Leu	U

(contd.)

First position	Second position				Third position
	G	A	C	U	
U	Trp	STOP	Ser	Leu	G
	STOP	STOP	Ser	Leu	A
	Cys	Tyr	Ser	Phe	C
	Cys	Tyr	Ser	Phe	U

The consequence is that different triplets correspond to the same amino acid. For example, both AAG and AAA code for lysine. On the other hand, three of the possible codons do not code for any amino acid and are used instead to signal the end of a gene. These special termination codons are identified in Table 1.2 with the word STOP written in the corresponding entry. Finally, we remark that the genetic code shown above is used by the vast majority of living organisms, but some organisms use a slightly modified code.

Transcription, Translation, And Protein Synthesis

Now let us describe in some detail how the information in the DNA results in proteins. A cell mechanism recognizes the beginning of a gene or gene cluster thanks to *a promoter.* The promoter is a region before each gene in the DNA that serves as an indication to the cellular mechanism that a gene is ahead. The codon AUG (which codes for methionine) also signals the start of a gene. Having recognized the beginning of a gene or gene cluster, a copy of the gene is made on an RNA molecule. This resulting RNA is the *messenger RNA,* or mRNA for short, and will have exactly the same sequence as one of the strands of the gene but substituting *U* for *T*. This process is called transcription.

The mRNA will then be used in cellular structures called ribosomes to manufacture a protein. Because RNA is single-stranded and DNA is double-stranded, the mRNA produced is identical in sequence to only one of the gene strands, being complementary to the other strand—keeping in mind that T is replaced by U in RNA. The strand that looks like the mRNA product is called the *ontisense* or *coding* strand, and the other one is the *sense* or *anticoding* or else *template* strand. The template strand is the one that is actually transcribed, because the mRNA is composed by binding together ribonucleotides complementary to this strand. The process always builds mRNA molecules from their 5' end to their 3' end, whereas the template strand is read from 3' to 5'. Notice also that it is not the case that the template strand for genes is always the same; for example, the template strand for a certain gene *A may be* one of the strands, and the template strand for another gene *B* may be the other strand. For a given gene, the cell can recognize the cor-responding template strand thanks to a promoter. Even though the reverse complement of the promoter appears in the other strand, this reverse complement is *not* a promoter and thus will not be recognized as such.

One important consequence of this fact is that genes from the same chromosome have an *orientation* with respect to each other: Given two genes, if they appear in the same strand they have the same orientation; otherwise they have opposite *orientation.* We finally note that the terms *upstream* and *downstream* are used to indicate positions in the DNA in reference to the orientation of the coding strand, with the promoter being upstream from its gene. Transcription as described is valid for organisms categorized as *prokaryotes.* These organisms have their DNA free in the cell, as they lack a nuclear membrane. Examples of prokaryotes are bacteria and blue algae. All other organisms, categorized as *eukaryotes,* have a nucleus separated from the rest of the cell by a nuclear membrane, and their DNA is kept inside the nucleus. In these organisms genetic transcription is more complex.

Many eukaryotic genes are composed of alternating parts called *introns* and *exons.* After transcription, the introns are spliced out from the mRNA. This means that introns are parts of a gene that are *not* used in protein synthesis. An example of exon-intron distribution is given by the gene for bovine atrial naturietric peptide, which has 1082 base pairs. Exons are located at positions 1 to 120, 219 to 545, and 1071 to 1082. Introns occupy positions 121 to 218 and 546 to 1070. Thus, the mRNA coding regions has just 459 bases, and the corresponding protein has 153 residues. After introns are spliced out, the shortened mRNA, containing copies of only the exons plus regulatory regions in the beginning and end, leaves the nucleus, because ribosomes are outside the nucleus. Because of die intron/exon phenomenon, we use different names to refer to the entire gene as found in the chromosome and to the spliced sequence consisting of exons only. The former is called *genomic* DNA and the latter *complementary* DNA or cDNA. Scientists can manufacture cDNA without knowing its genomic counterpart. They first capture the mRNA outside the nucleus on its way to the ribosomes. Then, in a process called *reverse transcription,* they produce DNA molecules using the mRNA as a template. Because the mRNA contains only exons, this is also the composition of the DNA produced. Thus, they can obtain cDNA without even looking at the chromosomes.

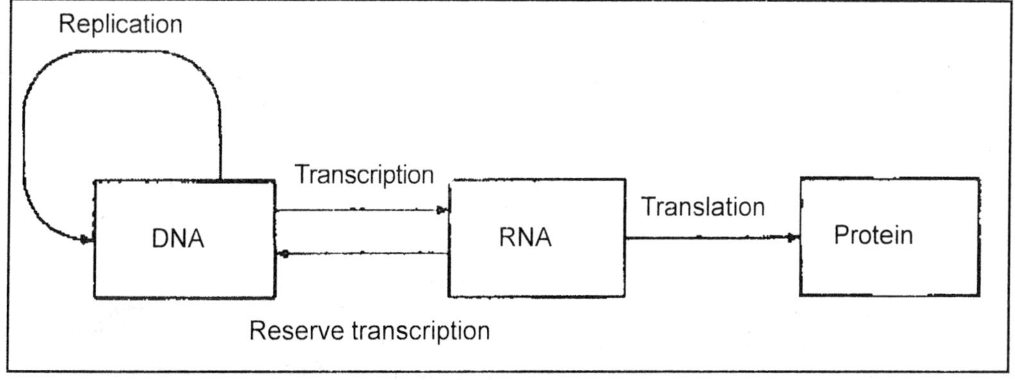

Fig. 1.9: Genetic information flow in a cell: the so-called central dogma of molecular biology.

Both transcription and reverse transcription are complex processes that need the help of enzymes. *Transcriptase* and *reverse transcriptase* are the enzymes that catalyze these processes in the cell. There is also a phenomenon called *alternative splicing*. This occurs when the same genomic DNA can give rise to two or more different mRNA molecules, by choosing the introns and exons in different ways. They will in general produce different proteins. Now let us go back to mRNA and protein synthesis. In this process two other kinds of RNA molecules play very important roles. As we have already mentioned, protein synthesis takes place inside cellular structures called ribosomes.

Ribosomes are made of proteins and a form of RNA called *ribosomal* RNA, or rRNA. The ribosome functions like an assembly line in a factory using as "inputs" an mRNA molecule and another kind of RNA molecule called *transfer* RNA, or tRNA. Transfer RN As are the molecules that actually implement the genetic code in a process called **translation**. They make the connection between a codon and the specific amino acid this codon codes for. Each tRNA molecule has, on one side, a conformation that has high affinity for a specific codon and, on the other side, a conformation that binds easily to the corresponding amino acid. As the messenger RNA passes through the interior of the ribosome, a tRNA matching the current codon—the codon in the mRNA currently inside the ribosome—binds to it, bringing along the corresponding amino acid (a generous supply of amino acids is always "floating around" in die cell).

The three-dimensional position of all these molecules in this moment is such that, as the tRNA binds to its codon, its attached amino acid falls in place just next to the previous amino acid in the protein chain being formed. A suitable enzyme then catalyzes the addition of this current amino acid to the protein chain, releasing it from the tRNA. A protein is constructed residue by residue in this fashion. When a STOP codon appears, no tRNA associates with it, and the synthesis ends. The messenger RNA is released and degraded by cell mechanisms into ribonucleotides, which will be then recycled to make other RNA. One might think that there are as many tRNAs as there are codons, but this is not true. The actual number of tRNAs varies among species. The bacterium *E. coli,* for in-stance, has about 40 tRNAs. Some codons are not represented, and some tRNAs can bind to more than one codon. The expression *central dogma* is generally used to denote our current synthetic view of genetic information transfer in cells.

Junk DNA and Reading Frames

As mentioned, genes are certain contiguous regions of the chromosome, but they do not cover the entire molecule. Each gene, or group of related genes, is flanked by regulatory regions that play a role in controlling gene transcription and other related processes, but otherwise intergenic regions have no known function. They are called "junk DNA" because they appear to be there for no particular reason. Moreover, they accumulate mutations, as a change not affecting genes or their regulatory regions is often not lethal and is therefore propagated to the progeny.

Recent research has shown, however, that junk DNA has more information content than previously believed. The amount of junk DNA varies from species to species. Prokaryotes tend to have little of it—their chromosomes are almost all covered by genes. In contrast, eukaryotes have plenty of junk DNA. In human beings it is estimated that as much as 90% of the DNA in chromosomes is composed of junk DNA. An aspect of the transcription process that is important to know is the concept of *reading frame,* A reading frame is one of the three possible ways of grouping bases to form codons in a DNA or RNA sequence. For instance, consider the sequence

TAATCGAATGGGC.

One reading frame would be to take as codons TAA, TCG, AAT, GGG, leaving out the last C. Another reading frame would be to ignore the first T and get codons AAT, CGA, ATG, GGC. Yet another reading frame would yield codons ATC, GAA, TGG, leaving out two bases at the beginning (TA) and two bases at the end (GC). Notice that the three reading frames start at positions 1, 2, and 3 in the given sequence, respectively. If we were to consider a reading frame starting at position 4, the codons obtained would be a subset of the ones for starting position 1, so this is actually the same reading frame starting at a different position.

In general, if we take starting positions i and j where the difference $j - i$ is a multiple of three, we are in fact considering the same reading frame. Sometimes we talk about six, not three, different reading frames in a sequence. In this case, what we have is a DNA sequence and we are looking at the opposite strand as well. We have three reading frames in one strand plus another three in the complementary strand, giving a total of six. It is common to do that when we have newly sequenced DNA and want to compare it to a protein database. We have to translate the DNA sequence into a protein sequence, but there are six ways of doing that, each one taking a different reading frame.

The fact that we lose one or two bases at the extremities of the sequence is not important; these sequences are long enough to yield a meaningful comparison even with a few missing residues. An *open reading frame,* or ORF, in a DNA sequence is a contiguous stretch of this sequence beginning at the start codon, having an integral number of codons (its length is a multiple of three), and such that none of its codons is a STOP codon. The presence of additional regulatory regions upstream from the start codon is also used to characterize an ORF.

Chromosomes

The complete set of chromosomes inside a cell is called a **genome**. The number of chromosomes in a genome is characteristic of a species. For instance, every cell in a human being has 46 chromosomes, whereas in mice this number is 40. Table 1.3 gives the number of chromosomes and genome size in base pairs for selected species. Prokaryotes generally have just one chromosome, which is sometimes a circular DNA molecule. On the other hand, in eukaryotes chromosomes appear in pairs (and for this reason the cells that carry them are called *diploid).*

In humans, for example, there are 23 pairs. Each member of a pair was inherited from each parent. The two chromosomes that form a pair are called *homologous* and a gene in one member of die pair corresponds to a gene in the other. Certain genes are exactly the same in both the paternal and the maternal copies. One example is the gene that codes for hemoglobin, a protein that carries oxygen in the blood. Other genes may appear in different forms, which are called *alleles.* A typical example is the gene that codes for blood type in humans. It appears in three forms: *A, B,* and *O*. As is well-known, if a person receives, say, the *A* allele from the mother and the *B* allele from the father, this person's blood type will be *AB*. Cells that carry only one member of each pair of chromosomes are called *haploid.* These are the cells that are used in sexual reproduction.

Table 1.3: Genome sizes of certain species. Apart from our own species, the organisms listed are important in molecular biology and genetics research.

Species	No. of Chromosomes (diploid)	Genome Size (base pairs)
Bacteriophage λ (virus)	1	5×10^4
Escherichia coli (bacterium)	1	5×10^6
Saccharomyces cerevisiae (yeast)	32	1×10^7
Caenorhabditis elegans (worm)	12	2×10^8
Drosophila melanogaster (fruit fly)	8	2×10^8
Homo sapiens (human)	46	3×10^9

When a haploid cell from the mother is merged with a haploid cell from the father we have an *egg cell,* which is again diploid. Haploid cells are formed through a process called *meiosis,* in which a cell divides in two and each daughter cell gets one member of each pair of chromosomes. It is interesting to note that despite the fact that all genes are present in all cells, only a portion of the genes are normally used (*expressed,* in biological jargon) by any specific cell. For instance, liver cells express a different set of genes than do skin cells. The mechanisms through which the cells in an organism differentiate into liver cells, skin cells, and so on, are still largely unclear.

Basic Mechanism

Having reviewed the basic mechanisms through which proteins are produced, it is tempting to view them in light of the so-called "genetic program metaphor." In this metaphor, the genome of an organism is seen as a computer program that completely specifies the organism, and the cell machinery is simply an interpreter of this program. The biological functions performed by proteins would be the execution of this "program".

This metaphor is overly simplistic, due in part to the following two facts:

- The "DNA program" in fact undergoes changes during transcription and translation, so that one cannot simply apply the genetic code to a stretch of DNA known to contain a gene in order to know what protein corresponds to that gene.
- Gene expression is a complex process that may depend on spatial and temporal context. For example, not all genes in a genome are expressed in an organism's lifetime, whereas others are expressed over and over again; some genes are ex-pressed only when the organism is subject to certain outside phenomena, such as a virus invasion. The opposite is also true: Genes that are normally expressed may be repressed because of outside stimuli. It is true that the expression of certain genes is essentially context-free, and this is what makes biotechnology possible. But this is by no means true for all genes. If we view gene expression inside a cell as a "computing process", we can say that, in the case of the human genome, there are more than 10^{18} such processes occurring and interacting simultaneously.

In view of these observations, it seems better to view an organism not as determined by its genome but rather as the result of a very complex network of simultaneous interactions, in which the genome sequence is one of several contributing factors.

SEQUENCING

In the scientific study of a genome, the first thing to notice is the different orders of magnitude that we have to deal with. Let us use the human genome as an example. The basic information we want to extract from any piece of DNA is its base-pair sequence. The process of obtaining this information is called **sequencing**. A human chromosome has around 10^8 base pairs. On the other hand, the largest pieces of DNA that can be sequenced in the laboratory are 700 bp long. This means that there is a gap of some 10^5 between the scale of what we can actually sequence and a chromosome size. This gap is at the heart of many problems in computational biology, in particular those studied in the form of fragment and physical mapping.

GENETIC MAPS

One particular piece of information that is very important is the location of genes in chromosomes. The term **locus** (plural **loci**) is used to denote the location of a gene in a chromosome. (Sometimes the word locus is used as a synonym for the word gene.) The simplest question in this context is, given two genes, are they in the same homologous pair? This can be answered without resorting to molecular techniques, as long as the genes in question are ones that affect visible characteristics, such as eye color or wing shape. We have to test whether the characteristics are being inherited independently. We will say that they are, or more technically, that they **assort** or **segregate** independently if an offspring has about a 50% chance of inheriting both characteristics from the same parent.

If the characteristics do assort independently, then the genes are probably not linked; that is, they probably belong to different chromosomes. Two genes carried in the same pair should segregate together, and the offspring will probably inherit both characteristics from the same parent. In fact, as is usually the case in biology, things are not so clear-cut: 100% or 50% segregation does not always happen. Any percentage in between can occur, due to crossing over. When cells divide to produce other cells that will form the progeny, new gene arrangements can form. We say that recombination occurs.

Recombination can happen because homologous chromosomes can "cross" and exchange their end parts before segregating. There are an enormous number of recombination possibilities, so what we re-ally see is that rates of recombination vary a great deal. These rates in turn give information on how far apart the genes are in the chromosome. If they are close together, there is a small chance of separation due to crossing over. If they are far apart, the chances of separation increase, the point that they appear to assort independently. The first genetic maps were constructed by producing successive generations of certain organisms and analyzing the observed segregation percentages of certain characteristics.

A *genetic linkage map* of a chromosome is a picture showing the order and relative distance among genes using such information. Genetic maps constructed from recombination percentages are important, but they have two drawbacks: (1) They do not tell the actual distance in base pairs or other linear unit of length along the chromosome; and (2) If genes are very close, one cannot resolve their order, because the probability of separation is so small that observed recombinant frequencies are all zero. Maps that reflect the actual distance in base pairs are called **physical maps**. To construct physical maps, we need completely different techniques. In particular we have to work with pieces of DNA much smaller than a chromosome but still too large to be sequenced directly. A physical map can tell the location of certain *markers*, which are precisely known small sequences, within 10^4 base pairs or so. Finally, for pieces of DNA that are on the order of 10^3 base pairs, we can use still other techniques and obtain the whole sequence. We have mentioned that current lab techniques can sequence DNA pieces of at most 700 bp; to sequence a 20,000-bp piece (in what is known as **large-scale sequencing**), the basic idea is to break apart several copies of the piece in different ways, sequence the (small) fragments directly, and then put the fragments together again by using computational techniques.

Specific Techniques

Some special laboratory techniques must be used to obtain maps and sequences. In this section we give an overview of such techniques. It is important to realize that lab techniques in molecular biology nearly always produce data that have errors. For this reason, most algorithms for problems in computational biology are useful only to the extent that they can handle errors. Such concern will appear throughout this book.

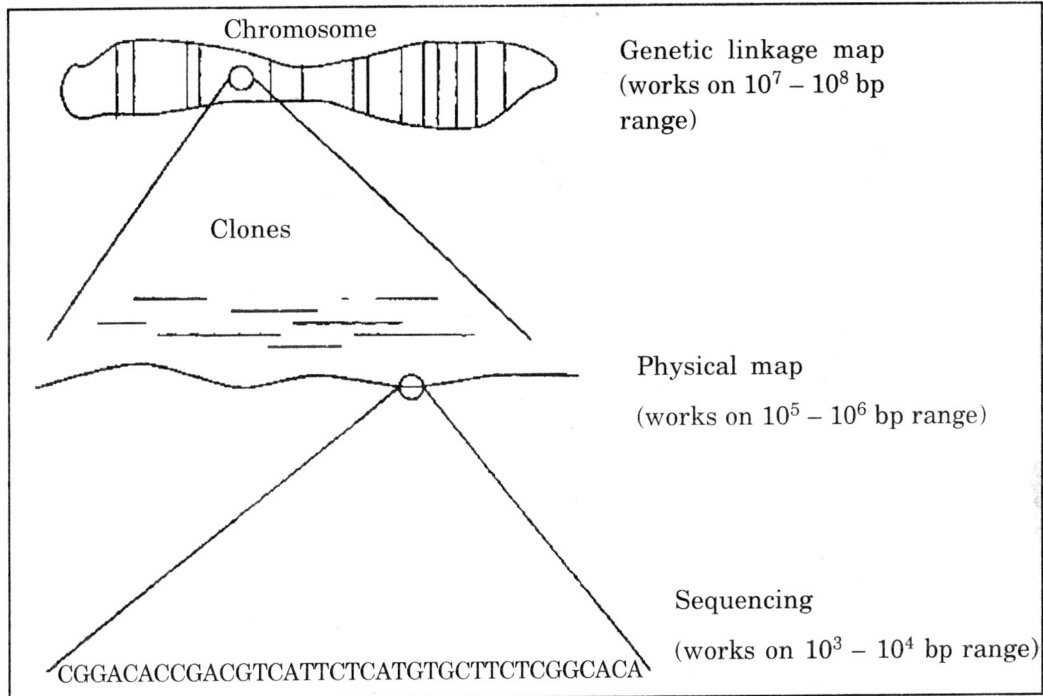

Fig. 1.10: The different levels at which a genome is studied

Viruses and Bacteria

We begin by briefly describing the organisms most used in genetics research: viruses and bacteria.

Viruses are parasites at the molecular level. Viruses can hardly be considered a life form, yet they can reproduce when infecting suitable cells, called **hosts**. Viruses do not exhibit any metabolism—no biochemical reactions occur in them. Instead, viruses rely on the host's metabolism to replicate, and it is this fact that is exploited in laboratory experiments. Most viruses consist of a protein cap (a *capsid*) with genetic material (either DNA or RNA) inside. Viral DNA is much smaller than the DNA in chromosomes, and therefore is much easier manipulate.

When viruses infect a cell, the genetic material is introduced in the cytoplasm. This genetic material is mistakenly interpreted by the cell mechanism as its own DNA, and for this reason the cell starts producing virus-coded proteins as though they were the cell's own. These proteins promote viral DNA replication and formation of new capsids, so that a large number of virus particles are assembled inside the infected cell. Still other viral proteins break the cell membrane and release all the new virus particles in the environment, where they can attack other cells. Certain viruses do not kill their hosts at once. Instead, the viral

DNA gets inserted into the host genome and can stay there for a long time without any noticeable change in cell life.

Under certain conditions the dormant virus can be activated, whereupon it detaches itself from the host genome and starts its replication activity. Viruses are highly specific; they are capable of infecting one type of cell only. Thus, for instance, virus T2 infects only *E. coli* cells; HIV, the human immunodeficiency virus, infects only a certain kind of human cell involved in defending the organism against intruders in general; TMV, the tobacco mosaic virus, infects only tobacco leaf cells; and so on **Bacteriophages**, or just *phages*, are viruses that infect bacteria. A bacterium is a single-cell organism having just one chromosome. **Bacteria** can multiply by simple DNA replication, and they can do this in a very short period of time, which makes them very useful in genetic research. The bacterium most commonly used in labs is the already mentioned *Escherichia coli*, which can divide in 20 minutes. As a result of their size and speed of reproduction, millions of bacteria can be easily generated and handled in a laboratory.

Cutting and Breaking DNA

Because a DNA molecule is so long, some tool to cut it at specific points (like a pair of scissors) or to break it apart in some other way is needed. The pair of scissors is represented by **restriction enzymes**. They are proteins that catalyze the hydrolysis of DNA (molecule breaking by adding water) at certain specific points called **restriction sites** that are determined by their local base sequence. In other words, they cut DNA molecules in all places where a certain sequence appears. For instance, *Eco*RI is a restriction enzyme that cuts DNA wherever the sequence GAATTC is found. Notice that this sequence is its own reverse complement, that is, $\overline{\text{GAATTC}} = \text{GAATTC}$. Sequences that are equal to their reverse complement are called *palindromes*. So, every time this sequence appears in one strand it appears in the other strand as well. The cuts are made in both strands between the G and the first A. Therefore, the remaining DNA pieces will have "sticky" ends, that is, their 5' end in the cut point will be four bases shorter than the 3' end; see Fig. 1.11. This favours relinking with another DNA piece cut with the same enzyme, providing a kind of "DNA cut and paste" technique very useful in genetic engineering for the production of recombinant DNA.

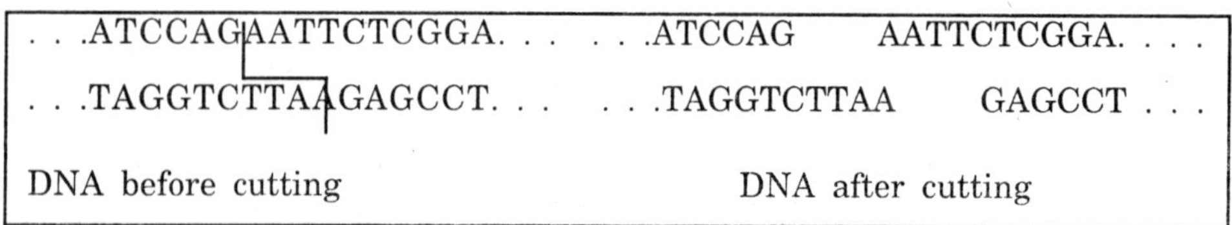

Fig. 1.11: Enzyme that cuts DNA, leaving "sticky" ends.

Some common types of restriction enzymes are 4-cutters, 6-cutters, and 8-cutters. It is rare to see an odd-cutter because sequences of odd length cannot be palindromes. Restriction enzymes are also called *endonucleases* because they break DNA in an internal point.

Exonucleases are enzymes that degrade DNA from the ends inwards. In bacteria, restriction enzymes may act as a defense against virus attacks. These enzymes can cut the viral DNA before it starts its damage.

Bacterial DNA on the other hand protects itself from restriction enzymes by adding a methyl group to some of its bases. DNA molecules can be broken apart by the *shotgun method*, which is sometimes used in DNA sequencing. A solution containing purified DNA—a large quantity of identical molecules—is subjected to some breaking process, such as submitting it to high vibration levels. Each individual molecule breaks down at several random places, and then some of the fragments are filtered and selected for further processing, in particular for *copying* or *cloning* (see below). We then get a collection of cloned fragments that correspond to random contiguous pieces of the purified DNA sequence. Sequencing these pieces and assembling the resulting sequences is a standard way of determining the purified DNA's sequence. This method can also be used to construct a cloning library—a collection of clones covering a certain long DNA molecule.

DNA Amplification

Another very important tool needed in molecular biology research is a *DNA copying process* (also called DNA *amplification*). We are dealing here with molecules, which are microscopic objects. The more copies we have of a molecule, the easier it is to study it. Fortunately, several techniques have been developed for this purpose.

DNA Cloning

To do any experiment with DNA in the lab, one needs a minimum quantity of material; one molecule is clearly not enough. Yet one molecule is sometimes all we have to start with. In addition, the material should be stored in a way that permits essentially unlimited production of new material for repetition of the experiment or for new experiments as needed. DNA cloning is the name of the general technique that allows these goals to be attained. Given a piece of DNA, one way of obtaining further copies is to use nature itself.

We insert this piece into the genome of an organism, a *host* or *vector*, and then let the organism multiply itself. Upon host multiplication, the inserted piece (the *insert*) gets multiplied along with the original DNA. We can then kill the host and dispose of the rest, keeping only the inserts in the desired quantity. DNA produced in this way is called recombinant.

Popular vectors include *plasmids*, *cosmids*, *phages*, and *YACs*. A *plasmid* is a piece of circular DNA that exists in bacteria. It is separated from and much smaller than the bacterial chromosome. It does get replicated when the cell divides, though, and each daughter cell keeps one copy of the plasmid. Plasmids make good vectors, but they place a limitation on insert size: about 15 kbp. Inserts much larger than this limit will make very big plasmids, and big plasmids tend to get shortened when replicated.

Phages are viruses often used as vectors. One example is phage λ., which infects the bacteria *E. coli*. Inserts in phage DNA get replicated when the virus infects a host colony. Phage λ normally has a DNA of 48 kbp, and inserts of up to 25 kbp are well tolerated. Inserts larger than this limit are impossible, though, because the resulting DNA will not fit in the phage protein capsule. However, if the entire phage DNA is replaced by an insert plus some minimum replicative apparatus, inserts of up to 50 kbp are feasible. These are called *cosmids*. For very large inserts, on the range of a million base pairs, a *YAC* (yeast artificial chromosome) can be used. A YAC is an extra, artificially-made chromosome that can be built by adding yeast control chromosomal regions to the insert and making it look like an additional chromosome to the yeast replication mechanism.

Polymerase Chain Reaction

A way of producing many copies of a DNA molecule without cloning it is afforded by the polymerase chain reaction (PCR). DNA Polymerase is an enzyme that catalyzes elongation of a single strand of DNA, provided there is a template DNA to which this single strand is attached. Nucleotides complementary to the ones in the template strand are added until both strands have the same size and form a normal double strand of DNA.

The small stretch of double stranded DNA at the beginning needed for polymerase to start its job is called a **primer**. PCR consists basically of an alternating repetition of two phases: A phase in which double stranded DNA is separated into two single strands by heat and a phase in which each single strand thus obtained is converted into a double strand by addition of a primer and polymerase action. Each repetition doubles the number of molecules.

After enough repetitions, the quantity of material produced is large enough to perform further experiments, thanks to the exponential growth of this doubling procedure. A curious footnote here is that Kary B. Mullis, the inventor of PCR (in 1983), realized it was a good idea because he "had been spending a lot of time writing computer programs" and was thus familiar with iterative functions and thereby with exponential growth processes.

Reading and Measuring DNA

How do we actually *"read"* the base pairs of a DNA sequence? Reading is done with a technique known as *gel electrophoresis*, which is based on separation of molecules by their size. The process involves a gel medium and a strong electric field. DNA or RNA molecules are charged in aqueous solution and will move to a definite direction by action of the electric field. The gel medium makes them move slowly, with speed inversely proportional to their size.

All molecules are initially placed at one extremity in a gel block. After a few hours, the smaller molecules will have migrated to the other end of the block, whereas larger molecules will stay behind, near the starting place. By interpolation, the relative sizes of molecules can be calculated with good approximation. In this kind of experiment, DNA molecules can be labelled with radioactive isotopes so that the gel can be photographed, producing a graphic record of the positions at the end of a run. This process is used in DNA sequencing, determination of restriction fragment lengths, and so on.

An alternative is to use fluorescent dyes instead of radioactive isotopes. A laser beam can trace the dyes and send the information directly to a computer, avoiding the photographic process altogether. Sequencing machines have been built using this technique. DNA or RNA bases can be read using this process with the following technique. Given a DNA molecule, it is possible to obtain all fragments from it that end at every position where an A appears. Similarly all fragments that end in T, in G, or in C can be obtained. We thus get four different test tubes, one for each base. The fragments in each tube will have differing lengths. For example, suppose the original DNA piece is the following:

GACTTAGATCAGGAAACT

The fragments that end in T are GACT, GACTT, GACTTAGAT, and the whole sequence itself. Thus if we separate these fragments by size, and we do it simultaneously but separately for all four test tubes, we will know the precise base composition of the original DNA sequence. Some errors may occur in reading a gel film, because sometimes marks are blurred, especially near the film borders. One other limitation is the size of DNA fragments that can be read in this way, which is about 700 bp.

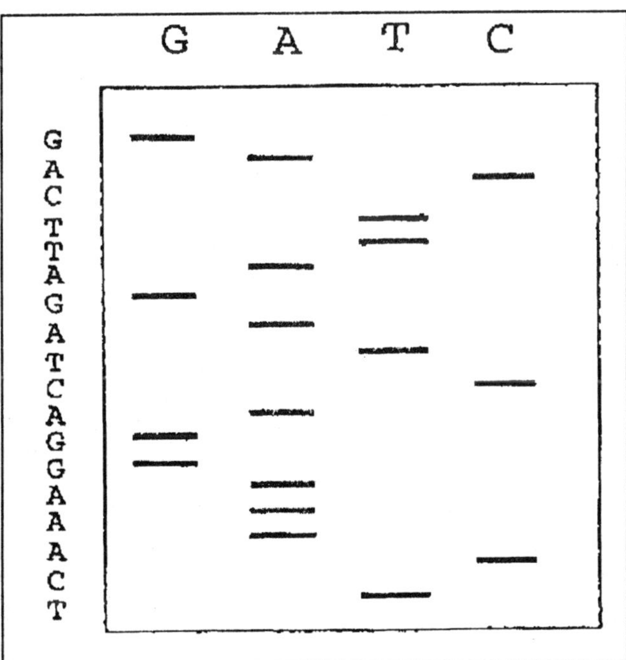

Fig. 1.12: Schematic view of film produced by gel electrophresis. Individual DNA bases can be identified in each of the four columns. Shorter fragments leave their mark near the top, whereas longer fragments leave their mark near the bottom.

The Human Genome Project

The Human Genome Project is a multinational effort, begun in 1988, whose aim is to produce a complete physical map of all human chromosomes, as well as the entire human DNA sequence. As part of the project, genomes of other organisms such as bacteria, yeast, flies, and mice are also being studied. This is no easy task, since the human genome is so large. So far, many virus genomes have been entirely sequenced, but their sizes are generally in the 1 kbp to 10 kbp range. The first free-living organism to be totally sequenced was the bacterium *Haemophilus influenzae*, containing a 1800 kbp genome. In 1996 the whole sequence of the yeast genome—a 10 million bp sequence—was also determined. This was an important milestone, given that yeast is a free-living eukaryote.

Mapping the human genome still seems re-mote, as it is 100 times bigger than the largest genome sequenced so far. One of the reasons why other organisms are part of the project is to perfect sequencing methods so that they can then be applied to the human genome. The cost of sequencing is also expected to drop as a result of new, improved technology. Another reason lies in research benefits. All species targeted are largely employed in genetic and molecular research. Some lessons have been learned. A large effort like this cannot be entertained by a single lab. Rather, a consortium of first-rate labs working together is a more efficient way—and perhaps the only way—of getting the work done. Coordinating these tasks is in itself a challenge.

On the computer science side, databases with updated and consistent information have to be maintained, and fast access to the data has to be provided. A major concern is with possible errors in the sequences obtained. The current target is to obtain DNA sequences with at most one error for every 10,000 bases. Another problem is that of the "average genome". Different individuals have different genomes (this is what makes *DNA fingerprinting* possible). A gene may have many alleles in the population, and the sequence of intergenic regions (which do not code for products) may vary from person to person. It is estimated that the genomes of two distinct humans differ on average in one in every 500 bases. So the question is: Whose genome is going to be sequenced? Even if one individual is somehow chosen and her or his DNA is taken as the standard, there is the problem of transposable elements.

It is now known that certain parts of the genome keep moving from one place to another, so that at best what we will get after sequencing is a "snapshot" of the genome in a given instant. Once the entire sequence is obtained, we will face the difficult task of analyzing it.

We have to recognize the genes in it and determine the function of the proteins produced. But gene recognition is still in its infancy, and protein function determination is still a very laborious procedure. Treatment of genetic diseases based on data produced by the Human Genome Project is still far ahead, although encouraging pioneering efforts have already yielded results.

SEQUENCE DATABASES

Thanks in part to the techniques described in previous sections, large numbers of DNA, RNA, and protein sequences have been determined in the past decades. Some institutional sequence databases have been set up to harbor these sequences as well as a wealth of associated data.. The rate at which new sequences are being added to these databases is exponential. Computational techniques have been developed to allow fast search on these databases. Here we give a brief description of some representative sequence databases.

GenBank

Maintained by the National Center for Biotechnology Information (NCBI), USA, GenBank contains hundreds of thousands of DNA sequences. It is divided into several sections with sequences grouped according to species, including:

- PLN: Plant sequences
- PRI: Primate sequences
- ROD: Rodent sequences
- MAM: Other mammalian sequences
- VRT: Other vertebrate sequences
- IVN: Invertebrate sequences
- BCT: Bacterial sequences
- PEG: Phage sequences
- VRL: Other viral sequences
- SYN: Synthetic sequences
- UNA: Unannotated sequences
- PAT: Patent sequences
- NEW: New sequences

Searches can be made by keywords or by sequence. The entry is divided into *fields*, with each field composed by a field identifier, which is a word describing the contents of the field, and the information *per se*. The en-tries are just plain text. One important field is *accession no.*, the accession number, which is a code that is unique to this entry and can be used for faster access to it. In the example, the accession number is M12174. Some entries have several accession numbers as a result of combination of several related but once distinct entries into one comprehensive entry. Other fields are self-explanatory for the most part.

This database is part of an international collaboration effort, which also includes the DNA DataBank of Japan (DDBJ) and the European Molecular Biology Laboratory (EMBL).

```
LOCUS          HUMRHOA    539 bp    nVRNA       PRI     04-AUG-1986
DEFINITION     Human ras-related rho mRMA (clone 6), partial cds.
ACCESSION      M12174
KEYWORDS       c-myc proto-oncogcne; raa oncogene; rho gene.
SOURCE         Human peripheral T-cell, cDNA to mRNA, clone 6.
ORGANISM       Homo sapiens
               Eukaryota; Animalia; Chordata; Vertebrata; Mammalia; Theria; Eutheria;
               Primates; Haplorhini; Catarrhini; Hominidae.
REFERENCE      1  (bases 1 to 539)
AUTHORS        Madaule, P.
JOURNAL        Unpublished (1985) Columbia U, 701 W 168th St, New York, NY 10032
REFERENCE      2  (bases 1 to 539)
AUTHORS        Madaule, P. and Axel. R.
TITLE          A novel ras-related gene family
JOURNAL        Cell 41, 31-40 (1985)
 MEDLINE       85201682
COMMENT        [2] has found and sequenced a family of highly evolutionarily conserved
               genes with homology to the ras family (H-ras, K-ras, N-ras) of oncogenes.
               [2] named this family rho (for ras homology). In humans at least three
               distinct rho genes are present. A draft entry and computer-readable copy
               of this sequence were kindly provided by P. Madaule (07-OCT-1985).
               NCBI gi: 337392
FEATURES       Location/Qualifiers
source         1. .539
               /organisiu^'Homo sapiens”
CDS            <1. .509
               /note="rho protein;  NCBI gi: 337393"
               /codon_start=2
BASE COUNT     105 a    180 c    172 g    82 t
ORIGIN         185 bp upstream of Hinfl site.
               1    cgogttcccc  gaggtgtacg  tgc  ....  tatgtggccg  acattgaggt
               61   ggacggcaag  caggtggagc  tgg  ....  ggccaggagg  actacgaccg
               121  cctgcggccg  ctctcctacc  egg  ....  atgtgcttct  cggtggacag
               181  cccggactcg  ctggagaaca  tec  ....  gaggtgaagc  acttctgtcc
               ......
               421  cgaggtcttc  gagacggcca  cgc  ....  cgctacggct  cccagaacgg
               481  ctgcatcaac  tgctgcaagg  tgc  ....  cgcgcctgcc  cctgccggc
//
```

Fig. 1.13: Typical GenBank entry. The actual file was edited to fit the page. In a real entry each sequence line has 60 characters, with the possible exception of the last line.

```
ID   ECTRGA    Standard; RNA; PRO; 75 BP.
XX
AC   M24860;
XX
DT   24-APR-1990 (Rel. 23,  Created)
DT   31-MAR-1992 (Rel. 31,  Last updated. Version 3)
XX
DE   E.coli Gly-tRNA.
XX
KW   transfer RNA-Gly.
XX
OS   Escherichia coli
OC   Prokaryota; Bacteria;   Gracilicutes; Scotobacteria;
OC   Facultatively anaerobic rods; Bnterobacteriaceae;
OC   Escherichia .
XX
RN   [1]
RP   1-75
RA   Carbon J., Chang S.,   Kirk L. L.;
RT   "Clustered tRNA genes in Escherichia coli: Transcription
RT   and processing";
RL   Brookhaven Symp. Biol. 26:26-36(1975) „
XX
FH   Key Location/Qualifiers
FH
FT   tRNA            1..75
FT              / no te= "Gly-tRNA"
XX
SQ   Sequence 75 BP; 13 A; 24 C; 19 G; 19 T; 0 Other;
     gcgggcatcg tataatggct attacctcag  . . .   tgatgatgcg ggttcgattc
     ccgctgcccg ctcca
```

Fig. 1.14: A typical EMBL entry, edited to fit the page. Actual entries have 60 characters per line in the sequence section, with the possible exception of the last line.

PIR1:CCHP
cytochrome. c - hippopotamus

Species	:	Hippopotamus amphibius (hippopotamus)
Date	:	19-Feb-1984 #sequence_revision 19-Feb-1984 #text_change 05-Aug-1994
Accession	:	A00008
		Thompson, R.B.; Borden, D.; Tarr, G.E.; Margoliash, E.
		J. Biol. Chem. 253, 8957-8961, 1978
Title	:	Heterogeneity of amino acid sequence in hippopotamus cytochrome c.
Reference number	:	A00008; MUID:79067782
Accession	:	A00008
Molecule type	:	protein
Residues	:	1-104 <THO>
Note	:	3-Ile was also found
Superfamily	:	cytochrome c; cytochrome c homology
Keywords	:	acetylated amino end; electron transfer; heme; mitochondrion; oxidative phosphorylation; respiratory chain

Residues	Feature
1	Modified site: acetylated amino end (Gly) #status predicted
14, 17	Binding site: heme (Cys) (covalent) #status predicted
18, 80	Binding site: heme iron (His, Met) (axial ligands) #status predicted

Number of residues = 104

Composition

6 Ala A	4 Gin Q	6 Leu L	2 Ser S
2 Arg R	8 Glu E	17 Lys K	8 Thr T
5 Asn N	14 Gly G	2 Met M	1 Trp W
3 Asp D	3 His H	4 Phe F	4 Tyr Y
2 Cys C	6 He I	4 Pro P	3 Val V

Mol. wt. unmod. chain = 11,530 Number of residues = 104

```
                5         10         15         20         25         30
1  G D V E K G K K I F V Q K C A Q C H T V E K G G K H K T G P
31 N L H G L F G R K T G Q S P G F S Y T D A N K N K G I T W G
61 E E T L M E Y L E N P K K Y I P G T K M I F A G I K K K G E
91 R A D L I A Y L K Q A T N E
```

Fig. 1.15: A typical PIR entry, edited to fit the page

```
HEADER    MYOGLOBIN (CARBONMONOXY)              19-JUL-95   IMCY
TITLE     SPERM WHALE MYOGLOBIN (MUTANT WITH INITIATOR MET AND
TITLE     2 WITH HIS 64 REPLACED BY GLN, LEU 29 REPLACED BY PHE
COMPND    MOL_ID: 1;
COMPWD    2 MOLECULE: MYOGLOBIN (CARBONMONOXY) ;
COMPND    3 CHAIN: NULL;
COMPND    4 ENGINEERED: YES;
COMPND    5 MUTATION: INS (MET 0), F29L, Q64H, N122D
SOURCE    MOL_ID: 1;
SOURCE    2 SYNTHETIC: SYNTHETIC GENE;
SOURCE    3 ORGAN! SM_SCIENTIFIC: PHYSETER CATODON;
SOURCE    4 ORGANISK_COMMON: SPERM WHALE?
SOURCE    5 EXPRESSIOJL.SYSTEM: ESCHERICHIA COLI
KEYWDS    HEME, OXYGEN TRANSPORT, RESPIRATORY PROTEIN
EXPDTA    X-RAY DIFFRACTION
AUTHOR    T. LI, G.N. PHILLIPS JUNIOR
REVDAT    1   07-DEC-95 IMCY    0
JRNL      AUTH  X . ZHAO, K . VYAS , B . D . NGUYEN, K . RAJARATHNAM,
JRNL      AUTH 2 G.N.LAMAR,T. LI, G,N. PHILLIPS JUNIOR, R. EICH,
JRNL      AUTH 3 J.S.OLSON,J.LING,D.F.BOCIAN
JRNL      TITL  A DOUBLE MUTANT OF SPERM WHALE MYOGLOBIN
JRNL      TITL 2 MIMICS THE STRUCTURE AND FUNCTION OF
JRNL      TITL 3 ELEPHANT MYOGLOBIN
JRNL      REF    J.BIOL.CHEM.                  V. 270 20763 1995
JRNL      REFN   ASTM JBCHA3  US ISSN 0021-9258          0071
REMARK    1
REMARK    2
REMARK    2 RESOLUTION. 1.7 ANGSTROMS.
REMARK    3
REMARK    3 REFINEMENT.
REMARK    3   PROGRAM                        X-PLOR
REMARK    3   AUTHORS                        BRUNGER
REMARK    3   R VALUE                       0.182
REMARK    3   RMSD BOND DISTANCES            0.020   ANGSTROMS
REMARK    3   RMSD BOND ANGLES               1.82    DEGREES
REMARK    3
REMARK    3   NUMBER OF REFLECTIONS          23187
REMARK    3   RESOLUTION RANGE          5.0 -1-7   ANGSTROMS
REMARK    3   DATA CUTOFF                   0.0      SIGMA(F)
REMARK    3
REMARK    3  DATA COLLECTION.
REMARK    3   NUMBER OF UNIQUE REFLECTIONS          25787
REMARK    3   RESOLUTION RANGE   INFINITY - 1.7   ANGSTROMS
REMARK    3   COMPLETENESS OF DATA                 95.8    %
```

Fig. 1.16: A typical PDB entry, partial header. The last columns containing the characters IMCY i, where i is the line number, are omitted. Some other editing was done to fit the page.

ATOM	1	N	MET	0	24.486	8	.308	-9	.406	1.00	37.00
ATOM	2	CA	MET	0	24.542	9	.777	-9	.621	1.00	36.40
ATOM	3	C	MET	0	25.882	10	.156	-10	.209	1.00	34.30
ATOM	4	O	MET	0	26.833	9	.391	-10	.078	1.00	34.80
ATOM	5	CB	MET	0	24.399	10	.484	-8	.303	1.00	39.00
ATOM	6	CG	MET	0	24.756	9	.581	-7	.138	1.00	41.80
ATOM	7	SD	MET	0	24.017	10	.289	-5	.719	1.00	44.80
ATOM	8	CE	MET	0	24.761	12	.009	-5	.824	1.00	41.00
ATOM	9	N	VAL	1	25.951	11	.334	-10	.816	1.00	31.10
ATOM	10	CA	VAL	1	27.185	11	.834	-11	.382	1.00	28.40
ATOM	11	C	VAL	1	27.330	13	.341	-11	.124	1.00	26.30
ATOM	12	O	VAL	1	26.444	14	.135	-11	.452	1.00	26.60
ATOM	13	CB	VAL	1	27.270	11	.547	-12	.912	1.00	29.30
ATOM	14	C01	VAT,	1	28.532	12	.207	-13	.526	1.00	28.60
ATOM	15	CG2	VAL	1	27.275	10	.038	-13	.163	1.00	29.70
ATOM	16	N	LEU	2	28.435	13	.739	-10	.500	1.00	23.00
ATOM	17	CA	LEU	2	28.691	15	.142	-10	.318	1.00	20.80
ATOM	18	C	LEU	2	29.289	15	.737	-11	.599	1.00	20.10
ATOM	19	0	LEU	2	30.129	15	.110	-12	.276	1.00	19.50
ATOM	20	CB	LEU	2	29.661	15	.356	-9	.134	1.00	20.90
ATOM	21	CG	LEU	2	29.036	15	.387	-7	.726	1.00	20.40
ATOM	22	GDI	LEU	2	28.556	13	.983	-7	.402	1.00	19.60
ATOM	23	CD2	LEU	2	30.058	15	.904	-6	.689	1.00	19.50
ATOM	24	N	SER	3	28.996	17	.003	-11	.826	1.00	18.80
ATOM	25	CA	SER	3	29.696	17	.733	-12	.852	1.00	19.40
ATOM	26	C	SER	3	31.096	18	.141	-12	.385	1.00	19.50
ATOM	27	0	SER	3	31.397	18	.121	-11	.174	1.00	19.10
ATOM	28	CB	SER	3	28.861	18	.954	-13	.223	1.00	20.60
ATOM	29	00	SER	3	29.019	19	.969	-12	.261	1.00	22.10
ATOM	30	N	GLU	4	31.947	18	.561	-13	.310	1.00	18.40
ATOM	31	CA	GLU	4	33.293	18	.956	-12	.937	1.00	19.00
ATOM	32	C	GLU	4	33.173	20	.215	-12	.047	1.00	19.20
ATOM	33	O	GLU	4	34.026	20	.457	-11	.206	1.00	19.10
ATOM	34	CB	GLU	4	34.135	19	.270	-14	.198	1.00	19.60
ATOM	35	CG	GLU	4	35.491	19	.937	-13	.932	1.00	21.10
ATOM	36	CD	GLU	4	36.537	19	.020	-13	.295	1.00	22.60
ATOM	37	OKI	GLU	4	36.355	17	.787	-13	.230	1.00	23.90
ATOM	38	OE2	GLU	4	37.569	19	.550	-12	.840	1.00	24.60
ATOM	39	N	GLY	5	32.182	21	.062	-12	.313	1.00	18.30
ATOM	40	CA	GLY	5	32.039	22	.287	-11	.532	1.00	19.10
ATOM	41	C	GLY	5	31.669	22	.025	-10	.056	1.00	18.20
ATOM	42	O	GLY	5	32.251	22	.638	-9	.140	1.00	18.60

Fig. 1.17: A typical PDB entry, coordinates. The last columns, containing the characters 1MCY i, where i is the line number, are omitted. A few blank characters in each line were removed.

EMBL

The European Molecular Biology Laboratory is an institution that maintains several sequence repositories, including a DNA database called the Nucleotide Sequence Database. Its organization is similar to that of GenBank, with the entries having roughly the same fields. In the EMBL database, entries arc identified by two-letter codes. The accession number, for instance, is identified by the letters AC. Code XX indicates blank lines. Both GenBank and EMBL separate the sequences in blocks of ten characters, with six blocks per line. This scheme makes it easy to find specific positions within the sequence.

PIR

The Protein Identification Resource (FIR) is a database of protein sequences cooperatively maintained and distributed by three institutions: the National Biomedical Re-search Foundation (in the USA), the Martinsried Institute for Protein Sequences (in Europe), and the Japan International Protein Information Database (in Japan).

PDB

The Protein Data Bank is a repository of three-dimensional structures of proteins. For each protein represented, a general information header is provided followed by a list of all atoms present in the structure, with three spatial coordinates for each atom that indicate their position with three decimal places.

Other Databases

Apart from those mentioned above, many other databases have been created to keep molecular biology information. Among them we cite ACEDB, a powerful platform prepared for the *C. elegans* sequencing project but adaptable to similar projects; Flybase, a database of fly sequences; and databases for restriction enzymes, codon preferential usage, and so on.

Chapter 2

DNA Computing

In this section we describe how to solve the Hamiltonian path problem in directed graphs with DAN molecules in a laboratory using a method whose number of laboratory procedures grows linearly with the number of vertices in the graph. Recall from the Hamiltonian path problem is NP-complete. This method is *not* a proof that P = NP, because at its heart the method is a brute-force algorithm executing an exponential number of operations. What enables the method to have a linear number of laboratory procedures is the fact that an exponential number of operations is done in *parallel* in the first such procedure. Formally, the problem to be solved is this. Given a directed graph G = (V, E), such that |V| = n and |E| = m, and two distinguished vertices s and t, verify whether the graph has a path (s, v_1, v_2, . . . , t) whose length (in number of edges) is n-1 and whose vertices are all distinct. We denote this problem by HDPP, the Hamiltonian directed path problem. A brute-force algorithm is very simple. Generate all possible paths with exactly $n - 1$ edges, and then verify whether one of them obeys the problem constraints. There are at most $(n - 2)!$ such paths, because for the first vertex on the path we can choose $(n - 2)$ vertices, for the second vertex on the path we have $(n - 3)$ choices, and so on.

In the method to be presented below paths are not generated as above. Rather, in the first step an enormous number of *random* paths are created, the fact that random paths are being generated makes possible the generation of each path independent of all the others,

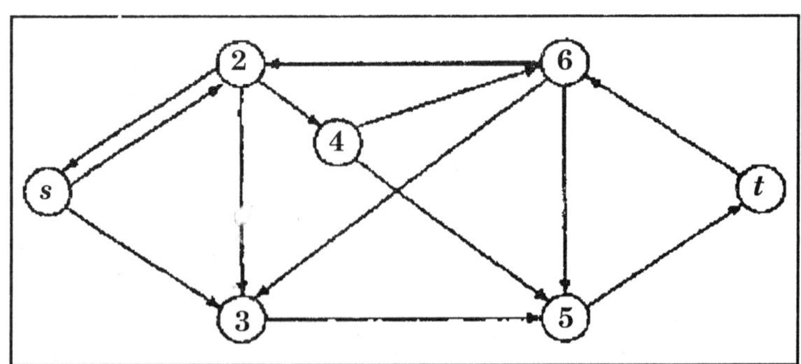

Fig. 2.1: A directed graph. An st-Hamiltonian path is (s, 2, 4, 6, 3, 5, t).

and hence all of them can be created simultaneously. On the other hand, this makes the algorithm probabilistic. The complete algorithm is as follows:

1. Generate random paths.
2. From all created in the previous step, keep only those that start at s and end at t.
3. From all remaining paths, keep only those that visit exactly n vertices.
4. From all remaining paths, keep only those that visit each vertex at least once.
5. If any path remains, return yes; otherwise return no.

We now describe how each of these steps can be implemented in the laboratory.

Random path

We assume that random single-stranded DNA sequences with 20 nucleotides are available, and that the generation of astronomical numbers (more than 10^{10}) of copies of short DNA strands is easy to do. Recall that $\bar{S}$ represents the reverse complement of DNA string S.

(i) **Vertex representation**: Choose at random n distinct single-stranded 20-character DNA sequences, and to each vertex v associate sequence S_v. For each such sequence obtain its reverse complement $\bar{S}_v$. Generate many copies of each $\bar{S}_v$ sequence in test tube T_1.

(ii) **Edge representation**: If $(u, v) \in E$, then build sequence S_{uv} by concatenating the 10-base suffix of S_u to the 10-base prefix of S_v. Suffixes and prefixes are distinguished by the 5'-3' direction of DNA strands. When $u = s$ the whole of S_s is used, and hence S_{sv} has 30 bases. Similarly for the case when v = t. Note that with this construction, $S_{uv} = S_{uv}$. Generate many copies of each S_{uv} sequence in test tube T_2.

(iii) **Path construction**: Pour T_1 and T_2 into T_3. We assume that in T_3 any binding between sequences can freely take place (that is, in T_3 many ligase reactions will take place).

By executing the three steps above we get many random paths. Why? We will explain by an example. Consider the sequences $\bar{S}_u$, $\bar{S}_v$, $\bar{S}_w$, S_{uv} and S_{vw}, for u, v, and w distinct vertices. Because many copies of these sequences are floating about in tube T_3, with high probability the 10-base suffix of one S_u sequence will bind to the 10-base prefix of one S_{uv} sequence (since one is complementary to the other); at the same time the 10-base suffix of the sane sequence S_{uv} binds to the 10-base prefix of one $\bar{S}_v$ sequence, whose 10-base suffix binds to the 10-base prefix of one S_{vw} sequence. And still at the same time the 10-base suffix of this S_{vw} sequence binds to the 10-base prefix of one $\bar{S}_w$ sequence. The final double strand thus obtained encodes path (u, v, w) in G. Let us make this clear by seeing what happens in the example of Fig. 2.1. That graph has $n = 7$ vertices. Suppose that the sequences chosen to represent vertices 2, 4, and 5 are the following:

S_2 = GTCACACTTCGGACTGACCT

S_4 = TGTGCTATGGGAACTCAGCG

S_5 = CACGTAAGACGGAGGAAAAA

The reverse complement of these sequences are:

$\bar{S}_2$ = AGGTCAGTCCGAAGTGTGAC

$\bar{S}_4$ = CGCTGAGTTCCCATAGCACA

$\bar{S}_5$ = TTTTTCCTCCGTCTTACGTG

We build edges (2, 4) and (4, 5) obtaining the following sequences:

(2, 4) = GGACTGACCTTGTGCTATGG

(4, 5) = GAACTCAGCGCACGTAAGAC

Finally, the path (2, 4, 5) will be encoded by the following double strand:

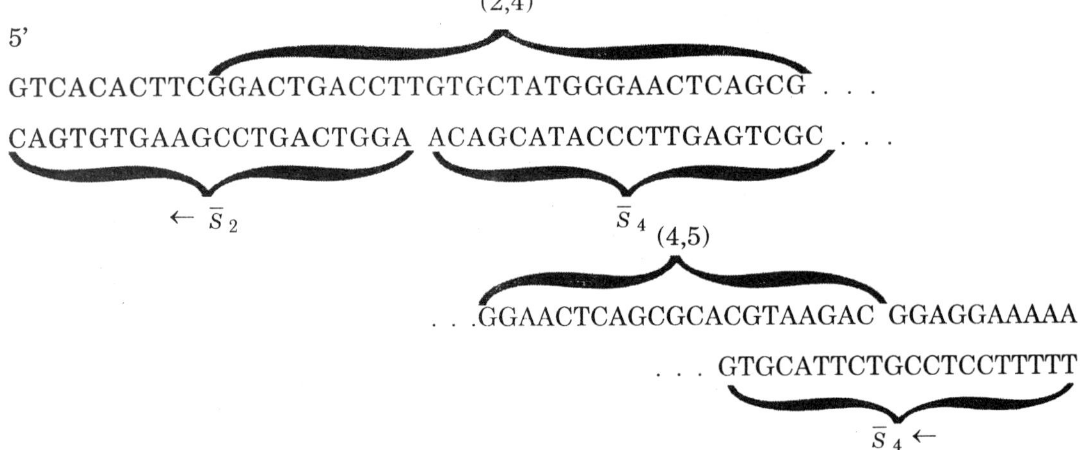

Further steps of the algorithm depend on the assumption that with high probability *all* possible paths in the graph with n vertices will be represented by at least one double strand. Note also that the number 20 was chosen so as to make it very unlikely that two sequences assigned to two different vertices would share long subsequences. Should this happen, we might have double-stranded sequences that would not encode any paths in the graph.

Further Steps

In the remaining steps we place the contents of T_3 through successive biochemical sieves that select paths with desired properties. Step (2) is implemented by a polymerase chain reaction using primers S_s and $\bar{S}_t$. Step (3) is implemented by separating double-stranded

DNA having exactly 20n bases, which therefore represent paths with exactly n vertices. Step (4) is implemented by retaining strands where $\bar{s}_u$ is present for each $u \in V$ different from *s* and *t*. In the final step we simply see what has remained through a conventional gel experiment. We can analyze this algorithm as follows. We need a number of operations proportional to the size of the graph in order to build it. After that we need to perform a number of operations that are proportional to the number of vertices in the graph, mainly because of step (4) above. In that step we have to check sequentially the presence of each vertex-representing sequence in the test tube.

SATISFIABILITY PROBLEM (SAT)

In this section we describe how we can use an approach similar to the one described in the previous section to solve the *satisfiability problem* (SAT). This is a very important problem in the theory of computing, since it was the very first one to be shown NP-complete. Moreover, a generalization of the algorithm described here leads to a general approach for solving any NP-complete problem directly. To explain the SAT problem, we need a few definitions. A *boolean variable* is a variable that can take either 0 or 1 as a value. Boolean variables may be put together in a formula using the operators $\vee$ (logical *or*), $\wedge$ (logical *and*), and parentheses. Furthermore we can also use a variable x in its negated form, $\bar{x}$. This means that if $\bar{x}$'s value is 1, then $\bar{x}$'s value is 0, and *vice versa*.

The actual way in which we use a variable is called a *literal*. It can be shown that any Boolean formula can be expressed by a collection of *clauses*, where each clause is a collection of literals connected only by the operator $\vee$, each clause is delimited by parentheses, and clauses are connected to each other by the operator $\wedge$ (we say that such formulas are in *conjunctive normal form*). An example of such a formula is $F = (x_1 \vee x_2) \wedge (\bar{x}_2 \vee x_3) \wedge (\bar{x}_3)$. We will consider formulas constructed using n variables, m clauses, and a fixed number of literal per clause. The SAT problem consists of determining whether there is an assignment of values to the variables that satisfies a given formula, in other words, an assignment that makes the formula evaluate to 1. In the above example, $x_1 = 1$, $x_2 = 0$, and $x_3 = 0$ satisfies F, whereas any other assignment does not. In general, a formula may have more than one satisfying assignment or it may have none. An algorithm to test whether a formula is satisfiable is this: Try out all possible values for the variables. If there are n variables, there are 2^n such assignments. A more systematic way of trying out all possible assignments, and the one actually used to solve SAT with DNA, is as follows:

1. Generate all possible 2^n assignments and place them in set S_0.
2. From all assignments in set S_0 keep only those that satisfy clause 1, and place them in set S_1.
3. From all assignments in set S_1 keep only those that satisfy clause 2, and place them in set S_2.

. . .

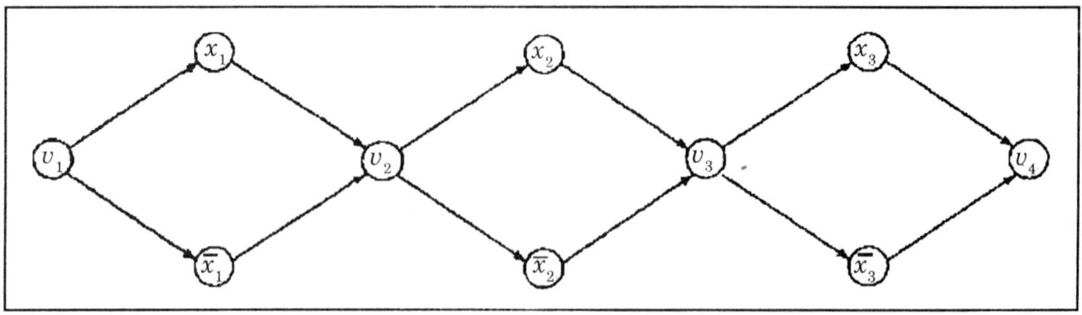

Fig. 2.2: An example of a graph whose (v_1 - v_n + 1)-paths encode Boolean variable assignments.

m. From all assignments in set S_{m-1} keep only those that satisfy clause m, and place them in set S_m.

After step m, perform one more operation: If set S_m is non-empty return yes; otherwise return no.

To implement this using DNA, we must have a way of encoding boolean variable assignments. We do this by using a graph. We will have one graph for each distinct number of variables; thus G_n will be the graph to be used in a problem with n variables. In Fig. 2.2 we show G_n for n = 3. Notice that it is a directed graph and that it has three kinds of vertices: the v_i vertices, $1 \leq i \leq n + 1$, the x_j vertices, and the $\bar{x}_j$ vertices, $1 \leq j \leq n$. Each possible assignment will correspond to one particular $(v_1 - v_{n+1})$-path in this graph in the following way. Starting from vertex v_1, we can go either to x_1 or to $\bar{x}_1$. Choosing x_1 means assigning 1 to variable x_1; choosing x_1 means assigning 0 to x_1. Then the path must go through v_2, and again we have a choice, which will correspond to the value assigned to variable x_2, and so on, thus a $(v_1 - v_{n+1})$-path in the graph encodes an n-bit number, which corresponds to one particular assignment of boolean values to n variables. Having determined this encoding, the next step is the construction of all paths in graph G_n, which we do in the same manner as was done for the HDDP (see previous section).

The starting point in our DNA SAT algorithm is thus a test tube T_0 containing an enormous number of sequences representing paths in G_n, and with high probability sequences representing *all* $(v_1 - v_{n+1})$-paths will be present in T_0. For the implementation of the next steps we need two kinds of operations on test tubes: *Detect* and *Extract*. *Detect* is an operation already employed in the HDDP problem (step 5 in that algorithm); it simply reports whether a test tube has any DNA sequence at all. The *Extract* operation is similar to the operations employed in the HDDP algorithm, in the sense that we use it to extract sequences having a certain property. Here we are interested in only one particular property: We apply *Extract* to test tube T in order to extract the DNA sequences in T that contain a given precise DNA substring of fixed length.

It is not required that all such sequences be extracted from T; thus it is not a problem if among the remaining sequences in T there are still some that contain the given substring. But it is a problem if we extract sequences that *do not* have the desired property, or if we fail to extract any relevant sequences even if T contains some. Even though these failures can happen in practice, in this section we assume *Extract* works as described. We will now describe how we deal with the first clause in our formula F. Let us assume that this clause is $(x_1 \vee \bar{x}_2)$. Consider the first literal in this clause, x_1. Because it is not negated, an assignment that makes $x_1 = 1$ is a candidate for the desired solution. Thus the corresponding n-bit number has a 1 in its first position (first from left to right). Similarly, an assignment that makes $x_2 = 0$ is another candidate.

By combining these two, we can say that only the n-bit numbers with a 1 in their first position, or with a 0 in the second position are the numbers that can possibly satisfy F; all others are out. This means that we should extract from T_0 all sequences encoding n-bit numbers that satisfy this requirement. Those sequences will then be the input for the treatment of the next clause. From the above description it is clear that we need the *Extract* operation in the following format: *Extract* (T, i, b), where T is a test tube, i is the index of a variable, and b is a Boolean value. We can now sketch more formally the steps that take care of the first clause, assuming that it has 2 literals:

```
// T0, T1, A1, A'1, A2 are test tubes
// v1 is the first literal
if v1 = x1 then
    A1 ← Extract (T0, 1, 1)
else // v1 = x1
    A1 ← Extract (T0, 1, 0)
A'1 ← T0 // the remainder
// v2 is the second literal
if v2 = x2 then
    A2 ← Extract (A'1, 1, 1)
else A2 ← Extract (A'1, 1, 0)
T1 ← A1 + A2 // We mix the tubes
```

Notice that when dealing with the second literal we extracted sequences from the remainder of the first extraction. This is essential to make the algorithm work. If the clauses were $(x_1 \vee \bar{x}_2)$ it should be clear that in tube T_1 we end up with all sequences encoding n-bit numbers that have a 1 in the first position (which were in tube A_1) and all sequences encoding n-bit numbers that have a 0 in the second position (which were in tube A_2). Using tube T_1 we are ready to proceed to the next clause, where the steps are similar to the ones described. This algorithm clearly executes $O(m)$ *Extract operations* and only one *Detect operation*.

```
Algorithm DNA Satisfiability
   input: test tube T_0 containing sequences that encode all n-bit numbers,
          and boolean formula F in conjunctive normal form
          (n variables and m clauses)
   output: YES if F is satisfiable, NO if not.
   // We use a fixed number of auxiliary test tubes A_l
          for i ← 1 to m do
   A'_0 ← t_{i-1}
   for j ← 1 to l_i do // l_i is the number of literals in clause i
          If v_j = x_j then
   A_j – Extract ( A'_{j-1}, j, 1)
Else A_j ← Extract ( A'_{j-1}, j, 0)
A'_j – A'_{j-1} // remainder
T_i ← A_1 + A_2 + . . . + A_i
If Detect (T_m) then return YES else return NO
```

Fig. 2.3: The DNA algorithm to solve the SAT problem.

Chapter 3

Genomic Experiment

Experimental biologists and biomedical scientists who are considering the use of microarrays in their research programmes have often asked us questions of the following form: :How do I best design an experiment with the following mutant mouse?", or, "We have a budget for 20 microarrays to analyze this system. How can we complete the experiment within budget?" Unlike the typical biomedical study design, many of these investigations that the experimentalists are contemplating are driven not by the testing of a particular hypothesis, but by the hope of generating hypotheses that will then lead to a directed and relevant experiment that will subsequently yield an interesting biological or biomedical insight. We find it even more challenging when we are asked secondary questions related to the subsequent microarray data analyses such as, "What is the right machine-learning technique or clustering technique to apply to a set of data to best obtain an answer about [some biological mechanism]?" The questioner often goes on to ask whether one particular software package works better than another, is more reliable, or whether it is worth its cost. We view these questions as well-intentioned but fundamentally flawed and ill-posed.

Although a particular machine-learning algorithm may be slightly superior to another in one set of circumstances, the key to a successful functional genomic investigation is to start with an experimental design that maximizes the possibility of observing gene expression patterns that are relevant to and informative of the biological aspect or question under investigation. If the experimental design is such, then *most* of the common machine-learning algorithms will reveal those patterns. So how does one proceed to an effective experimental design for the expression of thousands of genes in order to, hopefully, answer a biological question? We have found that it useful to frame our answer within the concepts of experiment design space and *expression space*. Let us start with the experiment design space.

EXPERIMENT DESIGN SPACE

Experiment design space is the space of potential combinations of stimuli or conditions to which one can subject the regulatory apparatus of the genome. These stimuli include exposure to pharmaceutical agents such as mitogens, environmental effects such as temperature differences, withdrawal of nutritional sources, over-expression of a particular gene such as a transcriptional factor in a transgenic animal, or conversely, an engineered mis-expression or "*knockout*" model. Other common experimental conditions include the passage of time, such as from a newborn to an old mouse; the observation of the environment (organisms form one particular condition vs. another); the observation of organismal processes (organisms with a particular morphology, or affected with a particular pathophysiologic condition vs. another); physical manipulation of the cell membrane; exposure to infectious agents; and changes in osmotic conditions, all of which can potentially affect the genomic regulatory mechanisms which maintain cellular existence and the regulation of the cellular response to the environment.

Experimental conditions can be represented as combinations of values on several scales:

(i) *Binary sales*, e.g., the presence or absence of a diagnosis.

(ii) *Continuous scales*, e.g., such as the gradients of exposure to a chemical, these can have value ranges, say, 10.0, 1.0, 0.1, 0.01, representing the chemical concentration or 0, 5, 10, 15, 20, representing time in minutes form the initial state.

(iii) *Unordered discrete scales*, e.g., the type of mutant mouse model created or which of the multiple mammalian muscle tissue types were sampled.

When different experimental conditions from these different scales are combined, then the experiment can be described as being located in a particular locus of the multidimensional experiment design space. Shown in Fig. 3.1 below is the simple example of an experimental space over three sets of conditions, in this case, the first dimension represents whether the mouse model being studies is an *irs*1 knockout, an *irs*2 knockout, or a wild-type mouse. The insulin receptor signaling proteins *irs*1 and *irs*2 are though to be part of the signaling cascade through which multiple cellular events are triggered upon binding of insulin to the cellular receptor.

The "knockout" is an engineered mouse in which a particular gene is no longer sufficiently expresses. The second dimension is the amount of insulin to which the mouse circulatory system is exposed. The third dimension represents the three types of mouse tissues being studied in this particular set of experiments, namely, the liver, brain, and fatty tissue. Each of these tissues has distinct metabolic responses to insulin exposure. Within this three-dimensional experiment design space, each set of possible experiments can be uniquely located. This notion obviously generalizes to the much larger number of possible dimensions in any experimental system. Note that the first and third dimensions have unordered discrete scales, whereas the second dimension is continuously scaled.

A functional genomic study might involve a trajectory across the experiment design space where the mouse models are studied across multiple insulin levels and multiple tissues. As the number of experimental dimensions increases, the number of experiments required to cover the experimental design space grows exponentially. However, the concept of an experiment design space by itself does not suffice to determine which kind or number of experiments is adequate to assess the genomic physiology of a tissue or system.

Expression Space

Now that we have defined the experiment design space, we define the *expression space* to be the space of potential expression values of all genes in a given genome. In the context of microarray measurements, this is a large *multi-dimensional space*—with each dimension corresponding to a single gene—and it can be immediately partitioned into several overlapping subspaces. For example, coordinated gene expression measurements made from normal muscle tissue under normal conditions can be represented by a subspace of expression space which will most likely have a non-empty inter-section with the subspace of coordinated gene expression measurements made from diseased muscle tissue. We often repeat mantra-like to ourselves the phrase "A cell is a cell is a cell", by which we mean that in order to survive as living and organized entities, most cells share a large repertory of biological processes, whether they be components of muscle tissue, liver tissue, or white blood cells. That is, the expression subspaces of most cell types have significant overlap. Expression space contains

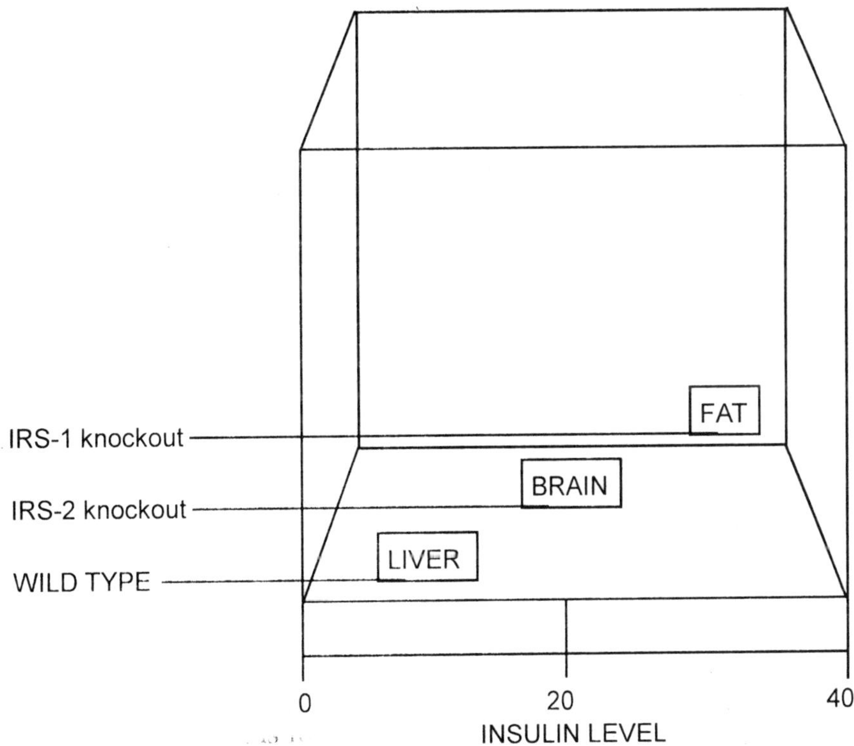

Fig. 3.1: Three experiments within the multi-dimensional representation of experiment design space. Experiment design space defined all the possible stimuli or conditions to which a particular biological system could be subjected. Shown here is an experiment design space that is concerned with insulin signaling in different tissues, in different mouse "knockout" models, with different levels of insulin.

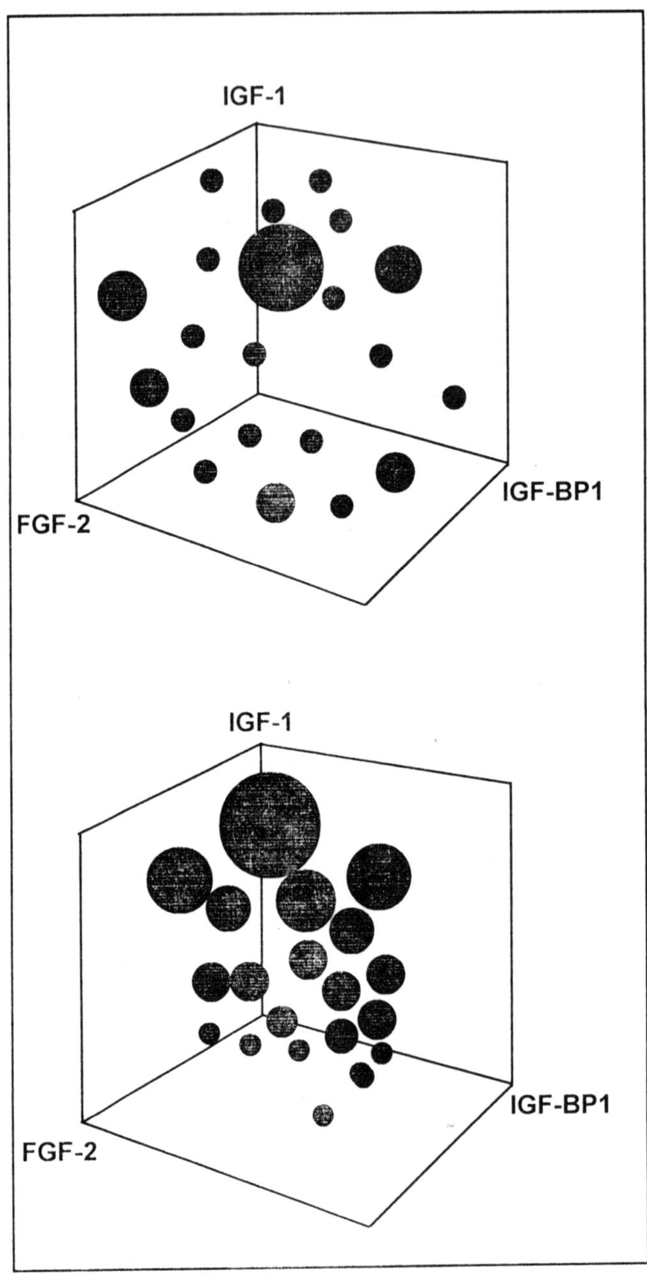

Fig. 3.2: **Expression space in which gene expression is loosely versus tightly coupled. If genes are tightly coupled in the expression space, they will tend to occupy a small subspace of expression space over any set of experiments. If, however, these genes are loosely coupled or even causally unrelated, then their expression levels will have little relation to one another and therefore tend to be scattered over a greater volume of the expression space.**

clues about the interactions between two or more clusters of genes. As an example, we start with the simple case of three arbitrary genes: insulin—like growth factor 1 (*IGF-1*), insulin—like growth factor binding factor 1 (*IGF-BP1*), and fibroblast growth factor 2 (*FGF-2*).

The regulatory relationship between these three genes is, in some tissues, that one gene partially determines the expression level of the other two. This will cause the three genes to move in a coordinated fashion across the range of their respective possible values, that is, if this is a particularly tight regulation system, then the expressions of this triplet will only be found in a small subspace of all possible loci of the expression space. If this were a loosely regulated or unregulated system, it is possible that the gene expression values might be found in all parts of the expression space. The totality of the potential loci that these genes can occupy constitutes the potential set of all coordinated gene expression levels, or expression space. The subset of the expression space wherein a cellular system has its expression under all stimuli is called the *transcriptome*. More precisely, the transcriptome encompasses the expression space of all the putative 30,000 genes in the human genome (or 100,000 gene products if current alternative splicing estimates are correct). Each point in the human transcriptome corresponds to the possible values that each gene's expression might attain under any physiological or pharmacological stimulus.

The meta-goal of the experimentalist/investigator then is to define the experiment design space that provides the best and most complete picture of the entire transcriptome or a subset of the expression space of interest. This is the point in experimental design where a great deal of domain-dependent knowledge, i.e., specific or *a priori* biological knowledge, is of significant utility, the investigator must ask herself or himself: "What is the minimal set of experimental conditions, i.e., the smallest subspace of the *experiment design space*, that will cause the largest set of relevant subspaces of the *expression space* to be sampled?" In order to answer this question, the investigator must first understand how each experimental condition could potentially affect the expression of potentially interesting genes and be able to quantify its effect. Second, the investigator must consider all possible subspaces or interactions between genes that are of interest.

For example, let us say that we are only interested in the interactions between *IGF-1* and *IGF-BP1* in the example earlier in the section and we wish to determine how the relationship between these two genes changes under both physiological and pharmacological conditions, then we would obtain those sets of experiments in the experiment design space which would most likely cause one or both of these genes to be expressed over the broadest range of their possible values so that all the evidence of the interactions would be captured, i.e., the subspace of experiments that covers the largest plane in the *IGF-1* and *IGF-BP1* expression subspace. A researcher might be interested in the signaling mechanisms of IGF-1 once it has bound to *IBF-BP1* and how it affects the liver's expression of other genes. These genes may code for many proteins involved in post-receptor signaling, including possibly *IGF-BP1* itself. With a finite budget and a finite source of tissue or animals, the investigator will have to choose between alternative strategies. A "*mutant screening*" strategy might be to assay expression levels in the liver over a set of knockout mice in which each element of the set constitutes a knockout of a different putative element of the *IGF-1* signalling pathway, whereas a "physiological stimulus" strategy, would be to take a single element of the set (one of the knockout strains or a wild-type animal) and subject it to a sequence of different concentrations of extracellular *IGF-1.*

Determining which of these alternatives provides the best exploration of the expression space will require an understanding of whether the range of mutants or physiological stimuli engages the portion of the genetic machinery that is relevant, and how close to the normal homeostatic behaviour of interest is the experiment intended to explore. The joint behaviour of two genes in which there exists a strong component of randomness in the relationship between the two genes—as evident form the low Pearson's correlation coefficient that is calculated between these two genes within the current range of expression levels. This is also summarized by the Pearson's correlation coefficient for the linear regression calculated between these two genes, which is calculated to be very low. However, if an additional stimulus is applied to the regulatory system such that gene 1 has a wider range of expression, then despite the randomness in the relationship between the two genes observed in the previous range of expressions, we begin to see what appears to be a linear relationship with a much larger Pearson's correlation coefficient.

If the expression level of gene 1 is perturbed to an even wider range, it becomes apparent that the relationship is not linear but, in fact, curvilinear, possible even a quadratic relationship. If the experiment space was incorrectly selected, then these and many other pairs of genes like it would be assumed to be poorly or randomly related when, in fact, with the right stimulus, a tight curvilinear relationship would be determined, and *vice versa.*

Calculating the Expression Space

In exploring the expression space, the investigator is constrained to the space of possible experiments. Informally, the goal of experiments should be to *maximally exercise the genome* bringing out the couplings or correlations between genes that are of greatest relevance to the biological process being studies. We have found the following "*watch metaphor*" to be of some use in explaining the interaction between expression space and experiment design space. We leave it to the reader to judge whether it is helpful, or simply a confusing distraction.

The Watch Metaphor

Suppose, that we were interested in discovering how an old analog watch of the sort illustrated in Fig. 3.6 worked. Simply watching the hands of the watch going around might allow us to generate some hypotheses about the mechanisms that generate this motion or behaviour. Empirically, we could derive the relationship between the movement of the minute hand and the hour hand, but the total sum of mechanisms that could explain the behaviour would be large. A more invasive investigation would be to take a hammer to the watch and examine the pieces that fall out of the watch's case.

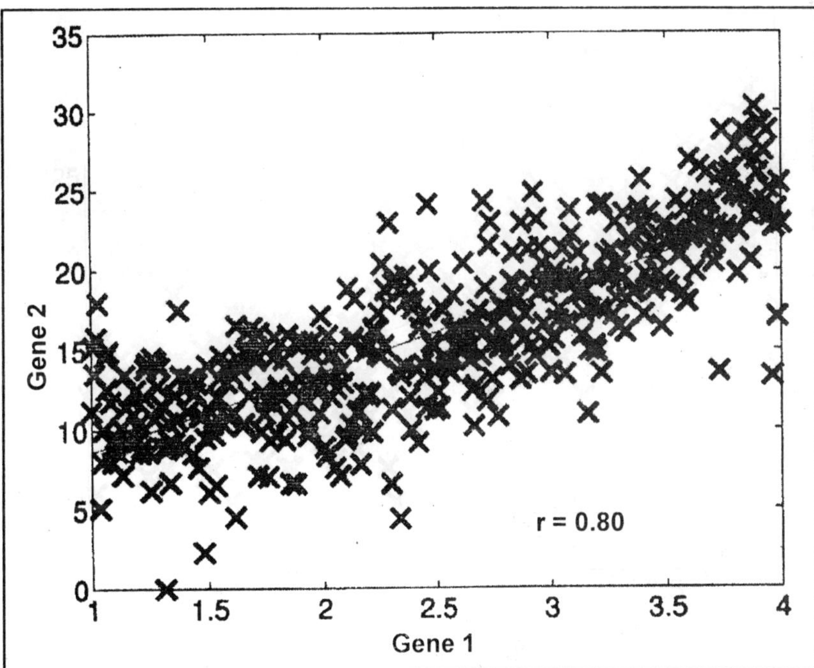

Fig. 3.3: Apparently random relationship in expression space between two genes.

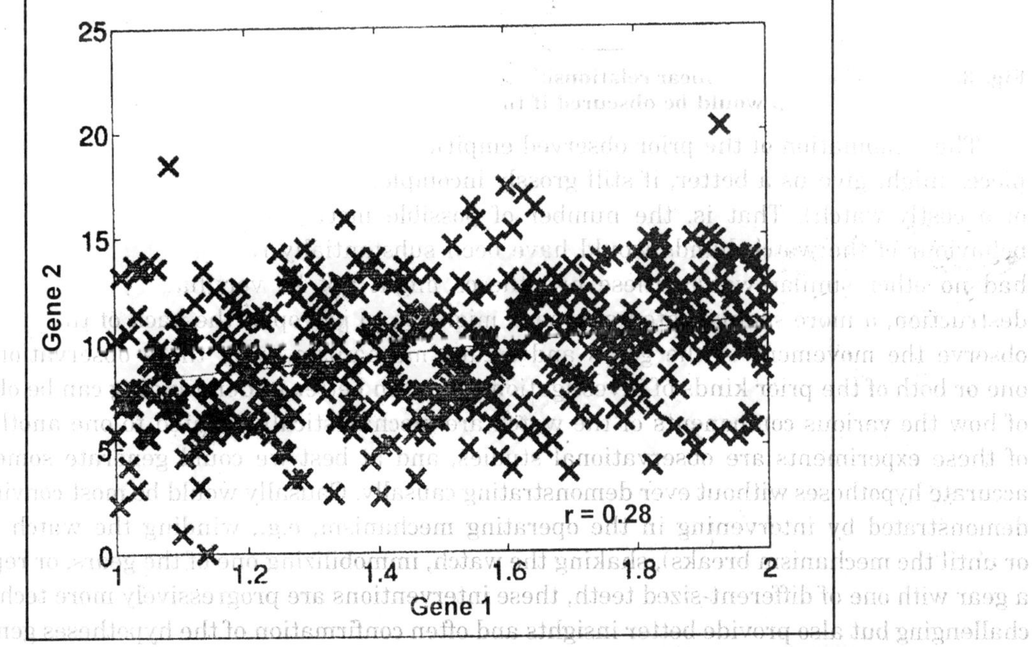

Fig. 3.4: Apparently linear relationship in expression space between two genes.

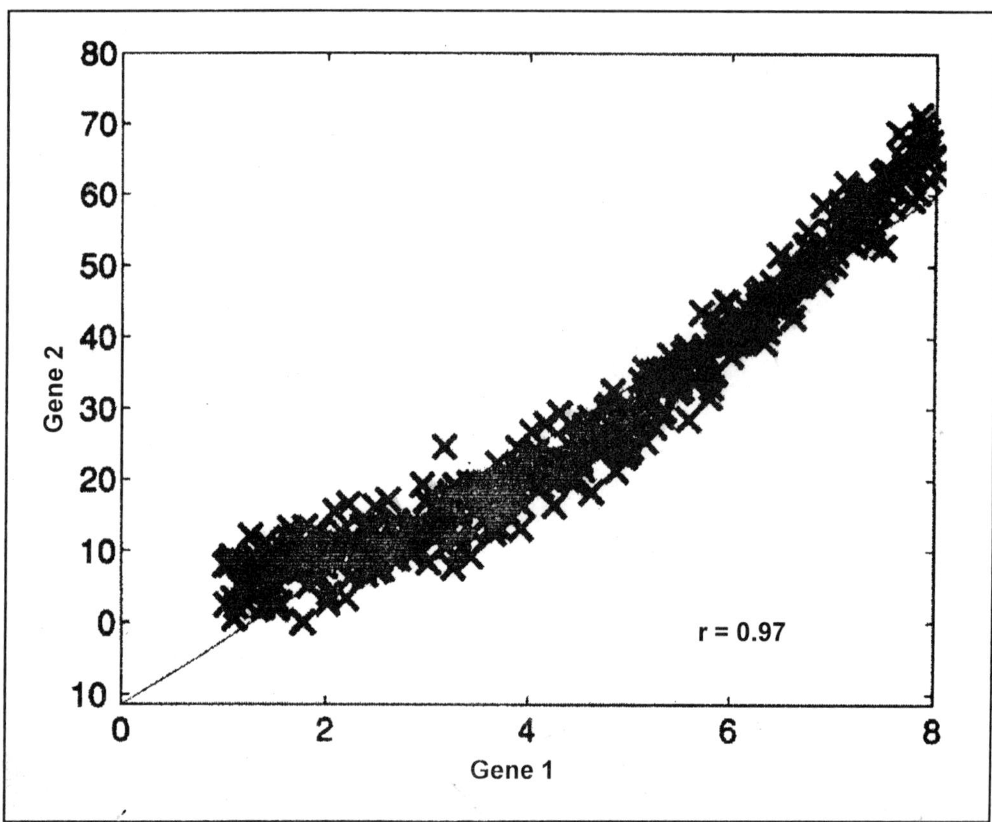

Fig. 3.5: Apparently curvilinear relationship in expression space between two genes. This curvilinear relationship would be obscured if the expression space had been insufficiently exercised.

The combination of the prior observed empirical behaviour and the examination of these pieces might give us a better, if still grossly incomplete hypotheses (not to mention the loss of a costly watch). That is, the number of possible mechanisms explaining the observed behaviour of the watch hands would have been substantially reduced. Furthermore, if one had no other similar watch, these hypotheses might not be vertifiable on account of its destruction, a more sophisticated approach might be to pry open the back of the watch and observe the movement of the gears and escape mechanism, with these observations, and one or both of the prior kinds of investigations, an even better understanding can be obtained of how the various components of the watch are mechanistically related to one another. All of these experiments are observational studies, and at best we could generate some fairly accurate hypotheses without ever demonstrating causally. Causally would be most convincingly demonstrated by intervening in the operating mechanism, e.g., winding the watch (partly or until the mechanism breaks), shaking the watch, immobilizing one of the gears, or replacing a gear with one of different-sized teeth, these interventions are progressively more technically challenging but also provide better insights and often confirmation of the hypotheses generated during the earlier observational studies.

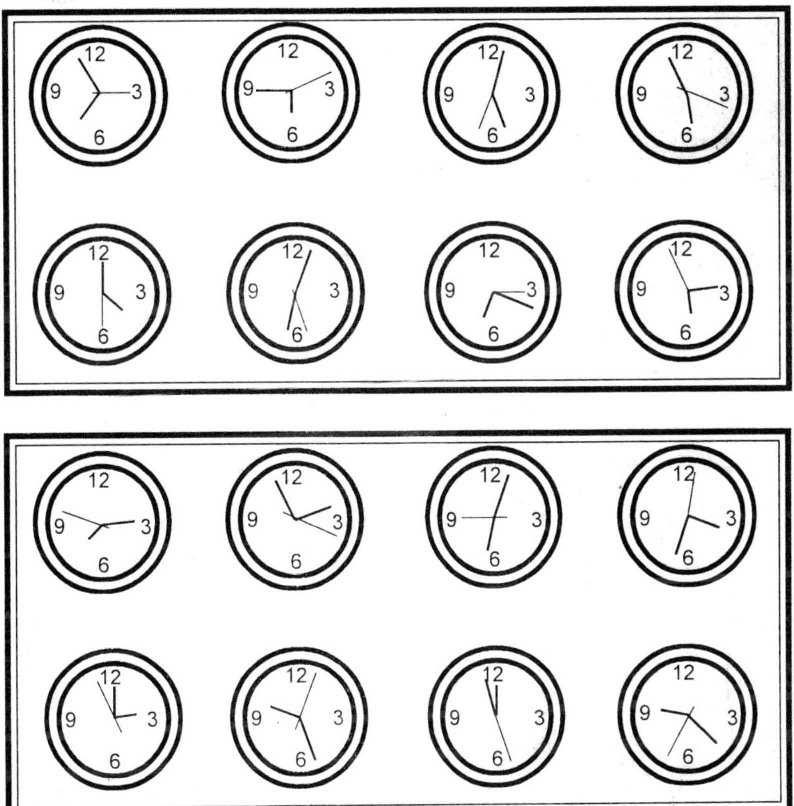

Fig. 3.6: **Noninvasive monitoring of mechanistic behaviour, by observing the concerted action of the watch hands, several competing weak hypotheses about the underlying mechanism can be generated. The more observations made at different times, the smaller the number of possible hypotheses generated.**

The analogy with expression studies, with the indulgence of the reader, can be made these describe different kinds of experiments—distinct points in the potential experiment space. Some of these experiments allow the components of the watch to working their usual fashion, i.e., within their usual "expression space," whereas others bring them into an operating mode or expression space in which they do not usually function, the question the watch investigator would have to ask himself or herself is what part of the expression space of the watch will best inform him or her of the normal function of the watch. It may be that smashing the watch is the only part of the experiment space that is cost effective or refeasible for the investigator, the smashed or overwound state of the watch may not accurately reflect the normal intact and possibly more interesting state of the watch's expression space. To take these analogies one step further, there are multiple arrangements of gears that could account for the observed interactions between these gears and their effect on the hour hand

and minute hand.

Similarly, although perturbations of the genome can provide insight into the interactions among the major genes involved in a given pathological pathway, they do not reveal the entire pathway. Rather they expose a significantly large family of possible mechanisms that could account for the observed behaviour. We address this issue in greater depth in the section on pathway re-engineering. Nevertheless, as suggested earlier, there are some misconceptions about what can be done on the basis of microarray expression studies implicit in some of the current genomic and bioinformatics literature.

Table 3.1: Exploring the experiment space with a watch

Experiment Space in Watch	Experiment Space Biological Analogy
Observing the movement of the hour, minute hands	Hypothesizing cellular function based on gross cellular behaviour
Smashing the watch, examining the pieces	Sequencing some of the genes in the cell or measuring a few proteins or RNA transcripts
Seeing all the gears working from the back of the opened watch	Parallel measurement of thousands of genes at a time Time series measurements
Partially winding up the watch	Physiological stimulus
Winding the watch until it is near breaking or until it breaks, thereby bringing the internal gears of the watch into behaviour they would not typically exhibit	Pharmacological stimulus that brings the gene expression patterns into an operating region outside the usual physiological range
Removing a gear	Creating an organism with a gene "knocked out"
Replacing a gear with one of different weight	Creating a transgenic organism

Misconceptions in "Reverse Engineering"

To date, there have been several efforts to try to reconstruct pathways from gene expression measurements. Given the success of recapitulating the first few steps of the glycolytic pathway from substrate measurements, it would appear plausible to do this, however, without careful consideration of the pathways that one is trying to reconstruct, there is significant risk of a methodological and metaphorical flaw in this analysis. It is important to realize that pathways could exist on several molecular and physiological levels.

The genes that are regulated in one pathway may play a role in another pathway. As an example, let us consider the genes coding for lactic acid dehydrogenase; The gene *LDHA* is expressed mostly in muscle and is located on chromosome 11p15.4, while *LDHB* is located on chromosome 12p 12.2 and is expressed mostly in heart. *LDHA* is known to have binding sites for NRE, NREBP/c-FOS, AP1, AP2, and other transcription factors in its promoter region. *LDHB* is thought to have a binding site for SP1. There is also likely to be a post-transcriptional level regulation of these two genes. Because of the differences in promoter regions, it is safe to assume that both LDHA and LDHB participate in their own gene

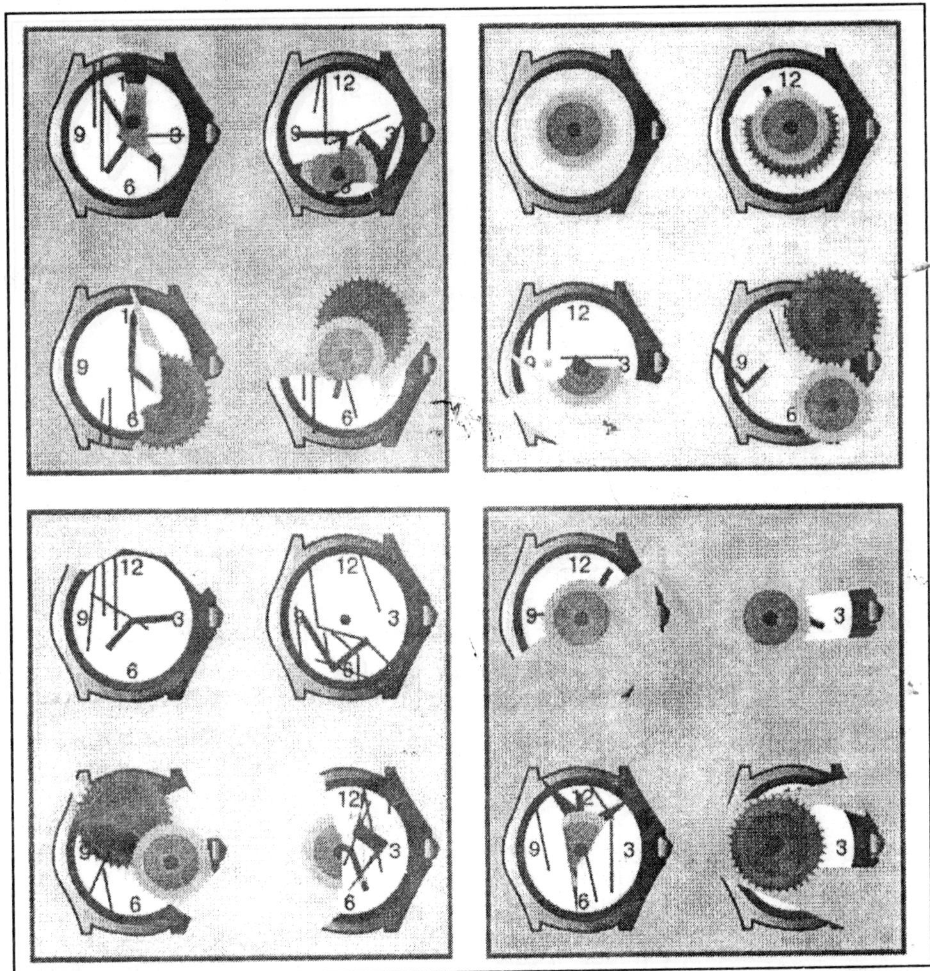

Fig. 3.7: Decomposition of the watch. An invasive exploration of mechanism well reveal most of the components of that mechanism. In the process of the invasive investigation some of the components will be damaged. Also, the relationship of these components inside a working watch must be inferred and is not directly on served because of the invasive nature of this investigation.

expression regulatory networks. Both LDHA and LDHB code for protein subunits, the final protein product is made up of four subunits.

Various combinations of the two types of subunits are assembled into five different lactic acid dehydrogenases, from LDH-1 (containing four LDHB subunits), to LDH-5 (containing four LDHA subunits). These five enzymes are found in a binomial distribution in mammals. Additionally, the final protein itself is an enzyme that assists in the conversion of lactic acid to pyruvate, called lactic acid dehydrogenase, and defined by the Nomenclature Committee of the International Union of Biochemistry and Molecular Biology as Enzyme commission

(EC) 1.1.1.27. Lactic acid and pyruvate happen to be substrates in the glycolytic pathway of anaerobic energy production. Thus, in this one simple example, one finds at least three pathways.

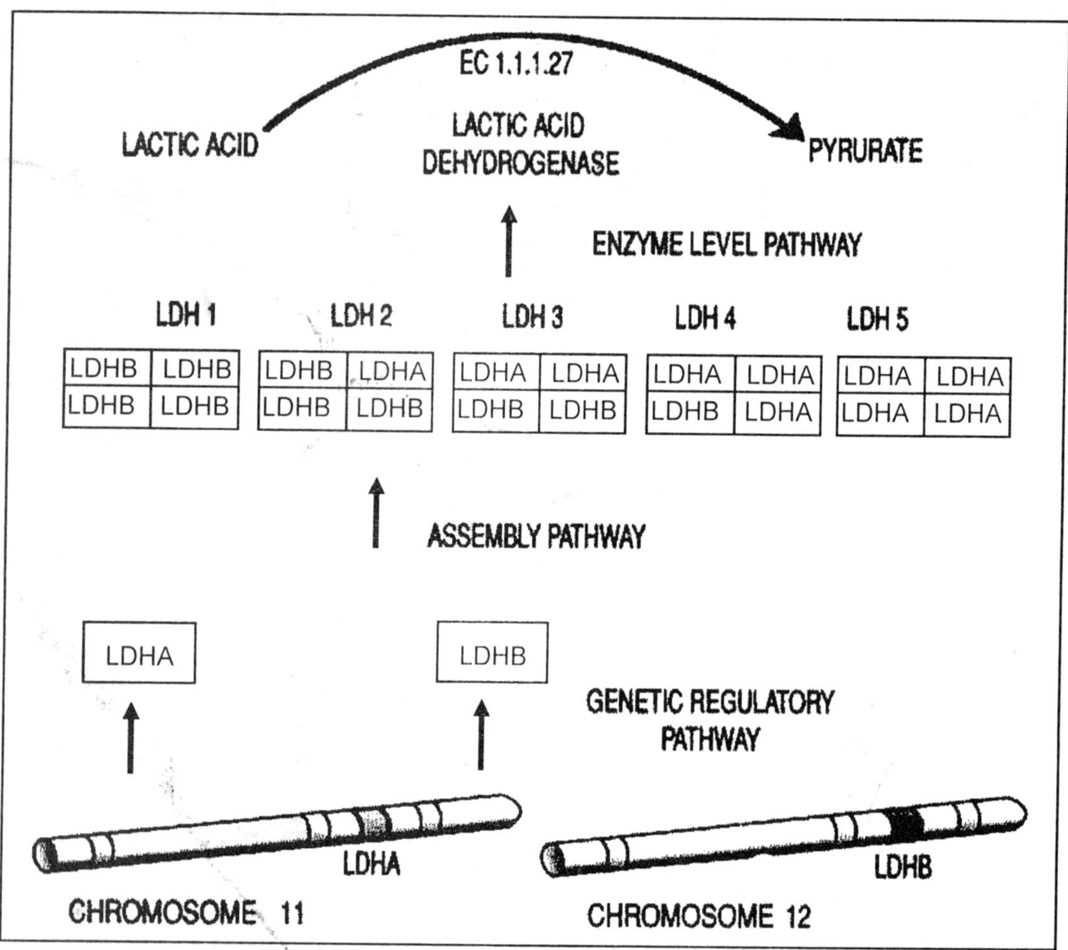

Fig. 3.8: **An example of why discovery of pathways solely using gene expression measurements. Is difficult. At least three "*pathways*" are involved in the conversion of lactic acid to pyruvate. The highest enzyme level pathway involves the action of an enzyme called lactic acid dehydrogenase, designated EC 1.1.1.27. However, when one traverses to a lower-level pathway, one learns that the role of this enzyme is performed by five separate protein products, which appear in a binomial distribution. When one traverses lower, through the assembly pathway, one learns this distribution is present because of the various possible combinations of *LDHA* and LDHB, two individual proteins subunits, of which four must be put together to make an active enzyme. Then, when one traverses even lower, through the genetic regulatory pathway, one learns that each of these subunits is regulated differently, and appears on different chromosomes. Not shown are the pathways involved in transcription, translation, movement of proteins to the appropriate locations, and movement of substrates into and out of the cells and organelles, and many other pathways.**

The operator at the *enzyme level pathway* participates as one enzyme in a major biochemical substrate pathway, that of glycolysis, that operator, however, is assembled stochastically in the *assembly pathway* form subunit components. Finally, those subunits are produced within their own *genetic regulatory pathways* involving a multitude of transcription factors. Considering the three layers of pathways, it should now become apparent that it may not be possible to "reconstruct" the glycolytic pathways simply from measuring multiple gene expressions. However, on a more optimistic note, given enough gene expression data, one should be able to reconstruct the first layer of these pathways, i.e., that of gene expression regulation and involving transcription factors and promoter regions.

"Maximally Exercising the Genome"

What constitutes a maximum perturbation will in part be defined by the goals of the investigator. If she or he wishes to understand the interactions between the genes under normal physiological conditions, then it may be that only relatively minor perturbations followed across time will be informative. Even though extreme perturbations are easier to measure, they might not accurately represent interactions that occur over the normal dynamic range of activities of the RNA species, structural proteins, and enzymes programmatically generated by the genome. There are, however, several circumstances when it may be advantageous to perturb the relationships between the various constituencies of the transcriptome much more violently and dramatically than is found under normal physiological or homeostatic conditions. An example of such a motivation might be to measure the effect of pharmacological doses, rather than physiological does, of a pharmaceutical.

The intended effect may lie outside the normal operating range of the genetic machinery, as is often the case in pharmacology. Other times, the goal may be to examine biological systems which naturally fall far outside the usual operating parameters of the genomic machinery. This includes a variety of disease states such as a malignancy or diabetes, where there are single or multiple massive perturbations of the usual homeostatic system, and therefore a perturbation of the model system that corresponds to the pathological condition would be appropriate, it is worthwhile considering some extreme examples to illustrate the importance of this decision. For example, if one is trying to understand the relationship between a set of genes thought to be involved in the effect of insulin signalling, it may not be particularly helpful to start with a sample population of human adipose tissue (fat) which is known to be insulin sensitive, but obtained from individuals who are all in the fasting state when insulin levels are low or absent.

There may be *insignificant* insulin exposure or stimulating effect of the insulin-responsive cellular machinery. Consequently, the experiment design subspace sampled will be too small to adequately explore the expression space of interest. Furthermore, if the sample population is ethnically homogeneous or, in the case of a mouse experiment, shares the same genetic background, then these samples will only inform us of the reproducibility and noise of the biology and the measurement system. In contrast, if the same population were exposed to varying degrees of insulin, then the behaviour of the genetic program in response to insulin might be better understood.

Alternatively, taking a population known to have varying degrees of insulin resistance would also provide a more diverse range or heterogeneity of interactions in insulin signalling, which would enable the investigator to explore the space of possible relationships between the genes. This would then allow the application of a machine-learning algorithm to elucidate these dependencies. If the degree of insulin resistance is clinically measured (e.g., as obtained in a carefully performed hyperinsulinemic euglycemic clamp) in close temporal proximity to the expression measurements, then the investigator will probably have the most accurate control of the experiment design space. The goal then is to stress or stimulate the cellular regulatory system, which in this instance is glucose homeostasis and insulin signaling, so that the genes "*show us*" how they interoperate under normal, and pathological or pharmacological circumstances. There are several other obvious variations of these kinds of stimulations such as:

1. Examine the expression profile in adipose tissue at different points in time after the glucose exposure to understand the timing of the genetic program. We should not expect all genes to act at the same time or rate, and therefore sampling the changes in the transcriptome across multiple time points at the appropriate frequency will be much more revealing than just one or two points in time.
2. Assess gene expression at different fixed concentrations of insulin.
3. Assess gene expression in ethnically diverse populations. The difference in genetic backgrounds of these populations will give a sense of how much variation in the signaling may be due to genes outside the set that is being studied in a particular expression experiment, this implies that we would need to hypothesize that changes in the genetic background affect the functioning and thus the functional genomics of the system. This presupposes that
   - (i) insulin production and insulin effect is the combined result of the activity of several genes;
   - (ii) polymorphisms in these genes are distributed heterogeneously across human populations;
   - (iii) such polymorphisms result in significant changes in gene expression or function, or both.
4. Explore the pharmacological effects of insulin. After a certain dosage level, increasing levels of insulin may no longer result in increased effects of glucose transport. However, increasing insulin may affect other physiological systems within the same cell, as, e.g., transporting ions across the cell membrane. Taking tissue samples at pharmacological concentrations of insulin and comparing their expression profile to those obtained with insulin concentrations in the physiological can range would allow

determination of genetic mechanisms that respond to the pharmacological dosing.

In summary of our discussion, we note that an appropriate set of (microarray) experiments is that in which the cellular response to the stimuli involved in the experiments will exercise the interaction between genes over the range of interest and relevance to the biologist-investigator. This critical aspect of the experimental design will trump any subsequent choice of bioinformatics analytic technique. Conversely, no bioinformatics analysis will be sufficient to extract functional genomic knowledge if the expression space is not adequately explored.

Microarray Data

After an investigator obtains his or her first microarray data sets, often the first question he or she will ask is: "Which of the observed up or down regulation of genes represents biologically significant changes in expression?" Here we provide a context and motivation for an answer that biologists often find discomfiting: *that very few genes are in fact significantly changed in expression in a way that is distinguishable from biological and measurement variation and noise.* We should emphasize at this point there is an important distinction between mathematical or statistical significance, and biological significant. The former has to do with analytically quantifying the difference between two or more different sample sets of numbers, i.e., microarray expression data, for which there exists standard statistical tests like the Student's *t* or Ç. These tests typically ignore the rich biological meaning and structure in these numerical representations of expression. The latter is a more complex matter which we shall discuss here. It is likely to be the case that there are thousands of genes with important biological outcomes effected by small relative changes in the expression levels.

In the present state of the microarray technology and with the typically small number of replicate experiments for any given condition, these small changes cannot be reliably and reproducibly distinguished form noise. This may mean that even though we measures tens of thousands of genes in a microarray experiment, we would only obtain hundreds of genes that we are reasonably convinced are involved in a particular biological system. On the other hand, not all statistically significant gene expression changes lead to a significant change in the biological or physiological state of a system. This answer is most disconcerting to those biologist-researchers who have had substantial experience in investigating the expression of only a few genes at a time. Their investigations typically measurements (all of which are carefully checked) have been made of the expression levels.

Consequently, the researchers have well-grounded ideas of what constitutes a biologically relevant or significant change in expression for these few genes. They do not have the benefit of this sort of knowledge for each of the thousands of genes that are represented on a microarray but nonetheless are uncomfortable with discarding from further analysis thousands of genes that appear to have large numerical changes in expression. In contrast, due to the large number of genes that can be measured in a single experiment, computer scientists and computational biologists are comfortable—perhaps too comfortable—with generating exhaustive lists of genes that are possibly involved or interacting in a particular biological process. This kind of exhaustive list generation has been quite common in publications of microarray experiments from 1997 until today. Unfortunately, such lists are not that helpful to the biologists who wish to determine which elements in these lists are worth pursuing, i.e., are biologically significant for their investigations. This is often the reason for questions of what constitutes a significant fold change, where significance can mean anything form "present in the tissue," "associated with the system being studied," "causative of the changes in the system being studied," or even simply meaning "worthy of further study." Thus, an analyst for a typical functional genomics study will simply draw two boundaries: one covering increases in expression and the other covering decreases in expression. Such an analyst will then declare that any gene found beyond these lines (the red points) is significant or relevant to the biological process being studied.

Unfortunately, there is no single threshold number or even function that we can provide that would generate a particular number of candidate genes that is worth pursuing for any arbitrary experiment and experimental design. Fortunately, there are well-founded decision analytic procedures that can be followed in order to come up with the appropriate number of candidate genes for any given investigation. The underlying motivation for picking a threshold is straightforward. The first question one has to consider is the cost of a *false positive*. A *false-positive* gene is a gene which, in reality, is not biologically significant to the process under investigation, but which the analytic technique deems to be significant, statistically or biologically. That ism what is the cost involved in the follow-up procedure to confirm the function of a gene in a particular biology process? Whether this confirmation process, or biological validation, involves quantitative PCR, *in situ* hybridization, the generation of a transgenic mouse, a transient-transfection assay, or transfecting a cell line, its cost in time and money will be substantial. This cost thus limits the number of genes that can be investigated.

The threshold that will then be picked will be determined in part by the disutility or cost of the biological validation step for false positives. The tens of thousands of genes present on any microarray now ensure that the number of false positives could potentially overwhelm the typical time and financial budgets of most (academic) research laboratories. Of course, there is the second and converse question of the cost of missing a *false negative* from the same system. A *false-negative* gene is a gene which, in reality, is biologically significant to the process under investigation, but which the analytic technique deems not to be significant, statistically or biologically.

There are likely to be several genes involved in any signaling pathway of interest. Not all genes in a pathway are equally amenable to biological validation. Furthermore, some of the chosen genes may be more suitable targets than others for diagnostic assays or therapeutic interventions. If the threshold picked for considering a gene to be significantly changed in expression excludes one of these targets, then the cost of that false negative will be quite high, e.g., a missed scientific opportunity or lost opportunity for commercial development. With this in mind, it becomes clearer how a threshold should be picked. First, the investigators should ask themselves how many false-positive and false-negative leads they can tolerate. Then they can conduct a series of replicate experiments and determine where they will have to draw the thresholds to attain the required sensitivity and specificity, so that design must be the decision analysis diagrammed in Fig. 3.9. This requires that the cost to one or one's enterprise of missing the one or more genes that are likely to be involved in the genetic regulatory pathway of interest be made explicit: Similarly, the cost of having to follow up on false leads must be estimated. These two costs define the principal disutilities that one is trying to minimize.

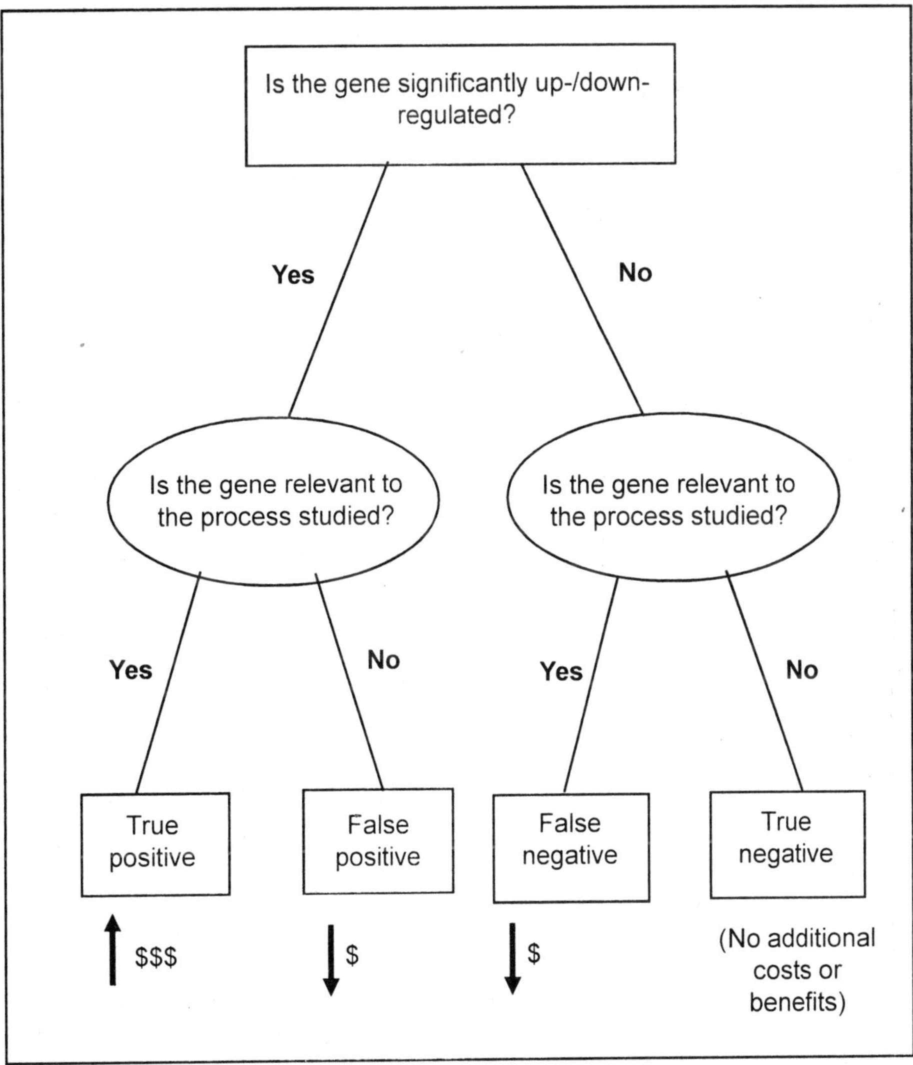

Fig. 3.9: A decision analytic procedure for picking a threshold for selecting genes from, a functional genomics experiment. Because of the large number of probes on current microarrays, it is all too easy to underestimate the cost and practical intractability caused by even a moderate false positive rate. Investigators are advised to perform this simple decision analysis in order to determine what false positive and false negative rate they can afford before they proceed with any experiments.

In order to complete this simple decision analysis, the probabilities corresponding to the sensitivities and specificities that are diagrammed will also have to be obtained. After this is complete, one will then know whether it is possible to engage in a productive high-throughput functional genomics strategy, or whether one has to increase the sensitivity or specificity of the measurement techniques, or whether one needs to increase one's biological validation budget, or all three. Given the current level of reproducibility of expression measurements in microarrays, most investigators will choose a high level of specificity to reduce the false-positive rate. This will inevitably lead to a high false-negative rate. At this point the biologists will find that among the false negatives are genes that are known to have changed expression (but did not meet the significance threshold computed by the bioinformatician) which will lead naturally to the following worry:

Many other genes of relevance to the biological system under investigation are being discarded form the analysis and follow-up. This worry is likely to be well-founded but given the state of the measurement technology, and the typical sample sizes employed, it cannot be easily remedied. It remains, that the low-hanging fruit, while only numbering in the hundreds, are likely, to shed new light on the processes studied. Already this represents several orders of magnitude more hypotheses to be tested as compared to the investigations of the pregenomic era. For those fruit higher up the tree, the above decision analysis suggests the need to wait for more accurate and cheaper microarray technologies.

It goes without saying that the preceding discussion is an oversimplified decision analytic procedure. Attempting to assess the utilities of false-negative findings will be particularly difficult for most investigators, but the constraints of realistic budgets, will be a driving factor in considering this analysis. However, the procedure we have described does have the merit of providing a rational basis for deciding upon the number of genes that one should seek to obtain from one's pipeline. Even a coarse decision model can be useful in avoiding unpleasant surprises after the initial excitement of obtaining a list of putatively interesting genes wears off. It should be noted that the raw specificities and sensitivities of the expression microarrays used in a functional genomics pipeline need not be the ones that are used in this decision analysis. The application of appropriate noise models over repeated experiments can lead to improved sensitivity and specificity.

The cost of such models is typically in the increased number of microarray experiments required to develop those models. Perhaps the best glimpse of the interactions between the various genetic components, and in particular their causal dependencies, is obtained through analysis of the dynamics of the system. This is elaborated considerably in the section on dynamics. Suffice it to say here that understanding the trajectory over time of various RNA component species rather than their absolute values at any given point in time provides far more information about the operation of the underlying system.

To return to our watch metaphor, observing how the gears turn inside the watch while we smoothly move the minute hand or the hour hand will give us much more information about how the gears move together than a few (unordered) snapshots in time of the different

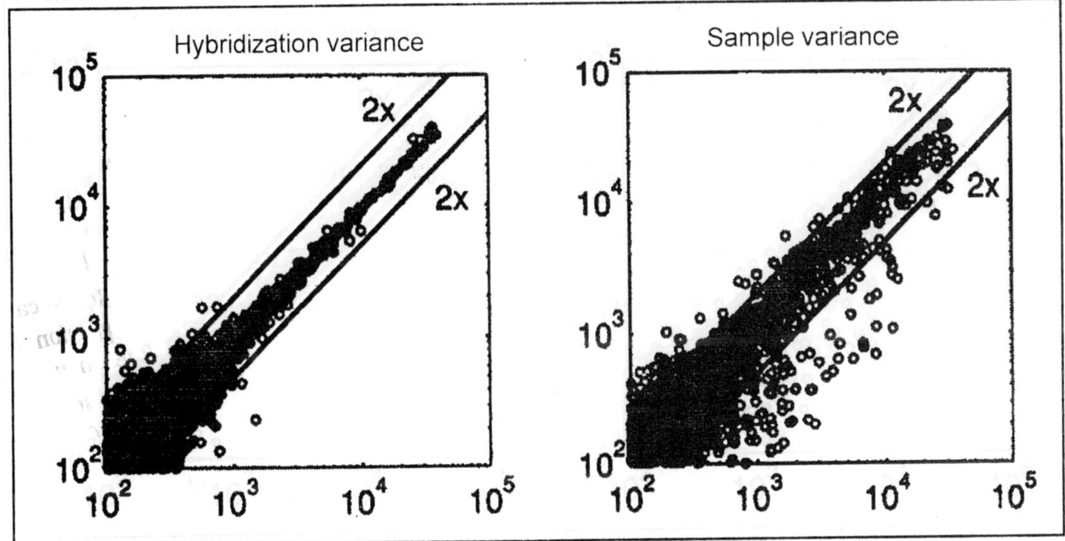

Fig. 3.10: **Typical uses of fixed fold thresholds, reproducibility scatter plots. In each of the experiments, samples were hybridized to identical oligonucleotide arrays containing probes for 6800 human genes. The message abundance in arbitrary units is plotted. Left pane: A single biotinylated RNA target was divided in two and each half hybridized to two arrays. Fifty-nine genes (0.8%) judged to be expressed "present" by the Affymetrix GeneChip software differed by more than twofold, and 0.3% differed by more than threefold. Middle panel: A single biotinylated target was hybridized to one array, the sample removed and then rehybridized to a second array. Of "present" genes 2.6% and 0.4% differed by more than twofold and threefold, respectively. Right panel: A single total RNA sample was converted to biotinylated cRNA in two independent labelling reactions, and the cRNA then hybridized to two arrays. Of the "present" genes 2.2% and 0.4% differed by twofold and threefold,**

positions of these hands on the face of the watch. In most analyses to date, even when expression data are obtained as a time series, the information buried in the specific ordering and timing of these measurements is rarely exploited. For example, many of the clustering and classification methods employed will generate the same answers even if the order of the measurements were shuffed. In fairness, the number of time points and the sampling intervals of many of the existing time-series expression data sets is minimal. They provide insufficient information for most of the standard armamentarium used to analyze time-oriented data.

As the price of microarrays falls, the quantity and quality of time-series experiments will undoubtedly increase, at which point many of the techniques for analyzing trends and periodicity that have been developed in statistics and the signal processing literature will become applicable. At that time, the temporal relationships between every single expressed gene with every other gene measured at all time points will provide a qualitatively improved set of insights into causal relationships in the genetic regulation of cellular physiology.

For example, cellular proliferation and differentiation during development.

All of these genes have been implicated in somatic growth and bone repair.

This involves giving the subject a continuous intravenous infusion of insulin and glucose, while maintaining a relatively constant blood glucose concentration by varying the rate of the glucose infusion.

This is particularly true given the noisy nature of most microarray-based expression measurement systems.

For the discussion here, we define "significant" as "worthy of further study."

Sensitivity can be described as (1 false negatives)/all positives; specificity as (1 false positives)/all positives.

GENE-CLUSTERING DOGMA

There is a central underlying assumption in all gene-clustering techniques for expression analysis. Simply put, the assumption is that genes that appear to be expresses in similar patterns are in fact mechanistically related. Furthermore, the corollary to this assumption is that although genes may distantly affect the function of other gene products, they fall into groups of more tightly regulated mechanisms. For instance, the genes that govern chromosome function or meiosis may be more tightly linked to each other than they are to the genes involved with another function, such as apoptosis. This has been the basis of our collective experience in biological investigations over the last century: that there are groups of proteins which interact together more closely than others.

Often they have been organized into pathways such as glycolysis, the Krebs cycle, and other metabolic pathways in which the gene products called enzymes have to work in concert. Other more obvious functional clusters are those of structural proteins which have to come together in a conserved and reproducible fashion in order to serve their purpose, whether they are the components of the ribosomal unit or the histoproteins which are essential to maintenance of the structure or chromatin. On this basis, if we can find genes whose expression patterns approximate one another, we can possibly impute that they are functionally clustered together. That is, they have functions that are related. Several important caveats are worth noting here. First of all, it remains unclear just how discrete the functional groupings of gene function are in the cellular apparatus. It may be that individual gene products have so many different roles under different circumstances that several of them partake of essential roles in significantly different functions. The second caveat is that the term "functionally related" is itself ill-specified. If the pattern of expression of one gene is similar to that of another, it could signify all kinds of relationships, ranging from "two genes having gene products that physically interact", to "one gene encoding a transcriptional factor of the other gene", to "two genes having different functions but similar promoter sequences", to "two genes both with promoter sequences bound by repressors which are knocked off when a nuclear receptor is activated, even though the two genes have widely disparate functions. "Of course there is a level of abstraction at which all genes are functionally related in their role of keeping the cell alive and producing whatever components are needed for the rest of the organism.

Below this level of abstraction, there are many alternative and by their nature, sloppy definitions of clustering. Therefore, we should be somewhat wary of the claim that similarity in expression corresponds to similarity in function. Nonetheless, it is a useful starting point for many analyses of a genome whose function remains by and large unknown at this time. For example, similarity could mean having similar patterns of change over time. It could mean similar absolute levels of expression at any given point in time, or it could mean perfectly opposite but well-choreographed patterns of expression. Just which dissimilarity measure is chosen for looking at patterns of expression will inuence the kind of functional clusters that we expect.

Learning

It has already been discussed why the computational techniques applied in the analysis of gene expression are qualitatively different form those of traditional biostatistics: the date sets are of high dimensionality and yet the number of cases are relatively small. Consequently, the number of solutions that could explain the observed behaviour is quite large. For this reason, the machine-learning community has recognized the potential role for their techniques specifically designed to explore high-dimensional spaces (such as those of voice or face recognition) and have also recognized the enormous need to apply these techniques to genomic data sets. To this effect, the first reaction of a computer scientist with a background in machine learning when he or she becomes aware of the new challenges created by genomic data sets is to pull out the tools of the standard armamentarium of machine learning, he or she then begins to explore informally, and subsequently revaluates formally the results obtained when these tools are applied to genomics data sets.

We provide a framework for the armamentarium in the genomic data mining chapter but here we only provide enough of an overview to discuss these tools with respect to experimental design. Two useful broad categorizations of the techniques used by the machine-learning community are *supervised learning techniques* and *unsupervised learning techniques*. These are also commonly known as classification techniques and clustering techniques, respectively. The two techniques are easily distinguished by the presence of external labels of cases. For example, labelling a tissue as obtained from a case of acute myelogenous leukemia (AML) or acute lymphocytic leukemia (ALL) is needed first before applying a supervised learning technique to create a method to learn which combinations of variables predict or determine those labels. In an unsupervised learning task, such as finding those genes that are co-regulated across all the samples, the organization or clustering of the variables operates independently of any external labels.

The kinds of variables (also known as *features* in the jargon of the machine-learning community) that characterize each case in a data set can be quite varied. Each case can include measures of clinical outcome, gene expression, gene sequence, drug exposure, proteomic measurements, or any other discrete or continuous variable believed to be of relevance to the case.

What kinds of questions are answered by the two types of machine learning? In supervised learning, the goal is typically to obtain a set of variables (e.g., expressed genes as measured on a microarray) on the basis of which one can reliably make the diagnosis of the patient, predict future outcome, predict future response to pharmacological intervention, or categorize that patient or tissue or animal as part of a class of interest. In unsupervised learning, the typical application is to find either a completely novel cluster of genes with a putative common (but previously unknown) function, or more commonly to obtain a cluster or group of genes that appear to have similar patterns of expression to a gene (i.e., they fall into the same cluster) already known to have an important well-defined function.

The goal there is to find more details about the mechanism by which the known gene works and to find other genes involved in that same mechanism either to obtain a more complete view of a particular cellular physiology or, in the case of pharmacologically oriented research, other possible therapeutic targets. Although the distinct goals of supervised versus unsupervised machine-learning techniques may appear rather obvious, it is important to be aware of the implications for study design. For example, an analyst may be asked to find classifiers between two types of malignancy, as was done in the investigation of AML and ALL. However, the lists of genes that reliably divide the two malignancies may have little to do with the actual pathophysiological causes of the two diseases and may not represent any particular close relationship of those genes and function. Why might this be? It is quite possible that small amounts of change of some gene products such as transcriptional activators and genes such as *p53* may cause large downstream changes in gene expression. That is, with only a subtle change, an important upstream gene may cause dramatic changes in the expression in several pathways that are functionally only distantly related but are highly influenced by the same upstream gene.

When applying a classification algorithm directly on the gene expression levels, the algorithm will naturally identify those genes which change the most between the two or more states that are being classified. That is, a study design geared toward the application of a supervised learning technique may generate a useful artifact for classification, diagnosis, or even prognosis, but it will not necessarily lead to valuable insights into the biology underlying the classes obtained. Let us consider the more general causes where gene expression values are not the only data type. For example, a given case can include several thousand gene expression measurements but also several hundred phenotypic measurements such as blood pressure, a laboratory value, or the response to a chemotherapeutic agent. Here again a clustering algorithm can be used to find those features that are most tightly coupled in the observed data. When designing an experiment that includes the various data types, it is worthwhile thinking ahead of time whether some kinds of features are more likely to cluster together, separately form the genomic data. That is, after application of a clustering algorithm the data set may reveal relationships between the non-genomic variables that are much more significant and stronger than any of those that involve gene expression or sequence.

While that is not necessarily a bad outcome, it will not help the investigator who is trying to understand the particular contribution of genetic regulation to the observed phenomenon. As an example of one looks at the effect of thousands of drugs on several cancer cell lines, then it should not be surprising if these drug effects were most tightly clustered around groups of pharmaceutical agents that were derived from one another through combinatorial chemistry. Similarly, phenotypic features which are highly interdependent, such as height and weight, will cluster together, the strength of these obvious clusters will often dominate those of heterogeneous clusters that contain phenotypic measurements as well as genes expression measurements. This suggests that careful use of feature reduction to only include those features that are non-redundant and lonely truly independent phenotypic measures for each case should be used.

Functional Genomics

Whereas the discussion above was motivated by the questions posed by out collaborators who are primarily biologists or clinicians, it parallels similar questions coming from those colleagues with a computer science background, one question is, how can we determine whether a particular methodology that we are applying is successful or not? Is one machine-learning algorithm better than another, or does one clustering method provide more robust clusters than another? In other words, how do we know how successful a particular functional genomics investigation is? What is the figure of merit that we are trying to obtain? The figure of merit is probably the most ascertainable for the case of a known classification. Take, for instance, the case of the task of the classifying whether a particular tissue belongs to one of two tissue types using a supervised learning technique. Suppose also that a test sample is available. Then, various classification algorithm can be run against one another and the sensitivity and specificity of these algorithms can be compared. More generally, the receiver operating characteristic (ROC) curve of each of these classification algorithms can be defined.

A good example of such a comparison is provided by Michael Brown *et al*. Several classification algorithms were employed and compared and ranked based on a weighted measure of true and false positives and true and false negatives. The best performance turned out to be in the application of support vector machines. The test that was used for this "bake-off" was the saccharomyces cerevisiae data set from Stanford involving expression data of 2467 genes across 79 different hybridization experiments. Because the classification is known in advance, the performance of these algorithms can be measured. However, there are several limitations which should be recognized.

The classification algorithms' performance only pertains to the particular population or set of experiments on which it was originally tested. That is, a classification algorithm working on a yeast data set may not necessarily work as well relative to other algorithms on distinguishing two different types of leukemia or two different types of diabetes mellitus in humans. Even less dramatically, the same algorithm trained on one set of patients with

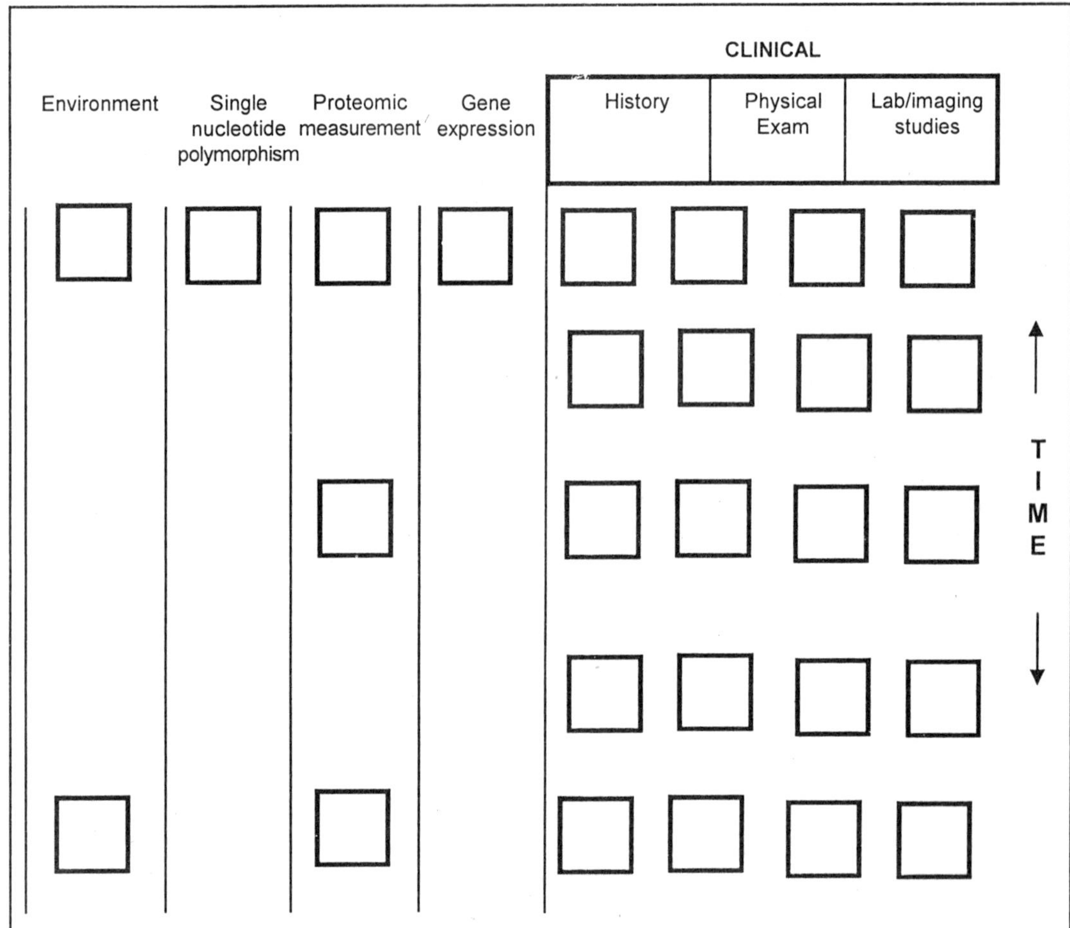

Fig. 3.11: Clustering more than only expression values. A full-edged functional genomic study with clinical relevance will involve multiple and quite heterogenous data types. The realities of clinical research and clinical care will ensure that there is a requirement for the handling of missing data.

leukemia may have a different relative performance compared to other classification algorithms on another set of leukemia patients drawn form a different population or obtained with a different ascertainment bias. For example, one set of patients may have been selected because of particularly refractory disease that caused them to be referred to a tertiary care hospital and another set of patients may have been treated in community hospitals. It is possibility of under-fitting or over-fitting the underlying classes of disease in these populations will differ. The figure of merit, therefore, in a classification experiment is the correct classification performance of an algorithm or measure for that particular population. Its applicability or merit for use in even related populations is uncertain or problematic.

Clustering Experiment

What is the figure of merit for a clustering analysis. That is, how can we evaluate the degree to which the cluster of genes obtained are in fact correct or relevant? For anybody who has performed a clustering procedure on microarray expression data, it will seem obvious that it is altogether too tempting when one obtains a cluster of genes through any machine-learning algorithm to come up with a post hoc justification for why those genes may have fallen or risen together. This temptation is all too evident in the publications of the past 3 years in which clusters of genes reported by investigators are described typically as falling into one of three different categories. The first is well-known published associations, either mechanistically verified or empirically known, that have been obtained through other more conventional techniques, second is associations that appear plausible to the authors, and presumably the reviewers, but have not been proven in the literature. The third is associations for which the authors are hard-pressed to find support.

For the biological investigator who wishes to further investigate the fundamental biology of genes and their function, the last two categories constitute a fairly unsatisfactory basis on which to invest several months, if not years, understanding the basis for the imputed associations or potential functional dependencies. The challenge then for the functional genomist using microarray data is to come up with means to validate the clusters obtained. There are two levels of validation. The first is a statistical or methodological validation with techniques such as permutation or cross-validation as described elsewhere in this book. These techniques can be used to ensure, within a specified degree of certainty, that for the particular data set being studies, the clusters are neither the result o serendipitous coordination of gene behaviours nor that the samples are apparently inadequate to estimate the reliable coordinated behaviour of gene expression.

An alternative method is to estimate the probability of particular clusters. This methodology is still only in its infancy and the relevant literature is only now beginning to be generated for functional genomics. As with other aspects of functional genomics, much larger data sets will allow more effective use of these methods. Ultimately, however, we should remember that the clusters or classifications that we obtain through machine-learning techniques are only reflections on the measurements made in a particular system. If we are to make broad claims about the empirical or scientific truth of the relationships or classes that we infer from expression date, we will have to submit analyses to the same kind of tests that have been developed for other experimental sciences with longer histories. Specifically, we need to at least come up with the microarray equivalent of Koch's postulates. In 1890, Robert Koch set out to develop criteria for judging whether a given bacterium was the cause of the disease.

It has been recognized there are several problems with Koch's postulates and there is a great deal of discussion and disagreement about this in the literature of the history of science. However, the postulates are a good first approximation of the requirement for validation of functional genomics experiments. Therefore the analogs to Koch's postulate in the domain of microarrays may be useful. The criteria describe the test for an inferred regulatory dependency obtained through microarray analysis.

Table 3.2: Koch's postulates

1. It (the suspected pathogen) should be present in every instance of the disease.
2. It should be isolated form the diseased host and grown in pure culture.
3. The specific disease must be reproduced when a pure culture of the agent is inoculated into a healthy susceptible host.
4. The same agent must be recoverable from the experimentally infected host.

Table 3.3: Functional genomics catalogs to Koch's postulates

1. If gene A is found to be correlated with gene B then this relationship should be reproducible through northern blots and other quantitative expression measurement techniques.
2. Furthermore, in all pathological conditions in which the gene is thought to play a critical role, the predicted level of that gene should be found at the right time and in the right part of the tissue during the disease process.
3. If a gene thought to be involved in a pathway is underexpressed or over expressed in a model system, then the process controlled by that pathway will be affected.

It is only with these biological tests that we can "rate' the performance of clustering algorithms- it is the only durable figure of merit. It is not surprising that very little of this kind of hypothesis testing has been reported in the literature. After all, for each gene in a cluster, there is implied a laborious sleuthing effort in which the bioinformatician and biologist must closely collaborate to verify the results using the analogs to Koch's postulates. If we are to avoid a functional genomics "meltdown" hinted at in the introductory chapter, then the interdisciplinary nature of the functional genomics pipeline will have to be substantively adopted. All gene associations or regulatory relationships suggested by bioinformatics techniques should undergo at least a minimal validation by the biologist. Conversely, experimental design without the involvement of the bioinformatician is likely to lead to wasteful, expensive, and unrewarding analyses.

In other words, there are several distinct and statistically distinguishable processes.

Well-known examples of these are the transcriptional factor such as sonic hedgehog (*shh*) which in some tissues and at some times are involved in cell proliferation and in others in cell differentiation processes.

Chapter 4

DNA Library in Forensic Science

To sequence a DNA molecule is to obtain the string of bases that it contains. In large-scale DNA sequencing we have a long target DNA molecule (thousands of bp) that we want to sequence. We may think of this problem as a puzzle in which we are given a double row of cards facing the table. We do not know which letter from the set (A, C, G, T) is written on each card, but we do know that cards in the same position of opposite strands form a complementary pair. Our goal is to obtain the letters using certain *hints*, which are (approximate) substrings of the rows. The long sequence to reconstruct is called the target. In the biological problem, we know the length of the target sequence approximately, within 10% or so.

It is impossible to sequence the whole molecule directly. However, we may instead get a piece of the molecule starting at a random position in one of the strands and sequence it in the canonical (5' → 3') direction for a certain length. Each such sequence is called a **fragment**. It corresponds to a substring of one of the strands of the target molecule, but we do not know which strand or its position relative to the beginning of the strand in addition it may contain errors.

5'.... □□□□□□□□.... 3'

3'.....□□□□□□□□.... 5'

Fig. 4.1: Unknown DNA to be sequenced.

By using the *shotgun method* we obtain a large number of fragments and then we try to reconstruct the target molecule's sequence based on fragment overlap. Depending on experimental factors, fragment length can be as low as 200 or as high as 700. Typical problems involve target sequences 30,000 to 100,000 base-pairs long, and total number of fragments is in the range 500 to 2000. The problem is then to deduce the whole sequence of the target DNA molecule. Because we have a collection of fragments to put together, this task is known as *fragment assembly*. We note that it suffices to determine one of the strands of the original molecule, since the other can be readily obtained because of the complementary pair cule.

In the remainder of this section we give additional details of a biological nature that are important in the design of fragment assembly algorithms.

Alignment

The best way of studying the issues involved is by looking at an example. Be aware, however, that real instances are much larger than the examples we present. Suppose the input is composed of the four sequences

ACCGT

CGTGC

TTAC

TACCGT

And we know that the answer has approximately 10 bases. One possible way to assemble this set is

– –ACCGT– –

– – – –CGTGC

TTAC – – – – –

– TACCGT– –

TTACCGTGC

Notice that we aligned the input set, ignoring spaces ate the extremities. We try to align in the same column bases that are equal. The only guidance to assembly, apart from the approximate size of the target, are the overlaps between fragments by *overlap* here we mean the fact that sometimes the end part of a fragment is similar to the beginning of another, as with the first and second sequences above. By positioning fragments so that they align well with each other we get a layout, which can be seen as a multiple alignment of the fragments.

BIOLOGICAL BACKGROUND

The sequence below the line is the **consensus sequence**, or simply **consensus**, and is the answer to our problem. The consensus is obtained by taking a majority vote among all bases in each column. In this example, every column is unanimous, so computing the consensus is straightforward. This answer has nine bases, which is close to the given target length of 10, and contains such each fragment as an exact substring, however, in practice fragments are seldom exact substrings of the consensus.

Anomalies

As mentioned above, real problem instances are very large. Apart form this fact, several other complications exist that make the problem much harder than the small example we saw. the main factors that add tot the complexity of the problem are errors, unknown orientation, repeated regions, and lack of coverage. We describe each factor in the sequel

Errors

The simplest errors are called *base call* errors and comprise base substitutions, insertions, and deletions in the fragments. Transpositions are also common but we can treat them as compositions of an insertion and a deletion or two substitutions. Base cell errors occurs in practice at rates varying form 1 to 5 errors every 100 characters. Their distribution along the sequence is not uniform, as they tend to concentrate towards the 3' end of the fragment. As we can see from the examples, it is still possible to reconstruct the correct consensus even in the presence of errors, but the computer program must be prepared to deal with this possibility and this usually means alogrithms that require more time and space.

Input:	Answer:
ACCGT	– – ACCGT– –
CGTGC	– – – – CGTGC
TTAC	TTAC– – – – –
TGCCGT	– TGCCGT– –
	TTACCGTGC

Fig. 4.2: In this instance there was a substitution error in the second position of the last fragment, where A was replaced by G. The consensus is still correct because of majority voting.

It is possible to find the best alignments between two sequences in linear time if there are no errors, and quadratic algorithms are needed to account for gaps. Apart form erroneous base calls, two other types of errors can affect assembly. One of them is the artifact of *chimeric* fragments, and the other is *contamination* by host or vector DNA. We explain each type in what follows.

Input:	Answer:
ACCGT	– – ACC-GT—
CAGTGC	– – – – CAGTGC
TTAC	TTAC– – – – –
TACCGT	– TACC-GT – –
	TTACC-GTGC

Fig. 4.3: In this instance there was an insertion error in the second position of the second fragment, base A appeared where there should be none. The consensus is still correct because of spaces introduced in the multiple alignment and majority voting. Notice that the space in the consensus will be discarded when reporting the answer.

Input:	Answer:
ACCGT	– – ACCGT – –
CGTGC	– – – – CGTGC
TTAC	TTAC – – – – –
TACGT	– TAC-GT – –
	TTACCGTGC

Fig. 4.4: In this instance there was a deletion in the third (or fourth) base in the last fragment. The consensus is still correct because of spaces in the alignment and majority voting.

Chimeric Fragments

Chimeric fragments, or *chimeras*, arise when two regular fragments from distinct parts of the target molecule join end-to-end to form a fragment that is *not* a contiguous part of the target. These misleading fragments must be recognized as such and removed form the fragment set in a preprocessing stage. Sometimes fragments or parts of fragments that do not have anything to do with the target molecule are present in the input set. This is due to contamination form host or vector DNA. The process of replicating a fragment consists of inserting it into the genome of a vector, which is an organism that will reproduce and carry along copies of our fragment.

In the end, the fragment must be purified from the vector DNA, and here is where contamination occurs—if this purification is not complete. If the vector is a virus, then the infected cell—generally a bacterial cell—can also contribute some genetic material to the fragment. Contamination is a rather common phenomenon in sequencing experiments. Witness to this is the significant quantity of vector DNA that is present in the community databases.

Scientists sometimes fail to screen against the vector sequence prior to assembly and submit a contaminated consensus to the database. As with chimeras, the remedy for this problem is to screen the data before starting assembly. The complete sequences of vectors commonly used in DNA sequencing are well known, and it is not difficult to screen all fragments against these known sequences to see whether a substantial part of them is present in any fragment, chimeric fragment detection leads to interesting alogrithmic problems, but we will not detail them any further in this book. Pointers to relevant references are given in the bibliographic notes.

Input:	Answer:
ACCGT	– – ACCGT – –
CGTGC	– – – – CGTGC
TTAC	TTAC – – – – –
TACCGT	– TACCGT – –
TTATGC	TTACCGTGC
	TTA – – – TGC

Fig.4.5: The last fragment in this input set is a chimera. The only way to deal with chimeras is to recognize and remove them form the input set before starting assembly proper:

Unknown Orientation

Fragments can come from any of the DNA strands and we generally do not know to which strand a particular fragment belongs to. We do know, however, that whatever the strand the sequence read goes from 5' to 3'. Because of the complementarity and opposite orientation of strands, the fact that a fragment is a substring of one strand is equivalent to the fact that its reverse complement is a substring of the other. As a result, we can think of the input fragments as being all approximate substrings of the consensus sought either as given or in reverse complement. Figure 4.6 shows an assembly involving fragments in both orientations, initially unknown, but with no errors.

Input:		Answer:
CACGT	→	CACGT - - - - - - - - -
ACGT	→	- ACGT - - - - - - - - -
ACTACG	←	- - CGTAGT- - - - -
GTACT	←	- - - - - AGTAC - - -
ACTGA	→	- - - - - - - - - ACTGA
CTGA	→	- - - - - - - - - CTGA
		CACGTAGTACTGA

Fig. 4.6: Fragment assembly with unknown orientation. Initially we do not know the orientation of fragments. Each one can be used wither in direct or reverse orientation. In the solution, we indicate by an arrow the chosen orientation: → means fragment as is, ← means it reverse complement.

In practice, we have to deal with both errors and unknown orientation at the same time. Because the orientations are unknown, in principle we should try all possible combinations, which are 2^n for a set with n fragments. Of course, this method is unacceptable and is not the way it is done in an assembly program, but it does hint at the complexity introduced by the orientation issue.

Repeated Regions

Repeated regions or **repeats** are sequences that appear two or more times in the target molecule. Short repeats, that is, repeats that can be entirely covered by one fragment, do not pose difficulties. The worst problems are caused by longer repeats. Also, the copies of a repeat do not have to be identical to upset assembly. If the level of similarity between two copies of a repeat is high enough, the differences can be mistaken for base call errors. Remember that the assembler must be prepared to deal with errors, so there is usually some degree of tolerance in overlap detection. The kinds of problems that repeats cause are twofold. First, if a fragment is totally contained in a repeat, we may have several places to put it in the final alignment, as it may fit reasonably well in the several repeat copies.

Fig. 4.7: Repeated regions. The blocks marked X_1 and X_2 are approximately the same sequence.

One could argue that it does not matter where we put it, since in any copy of the same repeat the consensus will be approximately the same. But the point is that, when the copies are not exactly equal, we may weaken the consensus by placing a fragment in the wrong copy. Second, repeats can be positioned in such a way as to render assembly inherently ambiguous; that is, two or more layouts are compatible with the input fragments and approximate target length at equivalent levels of fitness. The first one features three copies of the same repeat, and the second has two interleaving copies of different repeats. The common feature between these two cases is the presence of two different regions flanked by the same repeats. In the first example, both B and C are flanked by X and X. in the second, both B and D are flanked by X and Y. so far wed issued direct repeats, namely, repeated copies in the same strand. However, inverted repeats, which are repeated regions in opposite strands, can also occur and are potentially more dangerous. As few as two copies of a long, inverted repeat are enough to make the instance ambiguous.

Coverage and Mean Coverage

Another problem is lack of adequate coverage. We define the **coverage** at position i of the target as the number of fragments that cover this position. This concept is well defined, but it is impossible to compute it because we do not know the actual positions of the fragments in the target. Even after assembly, all we have is our best guess about these positions. We can compute the *mean coverage*, though, by adding up all fragment lengths and dividing by the approximate target length. If the coverage is equal to zero for one or

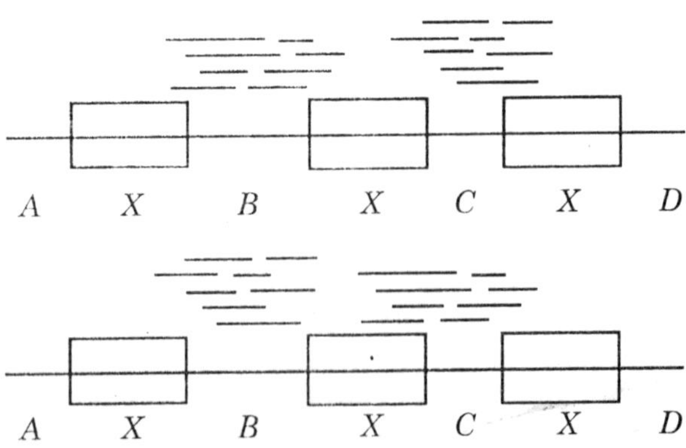

Fig. 4.8: Target sequence leading to ambiguous assembly because of repeats of the from XXX.

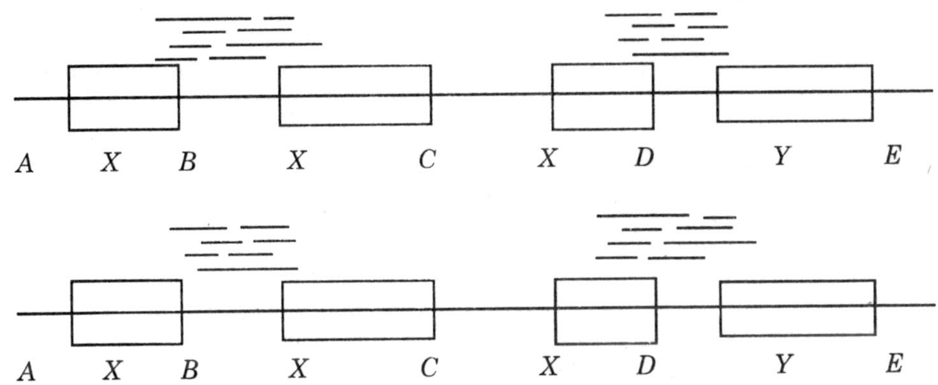

Fig. 4.9: Target sequence leading to ambiguous assembly because of repeats of the from XYXY.

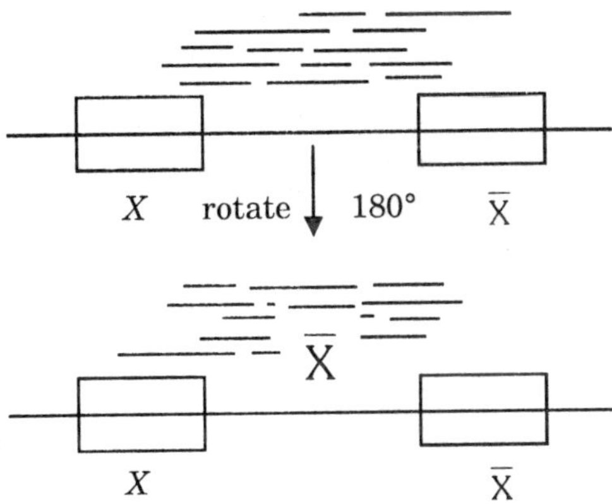

Fig. 4.10: Target sequence with inverted repeat. The region marked $\overline{X}$ is the reverse complement of the region marked X.

more positions, then there is not enough information in the fragment set to reconstruct the target completely. Figure 4.11 shows an example in which there is an uncovered area.

In such cases, the best we can hope to achieve is a layout for every one of the contiguously covered regions, called *contigs*. Lack of coverage occurs because the sampling of fragments is essentially a random process; it is therefore possible that some parts end up not being well covered or not covered at all. In general, we want not only one, but several different fragments covering any given point. The more fragments we have, the safer is our assessment of the consensus based on voting.

It is also desirable to have fragments from both strands covering a given region, for it has been observed in practice that certain kinds of errors occur consistently in one strand only. Insufficient coverage can usually be dealt with by sampling more fragments, but we

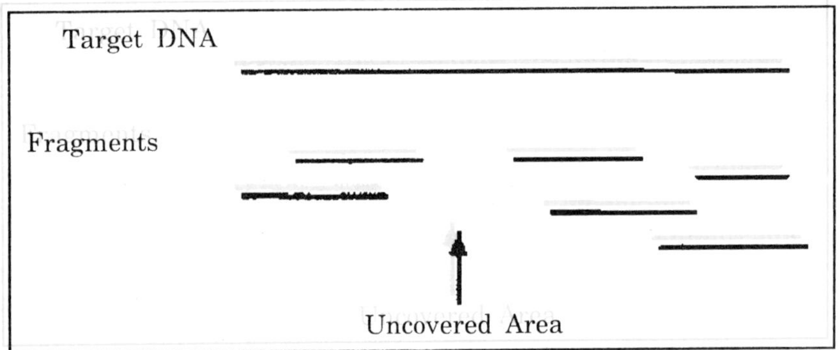

Fig. 4.11: Example of insufficient coverage.

must be careful with this approach. If everything is well covered except for a small portion of the target, sampling at random can be a very inefficient way of covering the gaps. An approach, called *directed sequencing* or *walking*, can be used in this case.

To avoid the formation of these little gaps, some researchers advocate sampling at very high coverage rates, collecting fragments whose combined length is enough to cover the target molecule 8 times or more. It is important to know how may fragments we need to generate in order to achieve a given coverage. The answer cannot be absolute, of course, since sampling is a random process, but reasonable bounds can be derived. We mention here an important result that is valid under certain simplifying assumptions. Let T denote the length of the target molecule. Assume that all fragments have about the same length l and that we can safely recognize overlaps of at least t bases. If we sample n fragments at random, the expected number p of *apparent contigs* is given by the formula:

$$p = ne^{-n(l-t)/T} \tag{4.1}$$

The term "apparent contig" refers to the fact that we are assuming that we are able to recognize overlaps only if they have size at least t. Thus some of the contigs we have may in fact overlap by less that t bases, so they are really parts of a longer true contig but "appear" to be two separated contigs to us. Notice that p approaches zero as the number n of fragments grows, which seems strange because we now there will always be at least one apparent contig. This happens because formula (4.1) is an approximation.

It can be used also in the context of DNA mapping, where the number of contigs is much larger, so a difference of one or two is negligible, the formula is useful in the sense that it provides a reliable ballpark in a wide variety of situations. A formula for the fraction of the target molecule covered by the fragments is also available. With the same notation as above, the fraction covered by exactly k fragments is given by

$$r_k = \frac{e^{-c}c^k}{k!} \tag{4.2}$$

where $c = nl/T$ is the mean coverage.

DNA Sequencing

We mention here a few supplementary and alternative approaches to DNA sequencing. We start with directed sequencing, a method that can be used to cover small remaining gaps in a shotgun project. In direct sequencing a special primer is derived form the sequence near the end of a contig, so that fragments spanning the region including the end of this contig and the continuation of it in the target are generated. These new fragments are then sequenced and give the sequence adjacent to the contig, thereby augmenting it. Continuing in this fashion, we can cover the gap to the next contig.

The problem with this approach is that it is expensive to build special primers. Also, the next step can be accomplished only after the current one, so the process is essentially sequential rather than parallel (but can be done for all gaps in parallel). Another technique that has become very popular is called *dual end sequencing*. In a shotgun experiment several copies of the target DNA molecule are broken randomly and short pieces are selected for cloning and sequencing. We recall from that these pieces are called *inserts* because they are inserted into a vector for amplification. Inserts sizes range from 1 to 5 kbp, but only about 200 to 700 bases can be directly read from one extremity to yield a fragment. However, it is also possible to read the other extremity if we have a suitable primer.

The two fragments thus obtained belong to opposite strands, and they should be separated in the final alignment by roughly the insert size minus the fragment size. This extra information is extremely useful in closing gaps, for instance. Sometimes the dual end is sequenced only if it is necessary to close a gap. Notice that dual end sequencing exploits the fact that inserts are usually larger than the portion read form them. A radically different approach from shotgun methods has been proposed recently. Called *sequencing by hybridization (SBH)*, it consists of assembling the target molecule based on many hybridization experiments with very short, fixed length sequences called probes. A hybridization simply checks whether a probe binds (by complementarity) to a DNA molecule. The idea is to design a DNA chip that performs all the necessary hybridizations simultaneously and delivers a list of all strings of length w, or w-mers, that exist in the target.

With current technology it is possible to construct such chips for probes of length up to eight bases. Larger probe sizes seem still prohibitive at this point. Important issues in SBH include the following. It is clear that not all target molecules can be reconstructed with a probe size as small as eight. One important problem then is to characterize the molecules that can actually be reconstructed with a given probe size. Another problem is that the hybridization experiments do not provide the number of times a given w-mer appears in the target, just whether an w-mer does appear. Clearly, it would help to have the extra information on how many copies do appear.

Finally, problems with errors in the experiments and orientation need to be tackled. One proposed strategy is to reduce the shotgun data to hybridization data by simply using all the w-mers of the fragments instead of the fragments themselves and to think of them as generated by hybridization experiments. However, this strategy throws away information, because it is generally impossible to recover the fragments form the w-mers, while it is possible to generate the w-mers from the fragments. Alogrithms for the SBH problem are not covered in this book, but references are given in the bibliographic notes at the end of this chapter.

Superstring

We will now examine formalisms for fragment assembly. We present in this section three models for the problem: shortest common superstring (SCS), *Reconstruction*, and *Multicontig*. Each throws light on the diverse computation aspects of the problem, although none of them completely addresses the biological issues. All three assume that the fragment collection is free of contamination and chimeras.

Shortest Common Superstring

One of the first attempts to formalize fragment assembly was through a string problem in which we seek the shortest superstring of a collection of given strings. Accordingly, this is called the Shortest Common Superstring problem, or SCS. Although this model has serious shortcomings in representing the fragment assembly problem- it does not account for errors, for instance-the techniques used to tackle the resulting computational problem have application in other models as well. It is worthwhile, therefore, to study these techniques. The *Shortest Common Superstring* problem is defined as follows:

PROBLEMS: Shortest Common Superstring (SCS)

INPUT: A collection $\mathcal{F}$ of strings.

OUTPUT: A shortest possible string S such that for every f $\in \mathcal{F}$, S is a superstring of f.

Example 4.1: Let $\mathcal{F}$ = {ACT, CTA, AGT}. The sequence S = ACTAGT is the shortest common superstring of $\mathcal{F}$. it obviously contains all fragments in $\mathcal{F}$ as substrings. To see that it is the shortest, notice that any string S' that has both u = ACT and v = AGT as substrings must have length at least 6. Moreover, with $|S'|$ = 6 we have just the concatenations uv and vu. But CTA is a substring of uv = S only.

In the context of fragment assembly, the collection $\mathcal{F}$ corresponds to the fragments, each one given by its sequence in the correct orientation, and S is the sequence of the target DNA molecule. Notice that the computational problem specifies that S should be a perfect superstring of each fragment, not an approximate superstring, so it does not allow for experimental errors in the fragments. Furthermore, the orientation of each fragment must be known, which is seldom the case. Finally, even in a perfect assembly project in which all these factors could be somehow controlled, the shortest common superstring may not be the actual biological solution because of repeated sections in the target DNA sequence, as the following example shows.

Example 4.2: Suppose that the target molecule has two copies of an exact repeat and that fragments are sampled as shown in Fig. 4.12. Notice that the repeat copies are long and contain many fragments. In this case, even if the fragments are exact substrings of the consensus, and even if we know their correct orientation, finding the shortest common superstring may not be what we want.

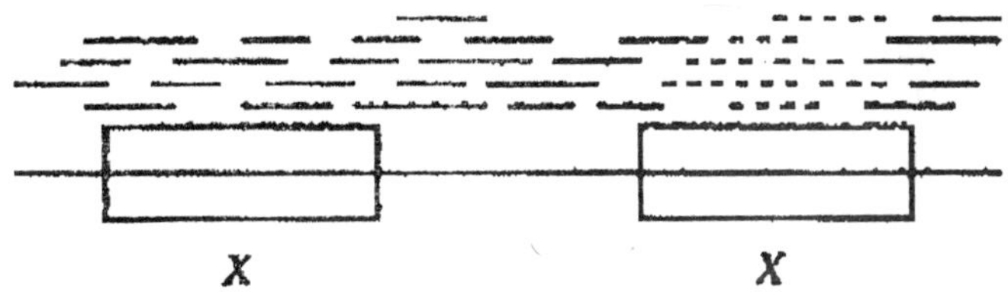

Fig. 4.12: Target sequence with long repeat that contains many fragments, this example shows that even with no errors and known orientation, the SCS formulation fails in the presence of repeats.

Indeed, Fig. 4.13 shows a different assembly, with a shorter consensus for the same fragment set. Observe that because the repeat copies are identical, superstring may contain only one copy, which will absorb all fragments totally contained in any of the copies. The other copy can be shorter, as it must contain only fragments that cross the border between X and its flanking regions. The shortened version of X is denoted by X' in this figure.

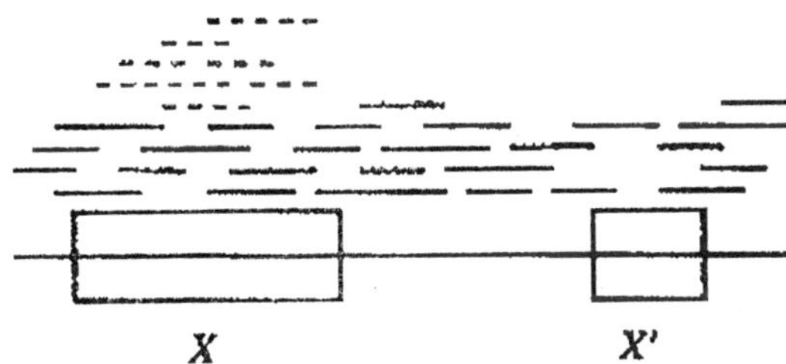

Fig. 4.13: Alternative assembly for the fragments in the pervious figure. This assembly leads to a shorter consensus, because all fragments totally contained in the rightmost copy (the dashed fragments) were moved to the leftmost copy, causing a decrease in length of the rightmost copy.

Notice that although shorter, this alternative assembly is poorer in terms of both coverage and linkage. The coverage is uneven, with many fragments spanning X and much fewer spanning X'. The linkage is poor because no fragment links the leftmost part of X' to its rightmost part. In fact, this consensus is the concatenation of two unrelated parts u and v, with u going form the beginning to about the middle of X', and v going from the middle of X' to the end. As far as shortest substrings are concerned, vu would be a perfectly equivalent consensus.

The Shortest Common Superstring problem is NP-hard, but approximation algorithms exist. However, given all the shortcomings of this model with respect to the real biological problem, such algorithms are primarily of theoretical interest.

Reconstruction

This model takes into account both errors and unknown orientation. To deal with errors, we need a few preliminary definitions. We can configure the basic dynamic programming sequence comparison algorithm to suit many different needs. Here we will use distance rather than similarity and a version that charges for gaps in the extremities of the second sequence only. The scoring system is edit distance; that is, one unit of distance is charged for every insertion, deletion, or substitution, except for deletion in the extremities of the second sequence, which are free of charge.

The distance thus obtained will be called *substring edit distance*, to distinguish it form the classical edit distance that charges for end deletions in both strings, we denote it by d_s and define it formally by the expression

$$d_s(a, b) = \min_{s \in S(b)} d(a, s),$$

where $S(b)$ denotes the set of all substrings of b and d is the classical edit distance. Notice that substring edit distance is asymmetric, that is, in general $d_s(a, b) = d_s(b, a)$.

Example 4.3: If a = GCGATAG and b = CAGTCGCTGATCGTACG, then the best alignment is an indicated in Figure 4.14 and the distance is $d_s(a, b) = 2$.

```
– – – – – GC-GATAG – – – –
CAGTCGCTGATCGTACG
```

Fig.4.14: Optimal alignment for substring edit distance, which does not charge for end deletions in the second string.

Let $\in$ be a real number between 0 and 1. A string f is an *approximate substring* of S at error level $\in$ when

$$d_s(f, S) \in |f|,$$

where f is the length of f. this means that we are allowed on average $\in$ errors for each base in f. For instance, if $\in$ = 0.05 we all allowed five errors per hundred bases. We are now ready for the definition of fragment assembly according to the **Reconstruction** model.

PROBLEM: Reconstruction

INPUT: A collection $\mathcal{F}$ of strings and an error tolerance $\in$ between 0 and 1.

OUTPUT: A shortest possible string S such that for every f $\in \mathcal{F}$ we have

$$min(d_s(f, S), d_s(\bar{f}, S) \leq \in |f|,$$

where $\bar{f}$ is the reverse complement of f.

The idea is to find a string S as short as possible such that either f or its reverse complement must be an approximate substring of S at error level $\in$. This formulation will assemble correctly all the examples except for the one involving a chimeric fragment. However, the general problem is still NP-hard. This is not surprising, though, as it contains the SCS as a particular case with $\in = 0$.

Multicontig Models

The *Multicontig model* adds a notion of good linkage to the answer. The previous models do not care about the internal linkage of the fragments in the layout—only the final answer matters. Thus, it is also necessary to accept answers formed by several contigs, and for this reason it is called "multicontig". We define first an error-free version of the *Multicontig model*. Given a collection $\mathcal{F}$ of fragments, we consider a multiple alignment, or layout, this layout must be such that every column contains only one kind of base. This is where the error-free hypothesis comes in. also, to contemplate orientation, we require that either the fragment or its reverse complement be in the alignment, but not both.

Let us call this layout $\mathcal{L}$. We number the columns form 1 to the length $|\mathcal{L}|$ of the layout. According to this numbering, each fragment f has a left endpoint $l(f)$ and a right endpoint $r(f)$, so that $| f | = r(f) - l\{f\} + 1$. We say that fragments f and g *overlap* in this layout if the integer intervals $[l(f) \,.\,.\, r(f)]$ and $[l(g) \,.\,.\, r(g)]$ intersect. The non-empty intersection $[l(f) \,.\,.\, r(f)]$ Ç $(l(g) \,.\,.\, r(g))$ is called the overlap between f and g.

The size of this overlap is just the size of the interaction. Because we are using integer intervals, all sets are finite. Among all the overlaps between fragments, we are interested in the most important ones, the ones that provide linkage. We say that an overlap $\{x...y\}$ is a *non-link* if there is a fragment if $\mathcal{F}$ that properly contains the overlap on both sides, that is, if the fragment contains interval $\{(x - 1) \,.\,.\, (y + 1)\}$. If no fragment has this property, the overlap is a *link*. The *weakest link* in a layout is the smallest size of any link in it. Finally, we say that a layout is a *t-contig* if its weakest link is at least as large as t. if it is possible to construct a t-contig from the fragments of a collection $\mathcal{F}$, we say that $\mathcal{F}$ *admits* a t-contig. We can now formalize the fragment assembly problem according to the *Multicontig model*. Given a collection of fragments $\mathcal{F}$ and an integer t, we want to partition $\mathcal{F}$ in the minimum number of subcollections C_i, $1 \leq i \leq k$, such that every C_i admits a t-contig.

Example 4.4: Let $\mathcal{F}$ = {GTAC, TAATG, TGTAA}. We want to partition $\mathcal{F}$ in the minimum number of t-contigs. If $t = 3$, this minimum is two, as follows:

```
– – TAATG                    GTAC
TGTAA – –
```

There is not way to make one single contig with all three fragments, so two contigs is the minimum possible in this case. Notice that GTAC is its own reverse complement.

If $t = 2$, the same solution is a partition in 2-contigs, since every 3-contig is a 2-contig. However, another solution exists now:

TAAATG – – – GTAC
– – – TGTAA

Here again it is impossible to assemble $\mathcal{F}$ into one t-contig.

Finally, if $t = 1$, we see that a solution with one t-contig exists:

TGTAA – – – – –
— TAATG – – –
– – – – – – – GTAC

A version contemplating errors can be defined as follows. The layout is not required to be error-free, but with each alignment we must associate a *consensus* sequence S of the same length, possibly containing space characters (–). The numbers $l(f)$ and $r(f)$ are still defined as the leftmost and rightmost columns of f in the alignment, but now we may have $|f| \neq r(f) + 1$. The image of an aligned fragment f in the consensus is $S[l(f) \,..\, r(f)]$. Given an error tolerance degree ϵ, we say that S is an ϵ-consensus for this contig when the edit distance between each aligned fragment f and its image in the consensus is at most $\epsilon|f|$. We are now ready for the formal definition.

PROBLEM: Multicontig

INPUT: A collection $\mathcal{F}$ of strings, an integer $t \geq 0$, and an error tolerance ϵ between 0 and 1.

OUTPUT: A partition of $\mathcal{F}$ in the minimum number of sub-collections $\mathcal{F}_i$, $1 \leq i \leq k$, such that every C_i admits a t-contig with an ϵ-consensus.

Notice that in the *Reconstruction* model a fragment is required to be approximate substring of S, but the particular place in S where it is aligned is not important. In contrast, in the *Multicontig* formulation we need to specify explicitly where each fragment should go.

The *multicontig* formalization of fragment assembly if NP-hard, even in the simplest case of no errors and known orientation. It contains as a special case the problem of finding a Hamiltonian path in a restricted class of graphs.

This formulation models errors, orientation, and lack of coverage but has no provision to use information on the approximate size of the target molecule. In addition, it partially models repeats, in the sense that it can satisfactorily solve some instances with repeats, although not all of them.

Algorithms

In this section we present two algorithms for the case of fragments with no errors and known orientation. One of them is known as the *greedy* algorithm, and many practical systems are based on the same idea, with additions that contemplate errors and unknown orientation. The other algorithm is based on the *Multicontig* model and is useful when we can obtain an acyclic overlap graph discarding edges with small weight. It should be mentioned that both algorithms are of little practical value in themselves, because of the restrictive hypothesis. Nevertheless, the ideal behind them can be useful in designing algorithms that deal with real instances of the problem.

Overlap Multigraph's

Common superstrings correspond to *paths* in a certain graph structure based on the collection $\mathcal{F}$. We can translate properties of these superstrings to properties of the paths. Because researchers have studies paths in graphs for a long time, many people feel more comfortable dealing with graphs, and relating a new problem to them is often a good idea. The overlap multigraph $\mathcal{OM}(\mathcal{F})$ of a collection $\mathcal{F}$ is the direçted, weighted multigraph defined as follows. The set V of nodes of this structure is just $\mathcal{F}$ itself. A directed edge from $a \in \mathcal{F}$ to a different fragment $b \in \mathcal{F}$ with weight $t \geq 0$ exists if the suffix of a with t characters is a prefix of b. In symbols, this is equivalent to

Suffix (a, t) = prefix (b, t),

Or

$\kappa^{|a|-t}\ a = b\kappa^{|b|-t}$,

Or

$(a\kappa^{t})b = a(\kappa' b)$,

where κ is the killer agent. Notice that we must have $|a| \geq t$ and $|b| \geq t$ for this edge to exist. Observe also that there may be many edges from a to b, with different values of t. This is why the structure is called a multigraph instead of simply a graph. Note that we disallow self-loops, that is, edges going from a node to itself. On the other hand, edges of zero weight are allowed. A word must be said about edge weights. These weights represent overlaps between fragments, and we use the overlaps to join fragments into longer strings, sharing the overlapping part.

This joining means that we believe that the two fragments come from the same region in the target DNA. Now this belief has to be based on solid evidence. In our case, this evidence is the size of the overlap. A short overlap provides weak evidence, whereas a long overlap gives strong evidence. When dealing with hundreds of fragments, each one a few hundred characters long, it is foolish to assume that two fragments share ends based on an overlap of, say, three characters.

For this reason, sometimes we impose a minimum acceptable amount of overlap, and throw away all edges with weight below this threshold. But, in principle, we will keep all the edges in the multigraph, including the zero weight edges. There are n(n – 1) such edges, because any two strings share a common part of zero characters.

Common Superstrings

The overlap multigraph is important because directed paths in it give rise to a multiple alignment of the sequences that belong to this path. A consensus sequence can then be derived from this alignment, providing a common superstring of the involved sequences. Let us investigate this process more closely. Given a directed path $\mathcal{P}$ in $\mathcal{OM}(\mathcal{F})$, we can construct a multiple alignment with the sequences in P as follows. Each edge $e = (f, g)$ in the path has a certain weight t, which means that the last t bases of the tail f or e are a prefix of the head g of e. thus, f and g can be aligned, with g starting t positions before f's end. Only edges with strictly positive weight were drawn. The collection F is (a, b, c, d), where

a = TACGA,

b = ACCC,

c = CTAAAG,

d = GACA.

Usually, we indicate paths by a list of the form vertex, edge, vertex, edge, . . ., vertex, with each edge connecting its two neighbouring vertices in the list. However, in this example there are no parallel edges, so, for simplicity, we will indicate the paths by just a list of the vertices involved. The path $P_1 = dbc$ leads to the following alignment:

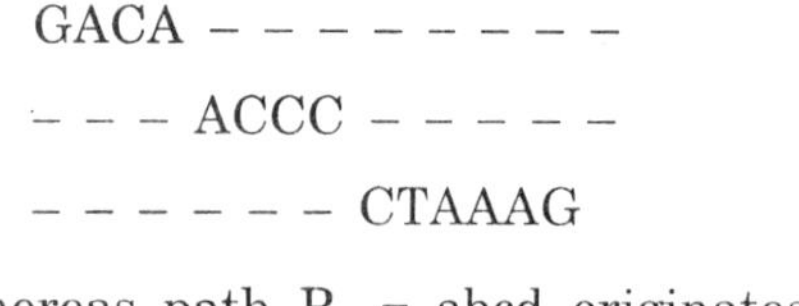

whereas path P_2 = abcd originates the following alignment.

TACGA – – – – – – – – – – – –

– – – – ACCC – – – – – – – – –

– – – – – – – – CTAAAG – – –

– – – – – – – – – – – – – – GACA

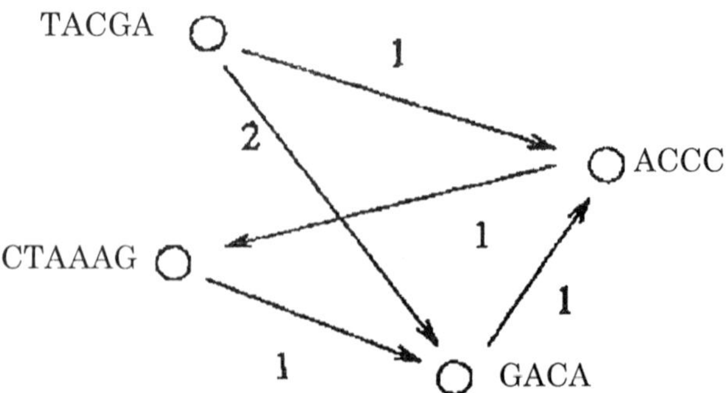

Fig. 4.15: Overlap multigraph with zero-weight edges omitted.

In general, let P be a path in $\mathcal{OM}(\mathcal{F})$, and let A be the set of fragments involved in P. paths by definition cannot repeat vertices, so we have exactly $|A| - 1$ edges in P. The common superstring derived form P as above will be called S(P).

The relationship between the total length of A, the path's weight, and the superstring length is given by the following equation.

$$\|A\| = w(P) + |S(P)|, \quad (4.3)$$

where $\| A \| = \Sigma_{a \in A} |a|$ is the sum of lengths of all sequences in A, and $w(P)$ is the weight of p. is this really true? Let us check it for some simple cases.

Start with a path with just one fragment and no edges. Then $A = \{f\}$, $\| A \| = |f|$, and $w(P) = 0$, since P has no edges. Furthermore, $S(P)$ is simply f. it is easy to see that Equation (4.3) holds in this case. Let us try something larger, suppose that $P = f_{1e1}f_{2e2} \ldots f_t$ and that the formula holds for P. what happens if we consider an additional edge? We have $P' = P_{et+1} f_{l+1}$ and

$$W(P') = w(P) + w(e_{l+1}) = w(P) + k,$$

If $w(e_{l+1}) = k$.

On the other hand, $S(P') = S(P)$ suffix $(f_{l+1}, |f_{l+1}| - k)$, because the first k characters of f_{l+1} are already present in $S(P)$. We need to concatenate the remaining suffix of f_{l+1} only. Then,

$$\begin{aligned} |S(P')| &= |S(P) + |f_{l+1}| - k \\ &= \|A\| - w(P) + |f_{l+1}| - k \\ &= \|A\| - w(P') \end{aligned}$$

and the formula holds. We have just proved Equation (4.3) by induction.

Another way of seeing it is by using the killer character. If $P = f_{1e1}f_{2e2} \ldots f_t$,

$$S(P) = f_1 k^{w(e_1)} f_2 k^{w(e_2)} \ldots k^{w(e_{t-1})} f_i$$

The right-hand side is not ambiguous in this case because

$$(f_1 k^{w(e_i)}) f_{i+1} = f_i (k^{w(e_i)} f_{i+1})$$

by definition of edge weight.

Computing the length,

$$|S(P)| + |f_1| - w(e_1) + |f_2| - w(e_2) + \ldots - w(e_{l-1}) + |f_t|$$

$$= \sum_{i=1}^{l} |f_i| \sum_{i=1}^{l-1} w(e_i)$$

$$= \| A \| - w(P).$$

So, every path originates a common superstring of the fragments involved. This is particularly important when we have a path that goes through every vertex. Such paths are called *Hamiltonian paths.* If we get a Hamiltonian path, we have a common superstring of A = $\mathcal{F}$ In this case Equation (4.3) becomes

$$|S(P)| + \|\mathcal{F}\| - w(P) \tag{4.4}$$

and, since $\|F\|$ is constant, that is, independent of the particular Hamiltonian path we take, we see that minimizing $|S(P)|$ is equivalent to maximizing $w(P)$.

Paths of Superstring

We have seen that every path corresponds to a superstring. Is the converse true? Not always, as the following example shows. For the fragment set that originated the graph in Fig. 4.15 the superstring

GTATACGACCCAAACTAAAGACAGGG

does not correspond to any path, and it easy to see why. Superstrings may contain unnecessary characters not present in any fragment. But shortest superstrings cannot waste any characters. So, do shortest superstrings always correspond to paths? The answer is affirmative, and we will prove this result in the sequel. We first observe that it is true for certain collections that are *substring-free*, defined below, and then generalize the result to any collection.

A collection $\mathcal{F}$ is said to be substring-free if there are no two distinct strings a and b in F such that a is a substring of b. the advantage of dealing with substring-free collections is that a partial converse of the path-to-superstring construction is easier to prove, as illustrated by the following result.

Theorem 4.1: Let $\mathcal{F}$ be a substring-free collection. Then for every common superstring S of F there is a Hamiltonian path P in $\mathcal{OM}(\mathcal{F})$ such that $S(P)$ is a subsequence of S.

Proof. The string S is a common superstring of $\mathcal{F}$, so for each $f \in \mathcal{F}$ there is an interval $[l(f) \,..\, r(f)]$ of S such that $S[l(f) \,..\, r(f)] = f$. For some fragments there may be many positions in which to anchor them. Never mind. Take one of them and fix it for the rest of this proof.

No interval of the form $[l(f) \,..\, r(f)]$ is contained in another such interval, because F is substring-free. These intervals have the following important property: All left end-points are distinct and all right endpoints are distinct, although there may be left end-points equal to right endpoints, moreover, if we sort the fragments in increasing order of left endpoint, the right endpoints end up in increasing order as well. Indeed, if we had two strings f and g in $\mathcal{F}$ with $l(f) \le l(g)$ and $r(f) \ge r(g)$, then the interval of g would be contained in the interval of f, which is impossible. So, we have the intervals sorted by increasing initial and final endpoints. Let $(f_i)_{i+1,\ldots,m}$ be this ordering. The fragments in this order make up the claimed path.

Consider the relationship between f_i and f_{i+1}, for each i such that $1 \le i \le m - 1$. If $l(f_{1+1}) \le r(f_i) + 1$, then there is an edge in $\mathcal{OM}(\mathcal{F})$ with weight $r(f_i) + 1 - l(f_{i+1})$. If l (f$_i$

$_{+1}) > r(f_i) + 1$, then take the edge of weight zero from f_i to f_{i+1}. The characters S[j] for $r(f_i) + 1 \leq j \leq l(f_{i+1}) - 1$ will be discarded, because they do not belong to any interval and are therefore superfluous as far as making a common superstring is concerned. These characters are in S but not in S(P), which is therefore a subsequence of S.

Corollary 4.1: Let F be a substring-free collection. If S is a shortest common superstring of F, there is a Hamiltonian path *P* such that $S = S(P)$.

Proof. By the previous theorem, there is such a path with *S(P)* a subsequence of *S*. But *S(P)* is also a common superstring, because P is Hamiltonian, because S is a shortest common superstring, we have $|S(P)| \geq |S|$ And therefore $S = S(P)$.

We now generalize this result to any collection. First, a few definitions are in order. A collection of strings $\mathcal{F}$ dominates another collection $\mathcal{G}$ when every $g \in \mathcal{G}$ is a substring of some $f \in \mathcal{F}$. For instance, if $\mathcal{G} \subseteq \mathcal{F}$, then $\mathcal{F}$ dominates $\mathcal{G}$. Two collections $\mathcal{F}$ and $\mathcal{G}$ are said to be *equivalent*, denoted $\mathcal{F} \equiv \mathcal{G}$, when $\mathcal{F}$ dominates $\mathcal{G}$ and $\mathcal{G}$ also dominates $\mathcal{F}$. Equivalent collections have the same superstrings, so this notion is important for us.

Lemma 4.1: Two equivalent substring-free collections are identical.

Proof. If $\mathcal{F} \equiv \mathcal{G}$ and $\mathcal{F}$ is substring-free, then $\mathcal{F} \subseteq \mathcal{G}$. To see that, consider $f \subseteq \mathcal{F}$. Since $\mathcal{F}$ is dominated by $\mathcal{G}$, there must be $g \in \mathcal{G}$ such that *f* is a substring of *g*. But $\mathcal{F}$ also dominates $\mathcal{G}$, so there is a superstring *h* of *g* in $\mathcal{F}$. By transitivity, f is a substring of h. But $\mathcal{F}$ is substring-free, so $f = h$, and, since *g* is "in between" them, $f = g = h$. It follows that $f = g \in \mathcal{G}$. Since *f* is arbitrary, we conclude that $\mathcal{F} \subseteq \mathcal{G}$. The same argument with $\mathcal{F}$ and $\mathcal{G}$ interchanged shows that $\mathcal{G} \subseteq \mathcal{F}$. Hence, $\mathcal{F} = \mathcal{G}$.

The next theorem helps us understand how to obtain a substring-free of collection form an arbitrary collection of fragments.

Theorem 4.2: Let $\mathcal{F}$ be a collection of strings. Then there is a unique substring-free collection $\mathcal{G}$ equivalent to $\mathcal{F}$.

Proof. Uniqueness is guaranteed by the previous lemma. Now let us turn to the existence. We want to show that for every $\mathcal{F}$ there is a substring-free collection $\mathcal{G}$ equivalent to $\mathcal{F}$. We do that by induction on the number of strings in $\mathcal{F}$.

For $|\mathcal{F}| = 0$, $\mathcal{F}$ itself is substring-free, so we are done. For $|\mathcal{F}| \geq 1$, $\mathcal{F}$ may or may not be substring-free. If it is, we are again done. If it is not, then there are *a* and *b* in $\mathcal{F}$ such that *a* is a substring of *b*. Remove a from $\mathcal{F}$, obtaining a smaller collection $\mathcal{F}' = \mathcal{F} - \{a\}$. By the induction hypothesis, $\mathcal{F}'$ has an equivalent substring-free collection $\mathcal{G}$. But it is easy to verify that $\mathcal{F} \equiv \mathcal{F}'$. Obviously $\mathcal{F}$ dominates $\mathcal{F}'$, which is a subset of $\mathcal{F}$. On the other hand, the only string that is in $\mathcal{F}$ but not in $\mathcal{F}'$ is *a*, which is covered by $b \in \mathcal{F}1$. Hence, $\mathcal{G}$ is also equivalent to $\mathcal{F}$ and we are done.

This results tells us that if we are looking for common superstrings, we might as well restrict ourselves to substring-free collections, because given any collection there is a substring-free one equivalent to it, that is, having the same common superstrings, incidentally, the proof of Theorem 4.2 also gives us a way of obtaining the unique substring-free collection equivalent to a given $\mathcal{F}$. Just remove from $\mathcal{F}$ all strings that ate substrings of other elements of $\mathcal{F}$.

Heaviest Path

We know that looking for shortest *common superstrings* is the same as looking of *Hamiltonian paths* of maximum weight in a directed multigraph. Moreover, because our goal is to maximize the weight, we can simplify the multigraph and consider only the heaviest edge between every pair of nodes, discarding other, parallel edges of smaller weight. Any path that does not use the heaviest edge between a pair of nodes can be improved, so it is not a maximum weight path. Let us call this new graph of the *overlap* graph of $\mathcal{F}$, denoted by $\mathcal{OG}(\mathcal{F})$. In an actual implementation we would not construct these edges one by one.

Using suffix trees, we can build the edges simultaneously, saving both space and time in the computer. The same structures can help in deciding which fragments are substrings of other fragments and may therefore be left out of the graph. The following algorithm is a "*greedy*" attempt at computing the heaviest path. The basic idea employed in it is to continuously add the heaviest available edge, which is one that does not upset the construction of a Hamiltonian path given the previously chosen edged. Because the graph is complete, that is, there are edges between every pair of nodes, this process stops only when a path containing all vertices is formed. In a *Hamiltonian path*, or in any path for that matter, we cannot have two edges leaving form the same node, or two edges leading to the same node.

In addition, we have to prevent the formation of cycles. So these are the three conditions we have to test before accepting an edge in our Hamiltonian path. Edges are processed in nonincreasing order by weight, and the procedure ends when we have exactly $n - 1$ edges, or, equivalently, when the accepted edges induce a connected subgraph. For each node, we keep its current indegree and outdegree with respect to the accepted edges. This information is used to check whether a previously chosen edge has the same tail or head as the currently examined edge. In addition, we keep the disjoint sets that from the connected components of the graph induced by the accepted edges. It is easy to see that a new edge will form a cycle if and only if its tail and head are in the same component. Therefore, this structure is needed to check for cycles. The disjoint sets are maintained by a disjoint-set forest data structure. We described the greedy algorithm in terms of graphs, but it can be implemented using the fragments directly. The algorithm corresponds to the following procedure being continuously applied to the collection of fragments, until only one fragment remains. Take the pair (f, g) of fragments with largest overlap k, remove the two fragments from $\mathcal{F}$, and add $fk^k\ g$ to $\mathcal{F}$. we assume that the collection F is substring-free.

This greedy strategy does not always produce the best result, as shown by the following example (Fig. 4.16).

In this graph, the edge of weight 3 is the first one to be examined and is obviously accepted because the first edge is always accepted. However, it invalidates the two edges of weight 2, so the algorithm is forced to select an edge of weight zero to complete the Hamiltonian path. The two rejected edges form a path of total weight 4, which is the heaviest path in this example. This example shows that the greedy algorithm will not always return the shortest superstring.

Is there an algorithm that works in all cases? Apparently not. We are trying to solve the SCS problem through the Hamiltonian path (HP) problem; but the HP problem is NP-complete. Perhaps we could have better luck trying another approach to solve the SCS problem. Unfortunately this is probably not the case either, since, as already mentioned, it can be shown that the SCS problem is NP-hard.

```
Algorithm Greedy
    Input: weighted directed graph OG(F) with n vertices
    Output: Hamiltonian path in OG(F)
    // Initialize
    for i ← 1 to n do
            in[i] ← 0 // how many selected edges enter i
            out[i] ← 0 // how many selected edges exit i
            MakeSet (i)
    // Process
    Sort edges by weight, heaviest first
    For each edge (f, g) in this order do
            // test edge for acceptance
            if in[g] = 0 and out [f] = 0 and FindSet(f) ≠ FindSet (g)
                    select (f, g)
                    in[g] ← 1
                    out[f] ← 1
                    Union (FindSet (f), FindSet (g))
            If there is only one component
                    break
            return selected edges
```

Fig. 4.16: Greedy algorithm to find Hamiltonian path.

Example 4.5: Suppose we have

$$\mathcal{F} = \{GCC, ATGC, TGCAT\}.$$

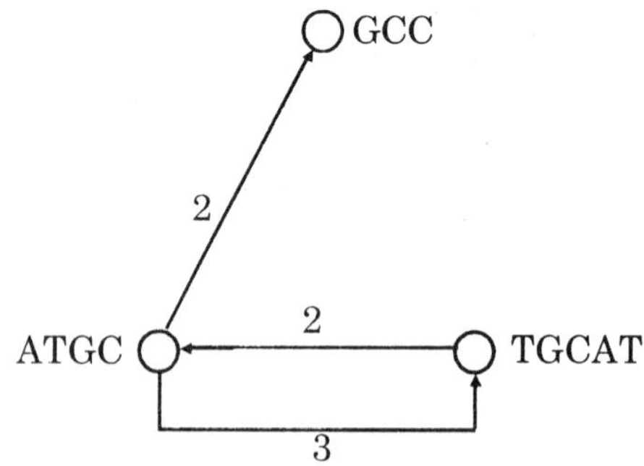

Fig.4.17: A graph where the greedy algorithm fails.

Good Sampling

The hardness results we mentioned in the previous section apply to an arbitrary collection of fragments. In this section we consider the problem of assembling fragments without errors and known orientation assuming that the fragments have been obtained from a "*good sampling*" of the target DNA. As shown in the sequel, this assumption enables us to derive algorithms that deliver the correct solution and are particularly efficient. What do we mean by a "*good sampling*"? Basically, we want the fragments to cover the entire target molecule, and the collection as a whole to exhibit enough linkage to guarantee a safe assembly. The definition of t-conitg. Our definitions here are based on similar concepts. Suppose that S is a string over the alphabet {A, C, G, T}. An *interval* of *S* is an integer interval $[i\ .\ .\ j]$ such that $1 \leq i \leq J + 1 \leq |S| + 1$. A *sampling* of *S* is a collection A of intervals of *S*. The sampling *A* covers *S* if for any i such that $1 \leq i \leq |S|$ we have at least one interval $[j\ .\ .\ k] \in A$ with $i \in [j\ .\ .\ k]$.

We say that two intervals α and β are *linked at level t* if $|\alpha \cap \beta| \geq t$. The entire sampling $\mathcal{A}$ is said to be *connected at level t* if for every two intervals α and β in $\mathcal{A}$ there is a series of intervals α_i for $0 \leq i \leq l$ such that $\alpha = \alpha_0$, $\beta = \alpha_l$ and α_i is linked to α_{i+1} At level *t* for $0 \leq i \leq l - j$. Note that we are using parameter t to measure how *strong* a link is. A sampling $\mathcal{A}$ will be considered good if it is connected at level t. Our goal is to study fragment collections coming from connected at level t samplings that cover a certain string *S*. The value of t is around 10 in typical, real instances, but in our study we will not fix it. We do assume that *t* is a non-negative integer, though.

The fragment collection *generated* by a sampling *A* is

$$S[\mathcal{A}] = \{\mathcal{A}\ [a]\ |\ a \in A\}$$

We say that a sampling $\mathcal{A}$ is *subinterval-free* if there are no two intervals $[i\ .\ .\ .\ j]$ and $[k\ .\ .\ .\ l]$ in $\mathcal{A}$ with $[i\ .\ .\ .\ j] \subseteq [k\ .\ .\ .l]$. The following observation about subinterval-free samplings will be coming into play again and again. In a subinterval free sampling, no two intervals can have the same left endpoint. Otherwise, whichever has the larger right endpoint would contain the other. An analogous argument holds for right endpoints. We consider also modified forms of our overlap graphs. Given a collection $\mathcal{F}$ and a nonnegative integer t define $\mathcal{OM}(\mathcal{F}, t)$ as the graph obtained form $\mathcal{OM}(\mathcal{F})$ by keeping only the edges of weight at least *t*. A similar construction defines $\mathcal{OG}(\mathcal{F}, t)$. This reduces the overall number of edges, keeping only the stronger ones.

We will use the symbol $\alpha \xrightarrow{w} \beta$ to mean that $|\alpha| \geq w$, $|\beta| \geq w$, and $l(\beta) + w - 1 = r(\alpha)$.

Similarly, $f \xrightarrow{w} g$ means that there is an edge from f to g in $\mathcal{OM}(\mathcal{F})$ of weight w.

Suppose now that we got such a collection *F* and would like to recover S. how can we do that? What are the good algorithms for it? It turns out that the answer depends strongly on the repeat structure of *S*. if *S* does not have any repeats of size t or more, then it is easy to reconstruct it, but if *S* has such repeats, the problem gets more complex, and in some cases S cannot be recovered without ambiguity. One thing that is independent of the repeat structure of S is the existence of a Hamiltonian path in the graph $\mathcal{OM}(\mathcal{F}, t)$. We state and prove this basic result here, before embarking on an analysis of the repeat structure of S. first, an auxiliary lemma.

Lemma 4.2: Let *S* be a string over {A, C, G, T} and *A* be a subinterval-free, connected at level t sampling of *S*, for some $t \geq 0$. If α is an interval in *A* such that there is another interval β in *A* with $l(\alpha) < (\beta)$, then β can be chosen so that

$$r(\alpha) + 1 - l(\beta) \geq t$$

Proof. We know *A* is connected at level t, therefore there is sequence $(\alpha_i)_{0 \leq i \leq l}$ of intervals in A such that $\alpha = \alpha_0$, $\beta = \alpha_l$, and $|\alpha_i \cap \alpha_{i+1}| \geq t$ for $0 \leq i \leq l - 1$.

Let *i* be the smallest index such that $l(\alpha) < l(\alpha_i)$. Because $\beta = \alpha_l$ Satisfies this property, we know that $i < l$. On the other hand, $i > 0$ because $\alpha = \alpha_0$ does not satisfy the property. Because *i* is the smallest index with the property, we have $l(\alpha_{i-1}) < l(a) < l(\alpha_i)$. It follows that $r(\alpha_{i-1}) < r(\alpha) < r(\alpha_i)$, since A is subinterval-free.

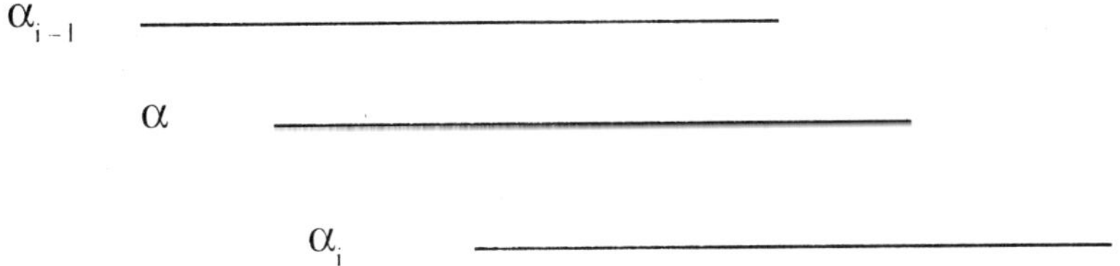

Fig. 4.18: Three mutually overlapping intervals.

We also know that $|\alpha_{i-1} \cap \alpha_i| \geq t$. But $|\alpha \cap \alpha_i| \geq ||\alpha_{i-1} \cap \alpha_i| \geq t$ (look at Fig. 4.18). So α_i is an interval in *A* such that

$$r(\alpha) + 1 - (\alpha_i) \geq t$$

Now, the Hamiltonian path existence result.

Theorem 4.3: Let S be a string over {A, C, G, T} and A be a subinterval-free, connected at level t sampling of S, for some $t > 0$. Then the multigraph $\mathcal{OM}(\mathcal{F}, t)$, where $\mathcal{F}$ is the fragment collection generated by A, admits a Hamiltonian path P. Furthermore, if A covers S, then P can be chosen so that $S(P) = S$.

Proof. Let $(\alpha_i)_{1 \leq i \leq n}$ be all the intervals in A stored by left endpoint. Because A is subinterval-free, this sequence is also sorted by right endpoint. Let fi = $S(\alpha_i)$ for $1 \leq i \leq n$. We claim that $f_i \overset{k_i}{\rightarrow} f_{i+1}$ for some $k_i \geq t$, so that the f_i in the above order form a Hamiltonian path.

Let α_i be any interval except the last one. By the previous lemma, there is another interval $\beta \in$ A such that $l(\alpha_i) < l(\beta)$ and $r(\alpha_i) + 1 - l(\beta) \geq t$. The sequence $(\alpha_i)_{1 \leq i \leq n}$ is sorted by left endpoint, so $l(\alpha_i) < l(\alpha_{i+1}) \leq l(\beta)$. (The intervals α_{i+1} and β may be the same.) Hence,

$$k_i = |\alpha_i \cap \alpha_{i+1}| = |[l(\alpha_{i+1}) \ldots r(\alpha_i)]| = r(\alpha_i) - l(\alpha_{i+1}) + 1 \geq r(\alpha_i) + 1 - l(\beta) \geq t.$$

This shows that an edge $f_i \rightarrow f_{i+1}$ really exists in $\mathcal{OM}(\mathcal{F}, t)$ and concludes the proof of the first part.

The second part, which says that S(P) = S, is a consequence of the following fact. Every item we have $\alpha \overset{w}{\rightarrow} \beta$, we also have $S[\alpha] \overset{w}{\rightarrow} S[\beta]$ and $S[\alpha]k^w S[b]$ $S[\alpha \cup \beta]$.

The presence of repeated regions, or repeated elements, in the target string S is related to the existence of cycles in the overlap graph. In this and following sections we make this relationship more explicit. A repeated element, or just a repeat in S, occurs when two different intervals of S give rise to the same substring. In other words, there are two intervals $\alpha \neq \beta$ such that $S[\alpha] = S[\beta]$. The size of this repeat is $|S[\alpha]|$, which is the same as $|S[\beta]|$. The repeat is *self-overlapping* when $a \cap b \neq \emptyset$, otherwise we have disjoint occurrences of the repeat.

Our next result shows that cycles in an overlap graph are necessarily due to repeats in S. the converse is not necessarily true; that is, we may have repeats but still an acyclic overlap graph. Again, a couple of intermediate results are needed before we can prove an important fact.

Lemma 4.3: If $\alpha \overset{w}{\rightarrow} \beta$ then $l(\alpha) \leq l(\beta)$ and $r(\alpha) \leq r(\beta)$

Proof. From $|\alpha| \geq w$ we have $l(\alpha) + w - 1 \leq r(\alpha)$. From $\alpha \overset{w}{\rightarrow} \beta$ we get $l(\beta) + w - 1 = r(\alpha)$. Therefore $l(\alpha) + w - 1 \leq r(\beta) + w - 1$, that is, $l(\alpha) \leq l(\beta)$. Similarly, $|\beta| \geq w$ implies $l(\beta) + w - 1 \leq r(\beta)$, or $r(\alpha) \leq r(\beta)$, in view of previous equations.

We need a few more definitions before proceeding. A false positive at level t is a pair of intervals α and β such that there is a $w \geq t$ with $S[\alpha] \overset{w}{\rightarrow} S[\beta]$ but $\alpha \overset{w}{\rightarrow} \beta$ does not hold., the name refers to the fact that we assumed that $S[\alpha]$ and $S[\beta]$ came from intersecting regions in the target molecule because their sequences overlap, but this is not the case. They look overlapping but they are not, so they constitute a false positive.

Lemma 4.4: The existence of a false positive at level t implies the existence of a repeat of size at least t.

Proof. Because $S[\alpha] \overset{w}{\rightarrow} S[\beta]$, we have $|\alpha| \geq w$ and $|\beta| \geq w$. If $l(\beta) = r(\alpha) - w + 1$, then we would have $\alpha \overset{w}{\rightarrow} \beta$, since $\alpha \neq \beta$ by hypothesis. Hence, $l(b) \neq r(a) - w + 1$.

The relation $S[\alpha] \overset{w}{\rightarrow} S[\beta]$ also implies that $S[\alpha'] = S[\beta']$, where α' is the last w elements of α and β' is the first w elements of β. Because $l(\beta) \neq r(\alpha) - w + 1$, α' and β' are different intervals giving rise to the same substring; that is, we have a repeat of size $w \geq t$.

Theorem 4.4: Let $\mathcal{F}$ be a collection generated by a sampling A of S. if $\mathcal{OG}(\mathcal{F}, t)$ has a directed cycle, then there is a repeat in S of size of least t.

Proof. Let $f_1 \rightarrow f_2 \rightarrow \ldots \rightarrow f_1 \rightarrow f_1$ be a directed cycle in $\mathcal{OG}(\mathcal{F}, t)$. Consider intervals $\alpha_1, \ldots \alpha_l$ such that $S[\alpha_i] = f_i$, for $1 \leq i \leq l$. We will show that at least one of the pairs (α_i, a_{i+1}), including (α_l, α_1) is a false positive, which implies the existence of a repeat in S by Lemma 4.4.

Suppose they are all true positives. Then $\alpha_i \rightarrow \alpha_{i+1}$ for $1 \leq i < l$ and $\alpha_l \rightarrow \alpha_1$. It follows that $l(\alpha_1) \leq l(\alpha_2) < \ldots < l(\alpha_l) < l(\alpha_1)$. So, all the $l(\alpha_i)$ are equal. An analogous observation holds for the $r(\alpha_i)$. But then $\alpha 1 = \alpha_2 = \ldots = \alpha_l$, which is impossible, since the f_i are distinct by hypothesis. Hence, there is at lest one false positive and the result is proved.

All these considerations lead to a satisfactory solution for fragment assembly when the target string S does not have repast.

Theorem 4.5: Let S be a string over {A, C, G, T}, A is a subinterval-free, connected at level t sampling of S that covers S, and $\mathcal{F} = S[A]$. If S has no repeats of size t or larger, then the graph $\mathcal{OG}(\mathcal{F}, t)$ has a unique Hamiltonian path P and $S(P) = S$.

Proof. We have seen that $\mathcal{OG}(\mathcal{F}, t)$ is acyclic (Theorem 4.4) and that $\mathcal{OM}(\mathcal{F}, t)$ has a Hamiltonian path P with $S(P) = S$.

To begin with, notice that $\mathcal{OM}(\mathcal{F}, t) = \mathcal{OG}(\mathcal{F}, t)$ if S has no repeats of size t or larger. This follows form the absence of false positives. It is clear that the vertex sets of OM(F, t) and OG(F, t) are the same. It is also clear that every edge of $\mathcal{OG}(\mathcal{F}, t)$ is also in $\mathcal{OM}(\mathcal{F}, t)$. What we need is to convince ourselves that all edges in $\mathcal{OM}(\mathcal{F}, t)$ are in $\mathcal{OG}(\mathcal{F}, t)$.

Let $f \overset{w}{\rightarrow} g$ be an edge in $\mathcal{OM}(\mathcal{F}, t)$. Then $w \geq t$. If this edge is not in $\mathcal{OG}(\mathcal{F}, t)$, this is due to the existence of another edge $f \overset{w}{\rightarrow} g$ with x>w. Let α and β be intervals in A with $S[\alpha] = f$ and $S[\beta] = g$. Since there are no false positives, we must have at the same time

$\alpha \overset{w}{\rightarrow} \beta$ and $\alpha \overset{w}{\rightarrow} \beta$, which is impossible. The weight in this edge is uniquely determined by $w = r(\alpha) - l(\beta) + 1$.

So, $\mathcal{OG}(\mathcal{F}, t)$ is an acyclic graph with a Hamiltonian path. It turns out that every time an acyclic graph has a Hamiltonian path this path is unique. To see why this is true, consider the nodes in the graph that have indegree zero (the *sources*). Because the graph is acyclic, it must have at least one source. But, having a Hamiltonian path, it can have at most one source: the starting point of this path. All other vertices have at least one incoming edge. We conclude that our graph must have exactly one source, and it must be the starting point of the Hamiltonian path. Now, if we remove this source form the graph, the remaining graph is still acyclic, and has a Hamiltonian path, because the vertex we removed is an extremity of the old *Hamiltonian path*. Proceeding in this way, we see that the Hamiltonian path is unique, and we know how to construct it: just keep removing sources form the graph. The algorithm just described is well known in computer science.

It is called *topological sorting* and, in its more general context, is used to find an ordering of nodes that is consistent with an acyclic set of edges in the following sense: For all edges $f \rightarrow g$ the node f appears before g in the ordering. If the graph has a Hamiltonian path besides being acyclic, then this ordering is unique, as we just saw. We end this section with an example where Greedy does not work, but there is an acyclic subgraph for a suitable threshold t, so that topological sorting will find the correct path. When we say that *Greedy* does not work, we mean that it does not do a good job regarding assembly, although it does do a good job finding a shortest common superstring. In this example the fragments come from a target sequence that does have repeats longer than t. this goes to show that the absence of long repeats is not necessary to yield an acyclic graph. It is merely sufficient, in mathematical jargon.

Example 4.6: Consider the following target string S and fragments *w, z, u, x,* and *y:*

S = AGTATTGGCAATCGATGCAAACCTTTTGGCAATCACT

w = AGTATTGGCAATC

z = AATCGATG

u = ATGCAAACCT

x = CCTTTTGG

y = TTGGCAATCACT

Choosing a threshold value of $t = 3$, we get the overlap graph. A greedy assembly will result in a shortest common superstring, formed by concatenation of the contig wk^9y with the contig zk^3uk^3x in any order. This solution has length 36, one character shorter than the target sequence, which is also a superstring. However, the solution given by the Hamiltonian path has better linkage.

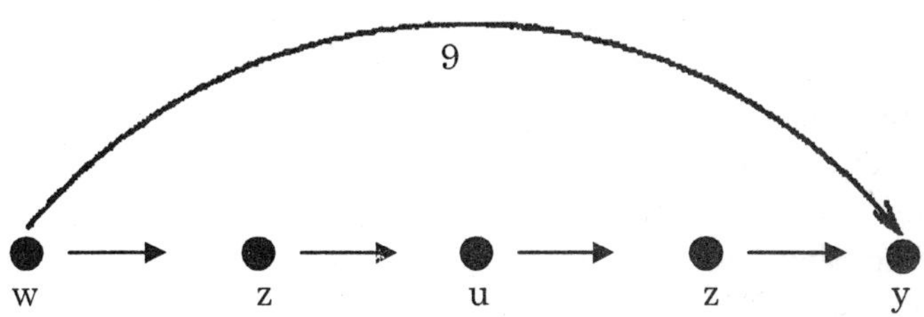

Fig. 4.20: Overlap graph for a given set of five fragments

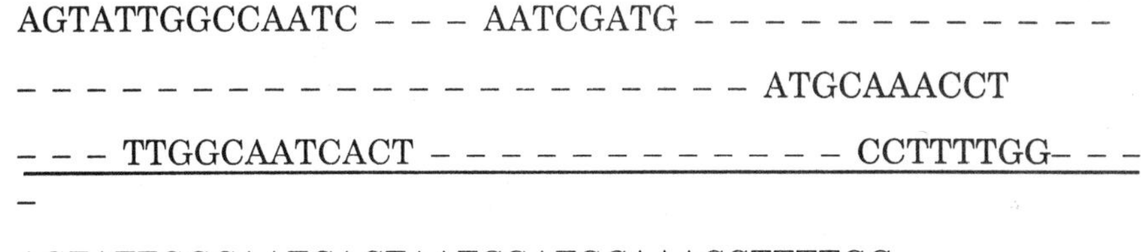

Fig. 4.21: A bad solution for an assembly problem, with a multiple alignment whose consensus is a shortest common superstring. This solution has length 36 and is generated by the Greedy algorithm. However, its weakest link is zero.

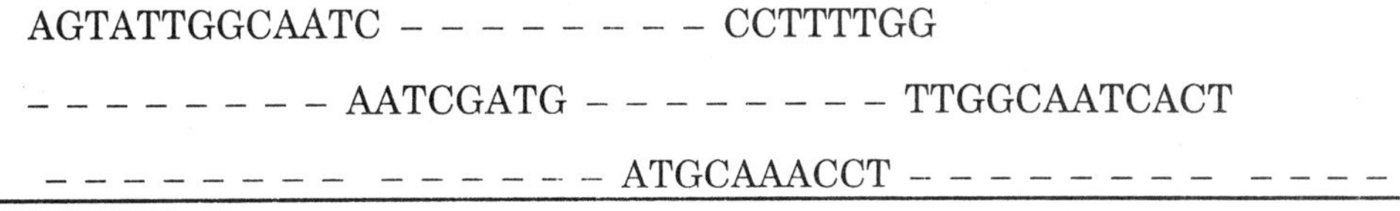

Fig. 4.22: Solution according to the unique Hamiltonian path. This solution has length 37, but exhibits better linkage. Its weakest link is 3.

FRAGMENT ASSEMBLY

We have seen that none of the formalisms proposed of *fragment assembly* are entirely adequate, so in practice we must often resort to heuristic methods. We review some of these in this section, starting by pointing out relevant properties of a good assembly layout. Fragment assembly can be viewed as a multiple alignment problem with some additional features. The most striking such feature is perhaps the fact that each fragment can participate with either the direct or the reverse-complemented sequence.

A second special feature is that the sequences themselves are usually much shorter than the alignment itself. This leads us to charge internal gaps and external gaps in the fragments differently, with internal gaps being much more penalized. If we charge both kinds of gaps the same or nearly the same penalty, then there would be no incentive to keep fragment characters together, potentially causing them to scatter along the entire length of the alignment. Moreover, the fact that fragments are short compared to the alignment forces us to look at other criteria, besides the score, to assess the quality of an alignment. These extra criteria include coverage and linkage. The latter measures how strongly the sequences are linked in terms of their overlap, now discuss these criteria in more detail.

Scoring

In each column of the multiple alignment we have a certain coverage. The ideal situation is when all sequences participating in a column have the same character there. In general, uniformity is good and variability is bad in a column. This suggests the due of a concept called *entropy* to measure the quality of a column. Entropy is a quantity that is defined on a group of relative frequencies, and it is low when one of these frequencies stands out form the others, and high when they are all more or less equal. In our context, the lower the entropy, the better. The frequencies we use are the frequencies of each character, where a character here can be an ordinary base A, C, G, T, or an internal space. Suppose that we have n sequences covering a certain column, and relative frequencies pA, pC, pG, pT, and p space for each character. The entropy of this distribution is defined as

$$E = -\sum_{c} p_c \log p_c$$

If $p_c = 1$ for a certain character c and all other frequencies are zero, then we have (assume $0 \log 0 = 0$)

$$E = 0.$$

If, on the other hand, all frequencies are equal, we have $p_c = 1/5$ for each c and then

$$E = -5 \times \frac{1}{5} \log \frac{1}{5} = \log 5,$$

which is the maximum value possible for five frequencies.

As described earlier that the sum-of-pairs scoring scheme when we discussed multiple alignments. This scheme combines in a single number both the overage and the uniformity of a column. The larger the coverage, the larger the sum-of-pairs value, provided the uniformity is high. Also, for constant coverage. The sum-of-pairs value increases with uniformity, in contrast, the entropy is independent of overage. It measures uniformity alone. Depending on the situation, a scientist may want to use sum-of-pairs, a combined measure, or the coverage and entropy separately.

Coverage

As already defined, the coverage of a set of fragments with respect to the target from which they were obtained. Here we define an analogous concept for a layout, a fragment covers a column *i* if it participates in this column either with a character or with an internal space. Let *i* (*f*) and *r* (*f*) be the leftmost position and the rightmost position, respectively, of a character of f (or of its reversed complement) in the layout. The previous definition means that fragment *f* covers column *i* if $l\ (f) < i < r\ (f)$. Having defined what the coverage is in each column, we may speak of *minimum*, *maximum* and *mean* coverage of a layout.

The meanings are as expected. If the coverage ever reaches value zero for a column *i*, then we do not really have a connected layout. Indeed, no fragment image span column *i*, so the parts of the layout to the left and to the right of *i* are in some independent, for instance, the right part could be placed to the left of the other, giving an alternative layout as good as the original one. This means that if we have more columns with zero coverage, any permutation of the intervening regions is acceptable. Each one of these regions corresponds to what we have called a contig. Coverage is also important to five confidence to the consensus sequence. If there are several fragments covering a column, we have more information on the particular base call for this column. It is also important to have enough fragments form both strands in the layout.

Linkage

Apart form the score and the coverage, the way individual fragments are linked in the layout is another determinant of layout equality, the coverage is high, but there are no real links between the fragments. Each block is composed of many almost exact copies of a certain sequence, but the blocks themselves do not seem related at all. Fragment images should have overlapping ends to show some evidence of linkage.

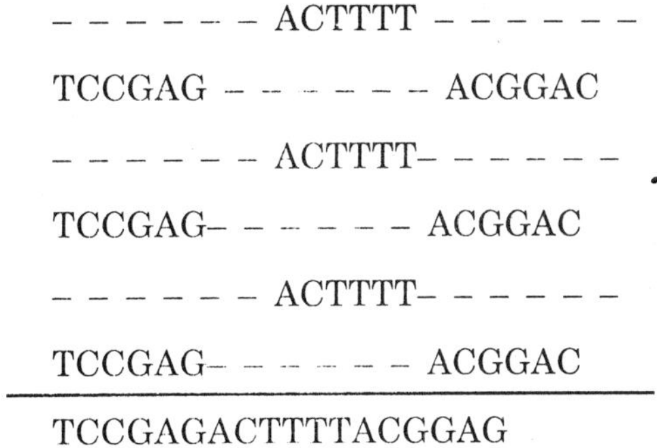

Fig. 4.23: Good coverage but bad linkage.

Assembly in Practice

A good assembly program must balance these requirement and find all solutions that satisfy them reasonably well. This is no easy task, and, coupled with the fact that no perfect formalisms exist, it makes algorithm design very difficult. As a result, practical implementations often divide the whole problem in three phases, each one using results produces by the previous phases, a typical division encompasses the following phases:

(1) Finding overlaps

(2) Building a layout

(3) Computing the consensus

Such partitioning makes the problem more manageable, because each phase can be treated more or less independently as a different problem. However, it becomes more difficult to understand the relationship between the initial input and the final output. Often, the methods used in these phases are heuristic, and no clear guarantees can be given with respect to solution quality, nevertheless, many existing implementations perform well in practice and it is worthwhile to know some of the techniques used.

Ecotonic Paths

The first step in any assembly program is fragment overlap detection. We need to try all pairs of fragments and their reverse complements to determine whether there is a prefix of a fragment that matches well with the suffix to another. In general, we will need all such pairs. Also, we are interested in fragments entirely contained in other fragments. All these comparisons are made within an error threshold. We mention here a way of comparing two fragments with the goal of finding such overlaps.

In a real problem we could apply this method to all pairs, including reverse complements, but this is seldom done, because it would take too long due to its being based on a quadratic time algorithm. However, it is one of the most reliable methods of assessing overlap. The comparison method consist in applying the dynamic programming algorithm, which is a variant to the basic algorithm. This variant does not charge for gaps in any extremity. Figure 4.23 represents the results of using such an algorithm for two fragments. The score system used in this comparison is 1 for matches, –1 for mismatches, and –2 for gaps. Notice that the algorithm, does not charge for the gaps after the first sequence or before the second one, so it was able to find a prefix-suffix similarity. The same algorithm would find an approximate containment.

Linear arrangement

In this section we briefly discuss one aspect of layout construction in a real assembly problem, namely, the issue of finding a good ordering of the fragments in a contig. Given such an ordering, and knowing that each fragment overlaps with the next one, a layout can be built using the methods of the next section. Unfortunately, there is no algorithm, that is simple and general enough, even restoring to heuristics. Therefore, rather than presenting an algorithm, we point out important considerations that should be taken into account when trying to construct good fragment orderings (Fig. 4.23).

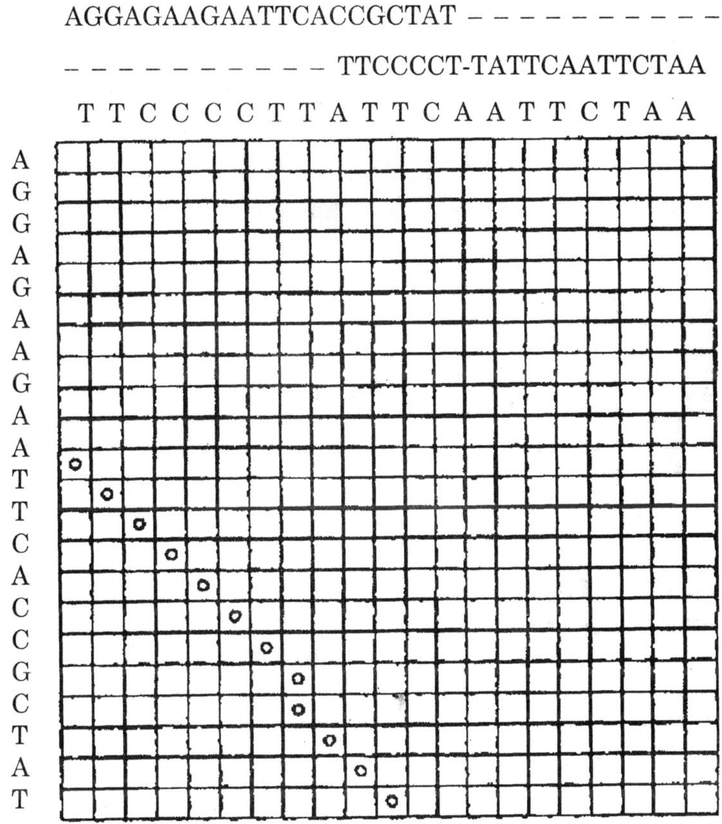

Fig. 4.24: An optimal semiglobal alignment and its corresponding path in the dynamic programming matrix. The algorithm is configured to find prefix-suffix or containment alignments, which are appropriate for fragment assembly.

In a real instance we have to deal with reverse complements, so it is necessary to extend the input collection $\mathcal{F}$ with the reverse complements of the fragments. Define

$$\bar{F} = \{\bar{f} \mid f \in F\}$$

and

$$DF = F \cup \bar{F}$$

In a real problem we have to worry about errors as well, and this means considering approximate matching, we will use the notation $f \rightarrow g$ in a slightly different sense from previously: It now indicates that the end of fragment f is approximately the same as the beginning of fragment g in some sense. We also assume that whenever $f \rightarrow g$ we also have $\bar{g} \rightarrow \bar{f}$, because it is reasonable to expect that whatever criterion is used to assess the similarity between two sequences, the same criterion would apply to their reverse complements. This can be shown as follows: If $f \rightarrow g$, we have that $f = uv$, $g = wx$, and v and w are similar. This implies that $\bar{g} = \bar{x}\bar{w}$, $\bar{f} = \bar{v}\bar{u}$ with $\bar{w}$ and $\bar{v}$ similar, which is equivalent to $\bar{g} \rightarrow \bar{f}$. We will use the notation $f \subseteq g$ meaning that f is an approximate substring of g. An argument analogous to the one just presented shows that $f \subseteq g \Rightarrow \bar{f} \subseteq \bar{g}$.

The graph we will be interested in is then $\mathcal{OG}(\mathcal{DF})$, with edges indicating approximate overlap. In addition it is important to know which fragments are approximate substrings of others.

Finding a good ordering of overlapping fragments is essentially equivalent to finding directed paths in $\mathcal{OG}(\mathcal{OF})$. These paths originate contigs. Notice that for any path

$$f_1 \rightarrow f_2 \ . \ . \ . \rightarrow f_k$$

in this graph there is a corresponding "complementary" path

$$\overline{f_k} \rightarrow \overline{f_{k-1}} \rightarrow \ldots \rightarrow \overline{f_1},$$

which gives rise to a contig with a consensus that is the reverse complement of the first path. This means that both strands of the target DNA are being constructed simultaneously. Ideally we would like to get a pair of complementary paths such that every non-contained fragment belongs to exactly one path in the pair. Contained fragments can be positioned afterward so their presence in the paths is not essential. Actually, forcing their presence may prevent the formation of a path. In rare occasions contained fragments may provide an important link, so we consider them optional fragments when building a path. Two obstacles may prevent the ideal goal of a pair of paths covering all non-contained fragments: lack of coverage and repeats.

A disconnected overlap graph is usually an indication of lack of coverage. Repeats cause the appearance of cycles in the overlap graph, if coverage is good, or may show up as edges $f \rightarrow g$ and $f \rightarrow h$ with g and h unrelated (g and h are unrelated if $g \not\rightarrow h$, $h \not\rightarrow g$, $g \not\subseteq h$ and $h \not\subseteq g$). This happens when f is a fragment near the end of a repeat X and g and h are fragments leading from X to distinct flanking regions in the target. A cycle involving both f and $\bar{f}$ for a certain fragment f indicates the presence of an inverted repeat. Another indication of repeated regions is unusually high coverage.

All fragments from all copies of a given repeat are piled together, causing coverage to increase to about the average value times the number of repeat copies (assuming an approximately uniform distribution of fragments along the target). In conclusion, there are four issues to keep in mind when building paths: (1) Every path has a corresponding complementary path; (2) it is not necessary to include contained fragments; (3) cycles usually indicate the presence of repeats; and (4) unbalances coverage may be related to repeats as well.

Construction

We briefly discuss in this section the problem of building a layout form a path in an overlap graph. This is straightforward if the overlaps are exact, but approximate overlap require extra care. We present here two techniques relate to alignment construction. The first one helps in building a good layout from a path in the presence of errors. The second one focuses on locally improving an already constructed layout. To introduce the first technique, we use the following example.

Example 4.7: Suppose we have a path $f \to g \to h$ with

f = CATAGTC

g = TAACTAT

h = AGACTATCC

The two optimal semiglobal alignments between f and g are

CATAGTC – – –   CATAGTC – – –

– – TAA–CTAT   – – TA – ACTAT

We do not have any reason to prefer one of them over the other, since their score is the same. However, when sequence h enters the scene, it becomes clear that the second optimal alignment (where the second A of g is matched to a T in f) is better:

CATAGTC – – – – –
– – TA-ACTAT – –
– – – AGACTATCC

CATAGACTATCC

The consensus S satisfies $d_s(f, S)$, = 1, $d_s(g, S)$ = 1, and $d_s(h, S)$ = 0. Had we aligned A with G, the voting in column 6 would have been ambiguous, with T, A, and – all tied with one occurrence each. In the consensus any choice other than A in this column would have led to a worse result, in the sense that the sum $d_s(f, S) + d_s(g, S) + d_s(h, S)$ would be greater then with the first solution. To avoid such problems, we use a structure that represents alignments distinguishing between bases that are actually matched to some other base, and bases that are "loose", as is the case of the second base A of sequence g in the above example.

We can implement this idea as follows. Each sequence is represented by a linked list of bases. When an alignment is formed, matched bases are unified and become the same node in the overall structure. If unsure about matching a particular base, we can leave it unmatched, for the sequences f and g. then, as new sequences are added, some of their bases will be unified according to suitable alignments, the structure can be seen as an acyclic graph, and a traversal in topological order yields a layout. Applying this technique, we get the best possible layout. When h is added, an arrow form G to the A node just below it will be added, as well as additional bases that belong to h alone.

Our second technique concerns improving a given layout, one common problem in practice is illustrated by the following example.

Example 4.8: Figure 4.25 shows two layouts for the same set of sequences. In the left-hand layout, the consensus ends up with two TS by majority voting, but four of the five sequences have just one T. The layout to the right is better, for its consensus agrees with the majority of the sequences. One idea to solve such problems is to perform a multiple alignment locally. We select a short range of consecutive columns that need doctoring, extract form each fragment its portion contained in these columns, and align the portions, for moderate values of the local coverage and width of the range the multiple alignment algorithm will run in a reasonable amount of time, despite its exponential complexity. A sum-of-pairs scoring scheme seems adequate here. In Example, the right-hand layout maximizes the sum-of-pairs score, assuming $p(a, b) = 1$ if $a = b$, $p(a, b) = 0$ if $a = b$, $p(a, -) = -1$, and $p(-, -) = 0$.

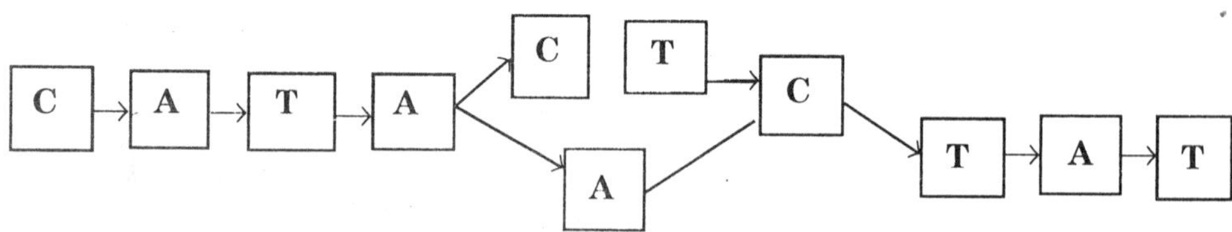

Fig. 4.25: Linked structure for approximate overlaps.

ACT-GG	ACT-GG
ACTTGG	ACTTGG
AC-TGG	ACT-GG
ACT-GG	ACT-GG
AC-TGG	ACT-GG
ACTGG	ACT-GG

Fig. 4.26: Two layouts for the same sequences. The layout to the right is better, but an assembly program might produce the left one due to ambiguities in pairwise alignments.

CONCLUDING REMARK

In this chapter we studied the fragment assembly problem, which appears in large-scale DNA sequencing. Basically, it consists of assembling in the correct order and orientation a collection of fragments coming from a long, unknown DNA sequence. We began by reviewing the biological setting of this problem, and how the biological factors influence the input. Then we presented attempts that have been made to capture the problem into a mathematical framework.

The proposed models fail to address all the issues involved. In particular, it is hard to deal with repeated regions and to take into account information about the size of the target sequence. Also, all models assume that the collection of fragments is free of chimeras and that there is no vector contamination, algorithms based on the models were also presented. Fragment assembly requires a particular kind of multiple alignment, also called layout, in which we have to worry about fragment orientation, coverage, and linkage.

We do not charge for end spaces when scoring this alignment. Because of the difficulties in finding a precise formulation for the problem, heuristic methods are often used in practice. The problem is usually broken into three phases: finding overlaps between fragments, building a layout, and computing the consensus. Programs based on this approach perform well in real instances without repeats. We presented an overview of such heuristics. New techniques, such as sequencing by hybridization, may radically change the way large-scale sequencing is done. If this happens, new algorithms will be required.

Chapter 5

Mapping Models

There are about 10^8 base pairs in a DNA of a single human chromosome. The techniques we have studied so far for sequencing are restricted to pieces of DNA with up to tens of thousands of base pairs. This means that whenever we sequence such a piece we will be looking at an extremely small part of a chromosome. It is as though we were viewing Earth from the moon with a telescope that enabled us to see features that are a couple of inches apart (a DNA base) but with a field of vision no greater than a mile (a piece of DNA with 15,000 bases. Such a telescope would not be a good instrument to study large-scale structures, such as mountain ranges, continents, and islands. To perceive those structures, we would need some other instrument. Analogously, molecular biologists use special techniques to deal with DNA molecules comparable in size to a chromosome. These techniques enable them to create maps of entire chromosomes or of significant fractions of chromosomes. In this chapter we examine computational techniques that can help biologists in the map-generation process.

CRITERIA FOR MAPPING

A *physical map* of a piece of DNA tells us the location of certain *markers* along the molecule. These markers are generally small but precisely defined sequences. Such a map helps molecular biologists further explore a genome. For example, suppose a certain stretch of DNA has been completely sequenced, giving us sequence S. If we know which chromosome S came from, and if we have a physical map of this chromosome, we could try to find one of the map's markers in *S*. If we succeed, we have located S in the chromosome. So the question now is. How do we create such maps? The first task is to obtain several copies of the DNA molecule we want to map (known as the target DNA). Each copy must then be broken up into several fragments, using restriction enzymes.

Mapping is done by carefully comparing the fragments obtained, in particular by observing overlaps. Note that in general each fragment is a DNA piece still too long to be sequenced, so we cannot determine overlap by sequencing and comparing fragments, as is done in fragment assembly. Instead, we obtain overlap information by generating fingerprints of the fragments. A fingerprint should describe part of the information contained in a fragment in some unique way, just like our fingerprints uniquely describe some part of ourselves. Two popular ways of getting fingerprints are restriction site analysis and hybridization. In restriction site mapping the aim is to locate the restriction sites of a given enzyme on the target DNA.

The restriction sites are thus the markers previously mentioned. The technique used for location is based on fragment length measurement; a fragment's length is thus its fingerprint. In hybridization mapping we check whether certain small sequenced bind to fragment. In hybridization mapping we check whether certain small sequences bind to fragments; the subset of such small sequences that do bind to the fragment become its fingerprint. The fingerprint of a fragment can also be given by hybridization information and fragment lengths, but in this chapter we consider each technique separately for clarity of exposition. In any case, by comparing fingerprints we try to determine whether the fragments overlap and thus determine their relative order (which was lost during the break-up process). Just as in other problems from molecular biology, possible lack of information and the presence of numerous experimental errors make the physical mapping problem especially hard.

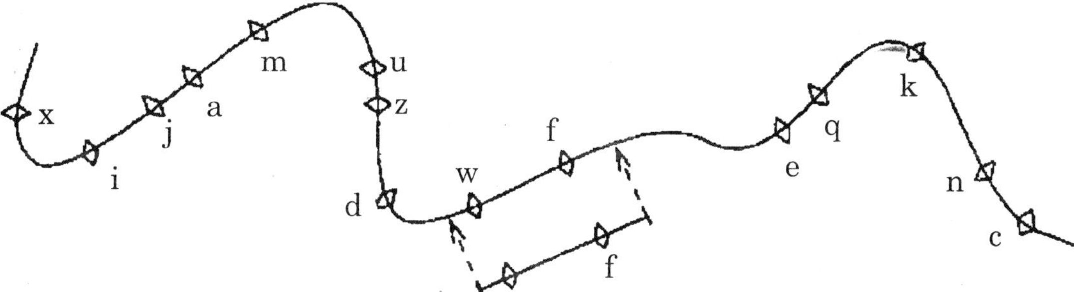

Fig. 5.1: Sketch of a physical map. In this figure a very long DNA molecule (e.g., a chromosome) is represented by the curved solid line. Markers along the molecule are represented by the diamond symbols. Each marker is distinct, and this is shown by the letters associated with each. Segment S represented a DNA fragment that has been sequenced, containing, for example, a gene. Thanks to the map, S can be located on the chromosome.

In particular, it may not be possible with a given collection of fragments to obtain one contiguous physical map. This may happen simply because the fragmentation process did not produce fragments covering certain sections of the target DNA. When this happens, the physical map pieces are called **contigs** (the same name used in fragment assembly). This is a feature common to all physical mapping processes. We now discuss each of the techniques mentioned above in more detail.

Limitations of Mapping

In this section we briefly present two possible techniques for mapping based on measuring lengths of fragments between restriction sites. One way to obtain fragments of many different sizes is to apply not one, but two restriction enzymes to the target DNA. We will explain this technique by an example. Suppose we have two restriction enzymes, A and B, each

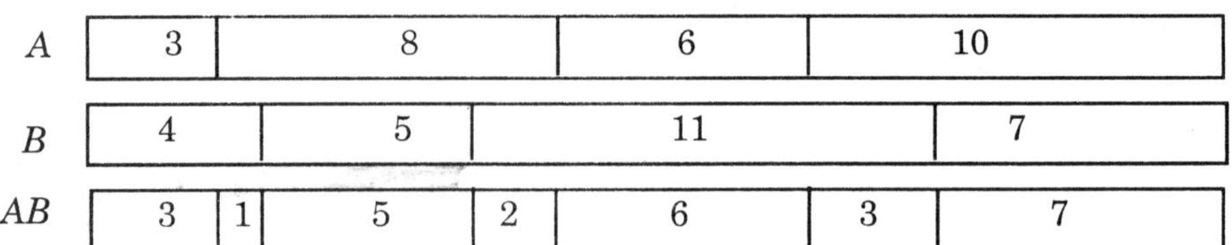

Fig. 5.2: Solution to a double digest problem.

recognizing a different sequence (restriction site). Applying A to one copy of the target DNA, we obtain fragments measuring 3, 6, 8, and 10 length units (for example, thousands of base pairs). Applying B to another copy of the target DNA, we obtain fragments measuring 4, 5, 7, and 11 units. Applying now both enzymes to a third copy of the target DNA we obtain fragments measuring 1, 2, 3, 3, 5, 6, and 7 units. The problem now is, How can we order these fragments in such a way that the order we determine is consistent with experimental results? This is known as the *double digest problem*, because the process of breaking a molecule by an enzyme is known as a digest.

A variant of the double digest approach is partial digest. Here we subject the target DNA to one enzyme only but perform many experiments on copies of the DNA, varying the time during which the enzyme acts on each copy.

By giving more or less time to the enzyme, more or less restriction sites will be recognized, thus yielding fragments of different lengths. Ideally the experiments should provide us with at least one fragment for every pair of restriction sites. We then try to pinpoint the location of the restriction sites by analyzing fragment lengths. For example, enzyme A recognizes three restriction sites. A partial digest experiment based on this enzyme would result in the following fragment sizes: 3, 11, 17, and 27 (fragments between the left endpoint and all other sites); 8, 14, and 24 (fragments between the first restriction site and those to its right); 6 and 16 (fragments between the second restriction site and those to its right); and 10 (last fragment). Several kinds of experimental errors may occur with digestion data. First, there is uncertainty in length measurement. This is done by gel electrophoresis, where there could be an error of up to 5%. Second, if fragments are too small, it may not be possible to measure their lengths at all. Third, some fragments may be lost in the digestion process, leading to gaps in the DNA coverage. As can be expected, all these errors bring complications to the mapping process. We will see later on that even when there are no errors the associated computational problems seem very hard.

A Clone Library

Recall that, in hybridization mapping, overlap information between fragments is based on partial information about each fragment's content. Before explaining what this partial information is, we note that in this context fragments are called clones, replicated using a technique called cloning. After each copy of the target DNA has been broken up (in a different way), and after each resulting fragment has been cloned, we obtain a collection of many thousands of clones (a clone library), each clone being typically several thousands of base pairs long. Partial Information about a clone is obtained through hybridization experiments. In these experiments we try to verify whether a small sequence known as a probe binds (or hybridizes) to the clone. If the binding does occur, this means that the clone contains the sequence complementary to the probe sequence.

We can do this experiment on a clone using several different probes. The clone's fingerprint will be the set of probes that successfully hybridized to the clone. Two clones that share part of their respective fingerprints are likely to have come from overlapping regions of the target DNA. For example, if we know that probes x, y, and z bound to clone A, whereas probes x, w, and z bound to clone B, we have good reason to believe that clones A and B overlap each other, for example (unless there are repeats; more on this later). Notice that with this kind of information we will not in general be able to tell the location of the probes along the target DNA, but only their relative order. We return to this important point further on.

Many kinds of errors may happen in hybridization experiments. First, a probe may fail to bind where it should; this creates a **false negative**. Second, a probe may bind to a site where it should not; this creates a **false positive**. Sometimes there is human misreading of experimental results, also resulting in false negatives or false positives. Errors may have appeared even before the hybridization itself. During the cloning process, two separate pieces of the target DNA may join and be replicated as if they were one single clone. This is known as a **chimeric clone**, and from it false inference about relative probe order can be made. Chimerism occurs frequently and is thus one of the most severe problems in hybridization mapping.

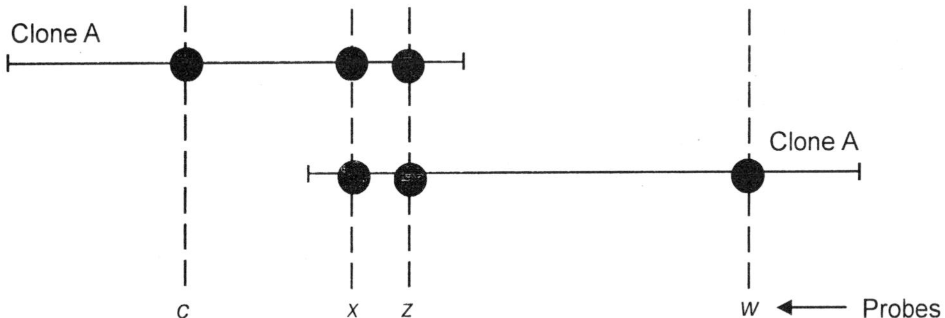

Fig. 5.3: An example of how four probes can hybridize to two clones.

Some estimates say that in many clone libraries between 40% and 60% of all clones are in fact chimeric. Another type of error in the cloning process is the deletion, in which an internal piece of a clone is lost, thus again joining two fragments that are not contiguous in the target DNA. Two other situations can cause problems to hybridization mapping. One happens when probes are not unique, meaning that there may be more than one site along the target DNA to which the probe can bind; these are the *repeats*. A technique called **Sequence Tagged Site** (STS) avoids this problem by generating probes that hybridize to unique sites along the target DNA. Another problematic situation is simply lack of data, as it may not be feasible to perform all hybridization experiments required. Pooling techniques have been developed to deal with this problem.

MODELS

Having looked at the various mapping strategies, we now review some of the corresponding mathematical models. Our aim is to explore the computational complexity of the mapping problems derived from these models. As we will see, mapping appears to be algorithmically hard, but it also brings up several interesting issues in problem solving that we discuss at the end of this section.

Fragmentary Models

Initially we will look at a model for the double digest problem, which we shorten to DDP. In this model we consider each fragment to be represented by its length. We will assume that the digestion process is perfect, meaning that lengths have no measurement errors, and all fragments are present. This is clearly unrealistic, but as we will see shortly even with these oversimplifications the resulting computational problem appears to be very hard. Placing the experimental results in the model framework, we get a collection of lengths positive integers) for each experiment performed. Digesting the target DNA by the first enzyme gives us collection A = $(a_1, a_2, \ldots, a_n)$. Some of these fragments may have the same length, which means that this collection is actually a miltiest. Similarly, from the second enzyme we get $B = \{b_1, b_2, \ldots, b_m\}$. Finally, from the digestion with both A and B we get the moltiset $O = \{o_1, o_2, \ldots, o_k\}$. Given these multisets, what is it that we want to find? We want a permutation π_A of the elements from A and a permutation π_B of the elements from B, such that the following can be achieved.

We plot the lengths from A on a line according to the order given by π_A. Similarly, we plot the lengths from B following π_B, but we do it on top of the previous plot. Due to overlap, we may now have several new subintervals. It should now be possible to establish a one-to-one correspondence between each resulting subinterval and each element from O. If this can be done, we will say that π_A and π_B are solutions to the DDP. Finding these permutations is an NP-complete problem, and here is the proof. First, it is not difficult to see that given a solution to the DDP it is easy to check whether it is a true solution. All we have to do is sort O and the plotted subintervals to determine whether we get identical results.

To complete the proof, we note that, as stated, the DDP is simply a generalization of the set partition problem, a well-known NP-complete problem. In the set partition problem we are given a set of integers $X = \{x_1, x_2, \ldots, x_i\}$ and want to know whether we can partition X into sets X_1 and X_2 such that the sum of all elements in X_1 is the same as those in X_2. This corresponds to a double digest problem in which A = X, B = {K/2, K/2}, and O = A, where

$K = \Sigma_{x \in X} x_i$.

In other words, the set partition problem is a DDP in which one of the enzymes produced only two fragments, and both of the same length. Considering that we are always seeking the true solution to the DDP, this problem has one other facet that makes this search a great deal harder.

The number of solutions may be exponential. To see this, consider the case where between two sites cut by enzyme A there are three sites b_1, b_2, b_3 cut only by enzyme *B*. In this case we will not be able to tell the order of fragments $[b_1, b_2]$ and $[b_2, b_3]$. In general, with a run of *k* restriction sites from one enzyme we will have (k − 1)! solutions. In fact, the situation is even worse than that because of the so-called coincidences. A coincidence happens when a restriction site for enzyme A is very close to a restriction site for enzyme B (they cannot exactly coincide; why?). When we have consecutive coincidences, the corresponding fragments for the A digestion and the B digestion will appear to have the same length (due to measurement inaccuracy). Therefore a run of n consecutive coincidences will also result in the possible solutions.

Taking this argument still further, it is possible to show that the number of solutions increases exponentially with the target DNA length. Thus we see that even with a very simple model double digestion poses serious computational problems. The NP-completeness of partial digest has not been proved, and in addition the problem seems to be more tractable for the following reason. In mapping by partial digest there are n restriction sites (including endpoints), and by experiment we obtain at least one fragment for every pair of sites, or $\binom{n}{2}$ fragments. When measured, these fragments yield a multiset of distances.

The problem is to determine the location of the sites on the line such that the set of distances between these locations is the same as the lengths obtained by the digests. What makes the problem easier is that researchers have shown that the maximum number of solutions is small, thus making partial digest, from a combinatorial point of view, a good alternative to double digest.

Finger Print Mapping

We can have a first look at the computational complexity of hybridization mapping (or more generally, fingerprint mapping) using simple models based on interval graphs. These models abstract away many of the complications of mapping; yet, simple as they are, they also give rise to problems that are NP-hard, as was the case with the double digest problem seen earlier. In these models we create graphs whose vertices represent clones and whose edges represent overlap information between clones. If we had complete and correct information about clone overlapping, the resulting graph would be an interval graph, and this is a key observation in these models. In the first model, we create two graphs. The first graph is $G_r = (V, E_r)$. If i, j) $\in E_r$, this means that we know for sure that clones i and j overlap. The second graph is $G_t = (V, E_t)$, where E_t represents known plus unknown overlap information (thus $E_r \subseteq E_t$). G_t is not necessarily complete, because we may know for sure that certain pairs of clones do not overlap and hence the corresponding edge would be left out of E_t. The problem we want to solve is this: Does there exist a graph $G_s = (V, E_s)$ such that $E_r \subseteq E_s \subseteq E_t$ and such that G_s is an interval graph? Notice that it is not necessarily the case that G_r or G_t are interval graphs (in which case a solution would be trivial).

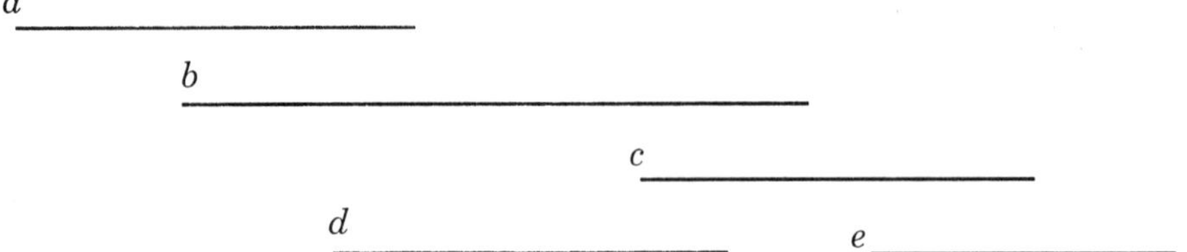

Fig. 5.4 : Example of clone overlapping patterns.

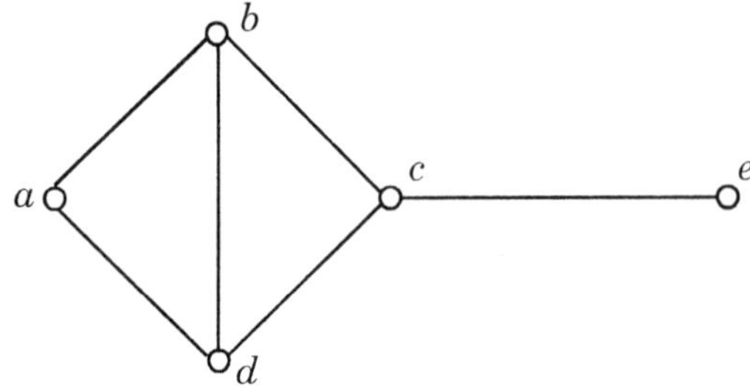

Fig. 5.5 : The interval graphs corresponding to Fig. 5.4.

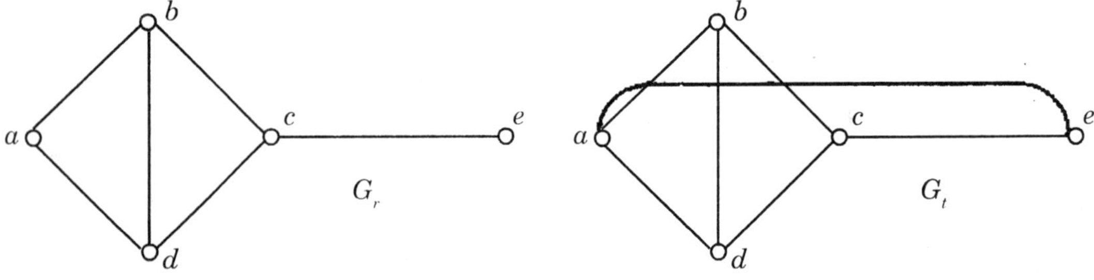

Fig. 5.6: Possible graphs G_r and G_t for the example from Fig. 5.4. None of them is an interval graph.

In the second model, we do not assume that the known overlap information is reliable. We construct a graph G = (V, E) using that information anyway. The problem we want to solve is this: Does there exist graph G' = V, E') such that E' $\subseteq$ E, G' is an interval graph and |E'| is maximum? The requirement that E' $\subseteq$ E means that, to get a solution for this problem, we may have to discard some edges of G, which would be interpreted as false positives. The solution is the interpretation that has the minimum number of false positives. In the third model, we use overlap information together with information about the source of each clone as follows. Because clones come from different copies of the same molecule, a reasonable idea is to label each clone with the identification of the molecule copy it came from.

We assume we have k copies of the target DNA, and different restriction enzymes were used to break up each copy. We now build a graph G = V, E), with known overlap information between clones. Because clones are labelled, we can think that the corresponding vertices are colored, using k colors (one colour for each molecule copy). The graph constructed will not have an edge between vertices of the same colour, because they correspond to clones that came from the same molecule copy and hence cannot overlap. WE thus say that the graph has a valid coloring. The problem we want to solve is this: Does there exist graph G' = (V, E') such that E $\subseteq$ E', G' is an interval graph, and the coloring of G is valid for G'? In other words, can we add edges to G transforming it into an interval graph without violating the coloring? We have just described three different models for hybridization mapping, and all are relatively simple. However, it turns out that all three problems described are NP-hard. We will not give the proofs here; they can be found in the references cites at the end of the chapter. We now proceed to describe more specialized models.

Clone Fingerprint

The models described in the previous section apply to hybridization mapping, but in principle they can be used in any situation where we can obtain some kind of fingerprint for each fragment. We now present a model that uses as a clone fingerprint the set of probes that bind to it. In this model we make the following assumptions:

(i) The reverse complement of each probe's sequence occurs only once along the target DNA ("probes are unique").

(ii) There are no errors.

(iii) All "clones × probes" hybridization experiments have been done.

The reader might think that another NP-hard problem is coming up. For once, this is not the case. As we will see the resulting problem is polynomial-time solvable, although some of its generalizations are indeed NP-hard. Assuming there are n clones and m probes, the experimental data will enable us to build an n × m binary matrix M, where entry M_{ij} tells us whether probe j hybridized to clone i (M_{ij} = 1) or not (M_{ij} = 0). Obtaining a physical map from this matrix becomes the problem of finding a permutation of the columns (probes) such that all is in each row (clone) are consecutive.

A binary matrix for which such a permutation can be found is said to have the consecutive 1s property for rows, or CIP for short. Below we also use the terms CIP permutation and CIP problem, with the obvious meanings. Verifying whether a matrix has this property and then finding a valid permutation is a well-known problem for which polynomial algorithms exist. However, the model is clearly too simple. If errors are present, even in the true permutation some rows may not have all is in consecutive positions. So why bother? The answer is that looking at the error-free case helps us to gain insights that are useful when designing algorithms for data with errors. In particular, we can use the following important assumption.

Laboratory researchers try to minimize the number of errors in their experiments. If their experiments were *perfect*, the resulting hybridization matrix would have the CIP. This means that any algorithm that tries to find or approximate the true column permutation should give solutions that minimize the number of situations that must be explained by experimental error. Hence, such algorithms should be able to find a CIP permutation if one exists. Note that even if a CIP permutation exists, we cannot claim that it is the true permutation. For one thing, there may be several such permutations for a given input matrix. How can we say which is the true one? The answer is that we cannot, in general. We can only say that all permutations that make all is consecutive likely candidates for the true permutation. For this reason an algorithm for the CIP problem should give us not one possible solution, but all of them.

In addition, given that errors do exist, it may even be the case, as noted above, that the true permutation is *not* a CIP permutation, even if there is one. On the other hand, how can we modify the model to account for errors? There are many possibilities. Instead of requiring exactly one block of consecutive is per row, we may try to find a column permutation such that in each row there are at most k blocks of consecutive 1s, where k's value could be 2 or 3. Or we may try to minimize the total number of blocks of consecutive 1s in the matrix. By doing this, we may expect that most rows will have just one block of consecutive 1s, while an occasional row might have four or five. Unfortunately, as already hinted above, these generalizations yield NP-hard problems. NP-hardness comes up again if we relax the assumption that probes must be unique, even if no errors are present.

Algorithmic Techniques

Before describing algorithmic techniques for obtaining physical maps, we would like to explore a bit the constraints forced upon us by the difficulty of the problem. These considerations give a new light on what it means to "solve" a problem using an algorithm, and should therefore be of interest outside physical mapping as well. The first consideration is that what we are really trying to solve is a real-life problem, not an abstract mathematical problem. And the real-life problem is this: We want to know what the true ordering of the clones is in the target DNA. It is a combinatorial problem:

There are finitely many possible orderings, and one of them is the one we want. We will try to discover this true ordering by means of abstract models that give rise to optimization problems. These problems may be themselves very hard, notwithstanding the fact that they are abstractions of a yet harder problem. For example, the optimization versions of all problems are NP-hard. In general we will not be able to find optimal solutions to these problems quickly; and even if we did solve one of them to optimally within a reasonable time, we will have no guarantee that the solution found corresponds to the true solution. Thus it looks like we are in a terrible predicament. However, the situation is not as bleak as it seems. First, since we have no guarantee that any optimal solution is the true solution, good approximate solutions may serve our purposes just as well. Second, we must remember that algorithms in molecular biology should be used in an iterative process. The scientist obtains experimental data and feeds it to an algorithm.

Based on the solution or solutions given by the algorithm, the scientist can make more experiments and obtain more data and thus use the algorithm again. At each step we may expect that "better" (or "closer to the truth") solutions will be found. These observations suggest desirable features of a mapping algorithm, which we now list:

(i) It should work better with more data, assuming that the error rate stays the same.

(ii) It should present a solution embedded in a rich framework of details, in particular showing how the solution was obtained, distinguishing "*good*" parts of the solution (groups of clones for which there was strong evidence for the ordering reported) from "*not so good*" parts. This greatly facilitates further experiments.

(iii) If several candidate solutions meet the optimization criteria, all of them should be reported. If to many solutions are reported, the optimization criteria may be too weak (or the input data may contain too many errors). Conversely, if no solutions are reported, the optimization criteria may be too strong.

We would also like to have some guarantee that the true solution is among the set of approximate/optimal solutions found by the algorithm. In general this will not be possible. Instead, we try to obtain a few different optimization functions and respective algorithms whose optima share properties with the true solution. The hope is that the true solution will lie in the intersection of the solution sets found by each algorithm with high probability. Note that instead of using several algorithms as suggested, we may try to design an algorithm that can optimize multi-objective functions. This is an important research area.

C1P Problem

In this section we present a simple algorithm that determines whether an $n \times m$ binary matrix M has the C1P for rows. The goal of the algorithm is to find a permutation of the columns such that in each row all 1s are consecutive. For simplicity we assume all rows are different (that is, no two clones have the same fingerprint) and that no row is all zeros (that is, every clone is hybridized by at least one probe).

Understanding the algorithm requires that we look first at the way rows can relate to each other. For this we need a definition.

Definition 5.1. For each row i of M, let S_i be the set of columns k where $M_{i,k}=1$.

With this definition we can now relate a row to another row. Given two rows i and j three situations can arise:

1. $S_i \cap S_j \emptyset$

2. $S_i \subseteq S_j$ *or* $S_j \subseteq S_i$

3. $S_i \cap S_j \neq \emptyset$,

and none of them is a subset of the other.

We can clearly deal separately with rows that fall into the first of the these situations, since the permutation of columns corresponding to the elements of S_i will not interfere with those from S_j. In such a case we place i and j into different components and deal with each component separately. Now, how about rows that have non-empty intersection? (We are abusing language a bit here and speaking of rows instead of their respective sets S.) Should they all be in the same component? Let us look at this case more carefully. Let us initially lump together in a component all rows that have non-empty intersection.

Suppose there is a row k in this component such that S_k is either a subset of or is disjoint from S_i for all $i \neq k$ in tis component. It is clear that we can leave row k out (and any others with the same property) while dealing with this component. Row k will thus be in its own component, possibly sharing it with other rows. What exactly these other rows are will be made clear. Based on the intuition just given, we give a formal way to determine when two rows belong to the same component. We build a graph G_C using matrix M. In

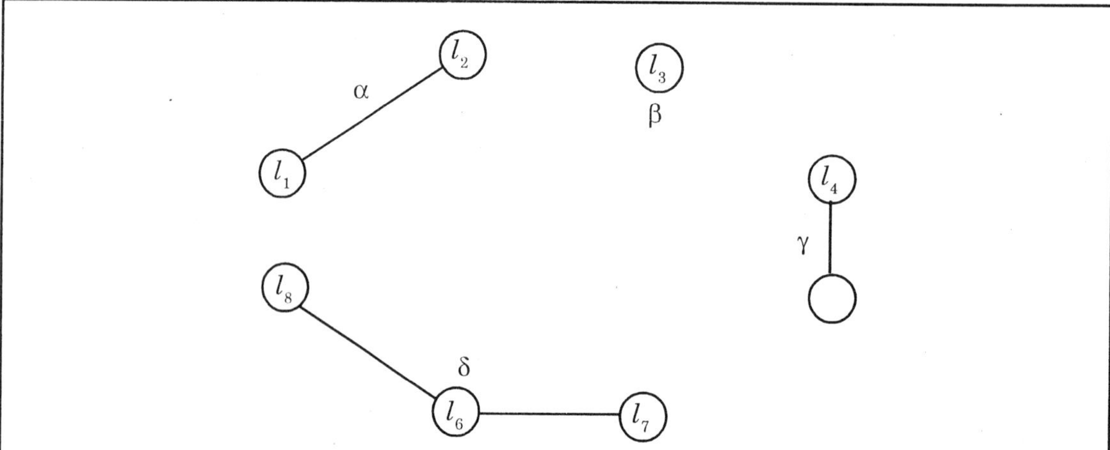

Fig. 5.7: Graph G_C corresponding to the matrix of Table 5.1. Connected components are indicated by Greek letters.

G_C each vertex will be row from M. There is an undirected edge from vertices i to j if $S_i \cap S_j \neq \emptyset$ and none of them is a subset of the other. The components we want will be the connected components of G_C. As an example, We can now sketch the algorithm:

Separate rows into components according to the rule above, permute the columns of each component, and then somehow joint components together. This leaves us with two subproblems: how to find the right permutation for one component, and how to join components together. Let us first work on the column permutation inside a component.

Table 5.1 : A binary matrix

	c_1	c_2	c_3	c_4	c_5	c_6	c_7	c_8	c_9
l_1	1	1	0	1	1	0	1	0	1
l_2	0	1	1	1	1	1	1	1	1
l_3	0	1	0	1	1	0	1	0	1
l_4	0	0	1	0	0	0	0	1	0
l_5	0	0	1	0	0	1	0	0	0
l_6	0	0	0	1	0	0	1	0	0
l_7	0	1	0	0	0	0	1	0	0
l_8	0	0	0	1	1	0	0	0	1

To explain this part of the algorithm, we shall make use of the example matrix shown in Table 5.2 (assume it is just a section of some matrix with more than three rows. Let us look first only at row l_1. If it were the only row in this component, all we would have to do is permute the columns such that all 1s become consecutive. Note that if there are k 1s in this row, we will have $k!/2$ possible permutations. We are not counting reversals as distinct solutions, because they do not matter for keeping 1s consecutive. In the case of row l_1 the 1s are in columns 2, 7, and 8. Then one set of possible permutations is 2, 7, 8, or 7, 2, 8 or 2, 8, 7. We can encode all these solutions by associating a set of possible columns for each element equal to 1 in a given row (we will call these sets the column sets). For just one row this set is set S defined above, and we describe the resulting configuration as follows:

Table 5.2 : A section of a binary matrix.

	c_1	c_2	c_3	c_4	c_5	c_6	c_7	c_8
l_1	0	1	0	0	0	0	1	1
l_2	0	1	0	0	1	0	1	0
l_3	1	0	0	1	0	0	1	1

$$\begin{array}{lccccc} & & \{2, 7, 8\} & \{2, 7, 8\} & \{2, 7, 8\} & \\ l_1 \rightarrow \ldots & 0 & 1 & 1 & 1 & 0 \ldots \end{array}$$

Now let us look at the following row, l_2. Because this row is in the same component as l_i it must be the case that l_2 has 1s in some columns where l_1 also has 1s, and some other 1s where l_1 does not. Now we have to place l_2 with respect to l_1 in a consistent manner.

We have two choices for this placement: Either put the *1*s of l_2 to the left of those of l_1 or to the right. This means placing column 5 either to the left or to the right of columns 2 and 7. Suppose we place 5 to the left. The possible solutions, encoded by the sets above each column, now become

		{5}	{2, 7}	(2, 7}	(8}		
$l_1 \rightarrow \ldots$	0	0	1	1	1	0	. . .
$l_2 \rightarrow \ldots$	0	1	1	1	0	0	. . .

If these were the only two rows in this component we would have arrived at two possible permutations for its columns: 5, 2, 7, 8 and 5, 7, 2, 8. If we had chosen to place 5 on the right, we would obtain the reverse of these two permutations. This means that the direction we choose to place the second row does not matter, and this is a key observation. Now let us consider the third row, l_3. Because it is in the same component as the other two, we know that in graph G_C we should find edge (l_3, l_1), or edge (l_3, l_2), or both.

In our example both edges are present. Let us now place l_3 with respect to l_2. It seems that again we have the choice of placing l_3 to the left or to the right of l_2. But now we have to take into account the relation between l_3 and l_1 as well. The way to do this is by considering the number of elements in the intersections between S_1, S_2, and S_3. Let $x \cdot y = |S_x \cap S_y|$, for any two sets x and y (the internal product of rows x and y). If $l_1 \cdot l_3 < \min(l_1 \cdot l_2, l_2 \cdot l_3)$, row l_3 must go in the same direction that l_2 was placed with respect to l_1. If $l_1 \cdot l_3 > \min(l_1 \cdot l_2, l_3)$, then we must place l_3 in the opposite direction used to place l_2 with respect to l_1 (we cannot have equality; why?). In either case, if the component is C1P, it must be the case that l_1 and l_3 are not properly placed, then the component is not C1P. The reason for this lies on the fact that the only choice we made up until now was in the placing of l_2 with respect to l_1, and we know that both possibilities (left or right) result in the same solutions up to reversal.

In our example above, $S_3 = \{1, 4, 7, 8\}$. Then 1.3 = 2, 1.2 = 2, 3.2 = 1, which means we have to put l_3 to the right of l_2, as in the following figure:

		{5}	{2}	{7}	{8}	{1, 4}	{1, 4}		
$l_1 \rightarrow \ldots$	0	0	1	1	1	0	0	0	. . .
$l_2 \rightarrow \ldots$	0	1	1	1	0	0	0	0	. . .
$l_3 \rightarrow \ldots$	0	0	0	1	1	1	1	0	. . .

We now know that the right permutation for columns 2, 5, 7, and 8 is 5, 2, 7, 8. Other columns may be determined as we consider more rows. But how do we handle additional rows? As we saw, in placing the third row we had no choice of placement. The same is true for all other rows in the same component. All we have to do for a new row k is find two previously placed rows i and j such that there exist edges (k, i) and (i, j) in G_C, and proceed as we did with the three-row case above. Note how the algorithm preserves possible permutations of a component in the sets above each column.

Considering that after placing the first two rows there are no other choices of placement for remaining rows, we see that this algorithm gives us all possible column permutations of a component having the C1P, up to reversal. In the example above, if the component had only three rows, the two possible permutations would be 5, 2, 7, 8, 1, 4 and 5, 2, 7, 8, 4, 1. As we have already mentioned, keeping track of all possible solutions is a desirable feature in the case of DNA mapping. An implementation of the preceding algorithm is simple Construct G_C and traverse it using depth-first search. When visiting a vertex invoke procedure Place. To check whether the algorithm has found a C1P permutation, we can verify the consistency of the column sets as we try each new placement, as indicated in the last line of procedure Place. If we attempt to place a row and we find out that one of its columns should go, say, to the right, and the existing placement tells us that it should go to the left, then the component, and hence the matrix, does not have the C1P.

```
Algorithm Place
    input: u, v, w, vertices of G_C=(V, E) such that (u, v) ∈ E and v, w) ∈ E.
    // v and w can have value nil
output: A placement for flow u, if possible
if v = nil and w = nil then
    Place all 1s of row u consecutively
else if w = nil then
    Left- or right-place the 1s of u with respect to the 1s of v
    Record direction used
else
if u . w < min(u . v, v . w) then
    Place u with respect to v using same direction used in v, w placement Record
    direction used
else
    Place u with respect to v using opposite direction used in v, w placement
    Record direction used
    Check consistency of column sets.
```

Fig. 5.8 : The procedure to place a row given two others.

Let us analyze the running time of this algorithm. Building graph G_C takes $O(nm)$ time. We then process the n rows as sketched above, spending $O(m)$ per row to check consistency of column sets. Total time is thus $O(nm)$.

Joining Components Together

We now tackle the second part of the algorithm, component-joining. For this we need another graph, G_M, which will tell us how components fit together. Each component of the original matrix M will be a vertex in G_M. A directed edge will exist between vertex α and vertex β if the sets S_i for all $i \in \beta$ are contained in at least one set S_j of component α (we are making no distinction between a row j and its corresponding set S_j). The graph G_M corresponding to the components of the matrix.

From the discussion in the beginning of this section we know that the containment relations between a set of β and sets of α are the same for all sets of β. Let us make this more clear. Suppose set S_i belonging to component β is contained in set S_j belonging to another component α. Because the rows i and j are in different components, there is no row k in component α such that S_i is not contained in S_k and $S_i \cap S_k \neq \emptyset$. In other words, S_i is contained in some sets of α and is disjoint from the others. We claim that the exact same containments and disjunctions hold for all other sets from β. Suppose not and let us analyze one such exception.

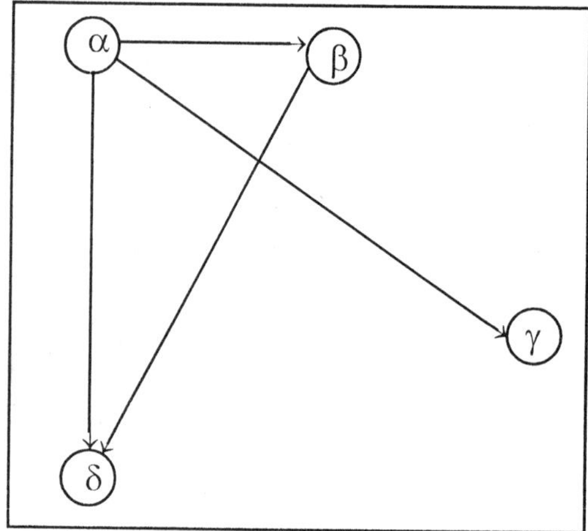

Fig. 5.9: Graph G_M corresponding to the components of the matrix

Suppose $i, l \in \beta$, $S_i \cap S_l \neq \emptyset$, $j \in \alpha$, and $S_i \subseteq S_j$, but $S_l \not\subset S_j$. In this case the only way l could be out of α is if $Sl \cap Sj = \emptyset$. But this cannot be, since S_i and S_l have at least one element in common and S_i is contained in S_j. We leave other cases to the reader. The joining of components depends on the way sets in one component contain or are contained in sets from other components. Intuitively, we should process first components that have sets that are not contained anywhere else. Because containment is given by edge direction, we have to process G_M in topological ordering. (We can only do this if G_M is acyclic. Is it? Yes, and this is not difficult to prove.) After having correctly permuted the columns of all components individually, we select all vertices in G_M without incoming edges, and freeze their columns.

We then take the next vertex in topological order. Suppose we are following edge (α, β). This means that we must find some sort of "reference column" in component α that will tell us how to place the rows of β. Choose the row l from β that has the left-most 1, and call the column where this 1 is cβ. We know that S_t is contained in some S_i of α but not in others. Find all rows from α that contain S_t, and find the left-most column where all such rows have 1s (and call this column cα). This is the reference column we are looking for, since we can now make cα and cβ one and the same. Let us illustrate this process with the matrix and the graph and let us consider its components in topological ordering. One such ordering is α, β, δ, γ. Here are the rows of component α with columns permuted so that its 1s are consecutive:

	{1}			{2, 4, 5, 7, 9}				{3, 6, 8}		
l_1 → . . .	1	1	1	1	1	1	0	0	0	. . .
l_2 → . . .	0	1	1	1	1	1	1	1	1	. . .

We will now join the next component, which is β, a singleton row:

	|		{2, 4, 5, 7, 9}		|	
l_3 → . . .	1	1	1	1	1	. . .

As should be clear, all we have to do is join l_3 with the rows from α, without any rearrangements:

	{1}		| {2, 4, 5, 7, 8}|					{3, 6, 8}		
l_1 → . . .	1	1	1	1	1	1	0	0	0	. . .
l_2 → . . .	0	1	1	1	1	1	1	1	1	. . .
l_3 → . . .	0	1	1	1	1	1	0	0	0	. . .

The next component is δ. Here are its rows, with consecutive 1s:

. . .	{9, 5}		{4}	{7}	{2}	. . .
l_6 → . . .	0	0	1	1	0	. . .
l_7 → . . .	0	0	0	1	1	. . .
l_8 → . . .	1	1	1	0	0	. . .

We can see that in component δ it is l_8 that has the leftmost 1 (in either column 5 or 9). In component α both rows l_1 and l_2 contain l_3, and it is the second column (representing a bunch of possible columns) that has all 1s. Therefore that second column should coincide with columns 9 or 5 from component δ. This is the result:

	{1}	{9, 5}	{4}	{7}	{2}		{3, 6, 8}			. . .
l_1 → . . .	1	1	1	1	1	1	0	0	0	. . .
l_2 → . . .	0	1	1	1	1	1	1	1	1	. . .

$l_3 \rightarrow \ldots$	0	1	1	1	1	1	0	0	0	...
$l_6 \rightarrow \ldots$	0	0	0	1	1	0	0	0	0	...
$l_7 \rightarrow \ldots$	0	0	0	0	1	1	0	0	0	...
$l_8 \rightarrow \ldots$	0	1	1	1	0	0	0	0	0	...

We finally glue in component γ. Its rows are

	{6}	{3}	{8}	
$l_4 \rightarrow \ldots$	0	1	1	...
$l_5 \rightarrow \ldots$	1	1	0	...

Joining with the rows above we obtain the final matrix with all 1s consecutive:

	{1}	{9, 5}		{4}	{7}	{2}	{6}	{3}	{8}
$l_1 \rightarrow \ldots$	1	1	1	1	1	1	0	0	0
$l_2 \rightarrow \ldots$	0	1	1	1	1	1	1	1	1
$l_3 \rightarrow \ldots$	0	1	1	1	1	1	0	0	0
$l_6 \rightarrow \ldots$	0	0	0	1	1	0	0	0	0
$l_7 \rightarrow \ldots$	0	0	0	0	1	1	0	0	0
$l_8 \rightarrow \ldots$	0	1	1	1	0	0	0	0	0
$l_4 \rightarrow \ldots$	0	0	0	0	0	0	0	1	1
$l_5 \rightarrow \ldots$	0	0	0	0	0	0	1	1	0

Notice that one of the column sets is not a singleton, indicating that we have found two solutions. In this particular case the reason is that columns 5 and 9 are identical. But in general there may be multiple solutions that do not involve just the permutation of identical columns. For example, 1, 2, 7, 4, 9, 5, 8, 3, 6 is another solution for this same matrix.

Multiple solutions may exist because G_M may allow different topological orderings, and this is compounded by the facts that each component may on its own have several solutions and each of them can be used in two ways (the permutation and its reversal). We now analyze the running time of the algorithm. Topological sorting of G_M takes time O(n+m). It is possible to preprocess the entries of M so that the queries needed when traversing G_M take constant time.

For example, we can store for each row the column where its leftmost 1 is. Such preprocessing takes at most O(nm). Therefore this is the time needed for joining components. Total time for processing each component was seen to be O(nm) as well, so this is the running time of the whole algorithm. Another algorithm exists that can find a C1P in time O(n+m+r), where r is the total number of 1s in the matrix. In addition, the algorithm encodes in a compact way all possible solutions. This algorithm, however, is considerably more complicated than the algorithm we presented in this section. References are given in the bibliographic notes.

MAPPING ANOMALIES

In the previous section we studied an algorithm that solves the consecutive 1s problem. We saw that such a problem is a good model of hybridization mapping when there are no errors and when probes are unique. If errors are present, some other approach is needed, and that is the subject of this section. Let us first examine the effect errors can have on a clones × probes binary matrix M. Suppose M is presented to us with the true column permutation. Given one row, if there are no errors, all its 1s will be consecutive. If a row corresponds to a chimeric clone, where two fragments were joined, we will see *two* blocks of 1s separated by some number of 0s (assuming no other errors are present in this row).

We will call a consecutive block of 0s bordered by 1s a *gap*. Notice that this is different from the use of the term *gap* in other chimeric fragment. If, on another row, there is a false negative, the corresponding 0 may separate two blocks of 1s, creating another gap, as shown below:

0 1 1 0 1 1 1 1 0 0

↑

A false negative

The gap will not be created if the probe was leftmost or right-most for this clone. Finally, a false positive may split a block of 0s in two, thus possibly creating yet another gap. In this way we see that there is a close correspondence between errors and gaps in the matrix.

Given the basic assumption that we want to avoid explaining gaps by experimental error as much as possible, a reasonable approach is to try to find a permutation where the total number of gaps in the matrix is minimum. Such an approach has the desirable property that, if there is a C1P permutation, it will have the minimum number of gaps. In other words, gap minimization can be seen as a generalization of the consecutive 1s problem. We have already mentioned that some extensions to the C1P are NP-hard. Such is the case also with the gap minimization problem just sketched. However, for this particular NP-hard problem, we can use many special techniques to get approximate solutions that we can expect to be reasonably good, as we show next.

Travelling Salesman Problem (TSP)

It turns out that gap minimization is equivalent to solving a well-known graph problem. This is the *travelling salesman problem* (TSP).

The input to this version of the TSP is a complete undirected weighted graph G. The vertices of G correspond to columns of the clones×probes binary matrix M; that is, they correspond to probes. For reasons that will soon become clear, we also have to add an extra column to M filled with zeros, and G must have the corresponding vertex. The weight on each edge of G is the number of rows where the two corresponding columns differ (this is also known as the *Hamming distance* between the rows). For example we have an example of a binary matrix, and we see the corresponding graph. We will now argue that a minimum-weight cycle in G corresponds to a column permutation in M with the least number of gaps.

Table 5.3 : A clones × probes matrix with added column p_6^*

	p_1	p_2	p_3	p_4	p_5	p_6^*
c_1	1	1	1	0	0	0
c_2	0	1	1	1	0	0
c_3	1	0	0	1	1	0
c_4	1	1	1	1	0	0

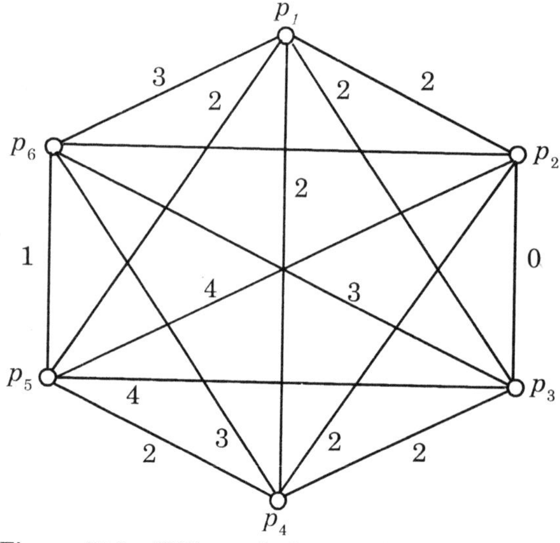

Figure 5.10 : TSP graph for matrix of Table 5.3.

To see this, note that given a permutation of the columns, a gap in a row means that at a certain point we have a transition from 1 to 0 and further on we have a transition from 0 to 1. Thus for each gap we have two transitions and each gap contributes exactly 2 to the weight of the cycle corresponding to the given column permutation. However edge weights may also be increased by *external transitions,* that is, transitions between elements in external (1 or m) columns, and these do not correspond to gaps. To ensure that every row has a pair of external transitions we include an extra column of zeros in column m + 1. Without such a column, cycles in the graph correspond to permutations where we allow consecutive

1s to wrap around in each row, and we do not want this to happen. The relationship between cycles and permutations now becomes.

cycle weight = number of gap transitions + 2n.

This means that for a given n minimizing cycle weight is the same as minimizing the number of gaps. What we have just done is to show how we can reduce the gap minimization problem to the TSP. It is well known that the TSP is an NP-hard problem, so in principle we have not accomplished much. However, a large array of techniques is available to solve or approximate traveling salesman problems, and these techniques can now be used in this context.

The existence of such techniques is not in itself enough to give us confidence that solving traveling salesman problems will give us the true probe permutation. What we need are *guarantees* that such solutions are, in some sense, close to the true solutions. Before that, let us go back to the gap minimization problem. We have defined above a function that, given an input matrix, returns the total number of gaps in the matrix. We have further argued that obtaining a permutation that has the minimum value for this function (or approximately minimum) will help us find the true column permutation. Since we do not have a guarantee that the true permutation will be among the solutions we find, we should look for other functions that might also be helpful.

The idea is that by carefully defining several such optimization functions and developing algorithms for them we will increase the likelihood of hitting upon the true solution. In particular, it is reasonable to expect that the true solution will be in the intersection of all solution sets. This will only be the case, however, if each function represents one property that true solutions do have or are likely to have. Here is an example of another optimization function.

One possible drawback of gap minimization is that in a permutation with a minimum value for this function one or a few rows may have many gaps, while others may have none. Having many gaps in one row is undesirable, since it would mean that one clone was subject to many more errors than other clones, which contradicts laboratory experience. Therefore we could try to minimize the number of gaps *per row*. The resulting graph problem is known as *the bottleneck travelling salesman problem,* which is also NP-hard.

Permutations

In this section we give a formal proof that the TSP approach outlined in the last section will give us, with high probability, the true permutation. The proof we present depends on two basic assumptions: that the number of probes is sufficiently large, and that the mapping process obeys a certain mathematical model. This model seems to be a good representation of what actually happens in large mapping projects. We describe the model next. First we assume that the DNA molecule we are dealing with is so long that we may think of it as in interval on the real line, extending from 0 to N. The clones are subintervals of this long interval, and we assume that all of them have the same length; for convenience each clone is one unit long. To simplify the exposition, we will speak of *clone permutations* rather than probe permutations.

The unit length assumption makes them equivalent. This means that we will be looking for consecutive 1s in columns, not in rows, and that each vertex in the associated TSP will correspond to a clone. We use each clone's left endpoint as a clone locator. We, of course, do not know the precise position of each clone along the molecule; and because we are dealing with hybridization, all we will be able to determine is relative clone order. A critical feature of this model is clone distribution along the target DNA.

We will assume that each of clone's position is an independent random variable, that clone locators are distributed uniformly over [0, N –1], and that the cones cover the interval [0, N] (that is, for every subinterval I of [0, N], there always exists at least one clone C such that $C \cap 1 \neq \emptyset$. Another important aspect of the model is probe distribution. We will *not* assume that each probe is unique; instead, we will assume that each probe occurs rarely along the target DNA. More formally, we will say that the occurrences of a given probe obey a Poisson process with rare γ. Moreover, the Poisson process of any one probe is independent of all the others. This part of the model lets us immediately obtain an expression for the probability of a specific probe hybridizing to a clone. The expression is

$$\Pr\{\text{a given probe occurs } k \text{ times in a given clone}\} = e^{-\lambda} \frac{\lambda_k}{k!} \tag{5.1}$$

This expression can be obtained from any textbook formula for Poisson processes, using the fact that our clones are unit length intervals. This completes the model description. We shall now argue that, given a clones x probes binary matrix, the row permutation given by solving the associated TSP is a good approximation to the true permutation, in the following precise sense: The probability that both permutations are the same tends to 1 as the number of probes increases. Note that the number of probes is *fixed* for a given instance of the problem.

We are just claiming that in larger and larger instances of the problem (and we are measuring size here by number of probes) the TSP permutation will be the same as the true permutation with higher and higher probability. To prove this claim, we must argue in terms of graph weights, or more appropriately, clone distances. As we mentioned, the weight of each edge of the graph associated with input matrix M is called the Hamming distance between its two endpoints (clones). We denote by h_{ij} the *Hamming distance* between clone i and clone j. We can also think of the *true* distance between clones. Denoting a clone's coordinates by l (left) and r (right) we can define this distance to be

$$t_{ij} = |l_j - l_i| + |r_j - r_i| = 2\ |l_j - l_i| \tag{5.2}$$

given that clones are all of the same size. Suppose now that we *knew* all true distances. Then it is clear that the largest such distance would give us the clones that are farthest apart, which is to say, the clones that occur at opposite ends of the interval [0, *N*]. The next largest such distance gives us another similar pair that occurs between the previous two, and so on. This means that given the true distance we are able to obtain the true clone permutation, which is not surprising. However, we have distances h_{ij} and not distances t_{ij}. But because we are trying to obtain only the true *relative* order of clones, it would suffice if we could say that, given and four clones i, j, r and s, $h_{ij} < h_{rz}$ implies that $t_{ij} < t_{rs}$ and *vice versa*. The reason is that some notion of *order* between cone distances was all we needed to place clones relative to each order using the true distances. If we prove that the probability that $h_{ij} < h_{rs} \Leftrightarrow t_{ij} < t_{rs}$ tends to 1 as the number of probes increases, we will have the result we need. Let us look at a pair of clones i and j, and let us find the probability that a certain probe p contributes to their Hamming distance. This will happen if probe p hybridizes to i but not to j or *vice versa*. The probability that probe p does *not* occur on clone j is $e^{-\lambda}$, since $k = 0$. On the other hand, the probability of probe p occurring at least one in clone i is the same as the complement of the probability of probe p not occurring in that section of i that does not overlap j. If this overlap is z_{ij}, we obtain the following result:

$$\Pr\{\text{p hybridizes to } i \text{ and not to } j \text{ or } vice\ versa) = p_{ij} = 2e^{-\lambda}(1 - e^{-\lambda(1-zij)}). \qquad (5.3)$$

This essentially means that there is a well-defined probability of this event happening. We can now take into account all probes, and each of them will have a certain probability of hybridizing to clones i and j. Assume we have m probes, and consider h_{ij}/m. This represents the mean contribution of each probe to the Hamming distance between clones i and j. (Note that a probe's contribution to the Hamming distance is either 0 or 1.) If the number of probes is large, we can invoke the law of large numbers and say that h_{ij}/m approaches p_{ij}. Or, for any fixed small positive real number $\in$, that

$$Pr\left\{\left|\frac{h_{ij}}{m} - p_{ij}\right| > \in\right\} \rightarrow 0$$

as $m \rightarrow \infty$. Note now that since clones are unit length, the true distance between two overlapping clones i and j with an intersection measuring z_{ij} is given by $t_{ij}=2 - 2z_{ij}$. This allows us to substitute t_{ij} for z_{ij}, effectively showing us that $t_{ij} > t_{rs}$ if and only if $p_{ij} > p_{rz}$, for pairs of clones i, j and r, s. But if this is the case, then we can say that

$$\Pr\{|t_{ij} - h_{ij}| > \in\} \rightarrow 0.$$

This implies that $h_{ij} < h_{rs} \Leftrightarrow t_{ij} < t_{rs}$,

which is what we wanted to prove.

Hybridization Graph

In this section we present some actual results of computational tests with algorithms for the hybridization mapping problem. Before presenting these results, however, we must make considerations regarding the way the results of such tests can be interpreted. Some of these considerations are valid for many problems in computational biology. Initially we shall look at the input data using another graph. By doing so it will become clearer what we can expect of any algorithm that tries of obtain the true probe permutation. We define the *hybridization graph H* as a bipartite graph (U, V, E) that is built using information from the hybridization matrix: Clones are the vertices of the U partition, and probes are the vertices of the V partition. There is an edge between two vertices if the corresponding probe hybridized to the corresponding clone. In Fig. 5.11 we can see the bipartite graph H that was built based on matrix M shown in Table 5.3, but excluding the all-zeros column.

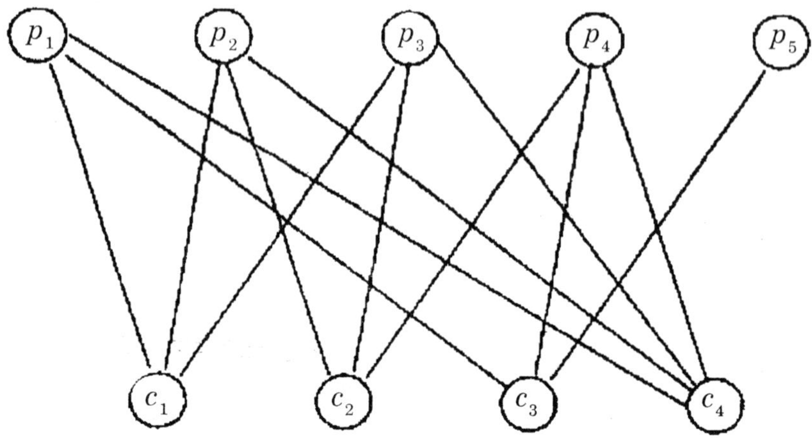

Fig. 5.11 : Hybridization graph h corresponding to hybridization matrix .

The first thing to notice is that H may not be connected, even if all entries in the hybridization matrix are correct. If this is the case, then no matter how good our algorithm is, we will not be able to tell the relative order between probes that belong to different connected components; the information to do so is simply not present in the hybridization matrix.

A connected component may be as simple as a singleton vertex, meaning that there is a probe that did not hybridize to any clone or a clone that was not hybridized to any probe. Another observation is that there may be *redundant probes,* or probes that hybridize to exactly the same set of clones. This could happen if the probes, although different, hybridize to parts of the target DNA that are close together. It could also happen if certain clones for that particular DNA stretch are missing, leaving some probes without any positive hybridization results. Connected components of H show up when we solve the corresponding consecutive 1s problem.

Redundant probes can also be easily seen in the hybridization matrix: They are columns that have exactly the same 1s and 0s. ut if there are errors we may get wrong information regarding the number and structure of connected components of H and whether a probe pair is redundant. We say that errors may *mask* these properties. So notice the difficult situation that errors create: The input matrix without errors may lack information necessary to find the true permutation (for example, the errorless H has several components), and we may have a lot of trouble just to recognize that lack of information (because our errorful H has only one component). Assuming matrix in Table 5.3 is errorless, probes p_2 and p_3 are redundant, but if there were a fast positive between clone c_3 and probe p_3 we would not be able to recognize that. Given this situation, it is clear that evaluation of a mapping algorithm is a difficult task (in addition to mapping itself!).

We will now take a look at how we can evaluate such algorithms assuming that we somehow know the correct answer to any mapping problem. This can be done, for example, if we use a computer program to generate artificial instances of mapping problems, simulating experimental errors. If such instances are faithful to real instances, it would be clear from the above discussion that the input matrix may lack information to enable an algorithm to determine the true probe order. Therefore we should try to evaluate how "close" the solution found by a particular mapping algorithm is to the true probe order. The question now becomes: How should we define closeness is this context? At the moment there is no consensus on how to do this. However, to give some idea of the performance of current mapping algorithms in practice, we will present as an example one definition that has appeared in the literature. It is a reasonable definition, but even if it becomes widely accepted, it may still undergo some refinements. We will measure a mapping algorithm by the fraction of *strong adjacencies* reported by it.

Strong adjacencies are defined in terms of the number by of blocks of consecutive 1s present in a hybridization matrix with a given probe permutation $\pi=p_1, p_2, ..., p_m$. We analyze the effect of *translocations,* which are operations that reverse the order of a set of consecutive probes. WE say that two adjacent probes p_i and p_{i+1} represent a strong adjacency if placing these probes apart by any translocation increases b in each row. When such an increase takes place we have some evidence (albeit not conclusive) that probes p_i and p_{i+1} should stay adjacent in all solutions.

Based on this concept, we can define the *strong adjacency cost* of a given permutation. This is given by the formula

$$100\left(\frac{1}{m-1}\sum_{0}^{m-2}\delta_i\right) \tag{5.4}$$

where $\delta_i = 1$ if p_i and p_{i+1} is a strong adjacency in the true permutation but these probes are not adjacent in the proposed permutation, and $\delta_i = 0$ otherwise. Note that the cost is given as a percentage. Good permutations should have low strong adjacency cost. With this definition we are finally able to give the reader an idea of algorithm performance. Table 5.4 presents the strong adjacency cost of two algorithms. One of them is "random": a random probe permutation is selected. The other is based on the TSP approach; this is how it works. Given the TSP graph the algorithm bids a cycle by choosing pairs of vertices and making them adjacent on a path. The pair (u, v) chosen at each iteration must fulfil the following conditions: If both u and v already belong to paths, and these paths are different, the paths can be joined only if u and v already belong to paths, and these paths are different, the paths can be joined only if u and v are endpoints in their respective paths (that is, they each have just one neighbor); and they must be the closest among all qualifying pairs.

After all vertices are on the same path, that path is closed forming a cycle. This solution is then submitted to another algorithm, which tries to improve the solution by applying another heuristic. In Table 5.4 we can see that the TSP-based approach (called "greedy") performs well compared to "random". The paper from which these results were obtained presents the performances of three other, much more sophisticated algorithms, and the results are similar to those shown above for "greedy TSP". This can be seen as a point in favor of the TSP approach, but in a sense it is yet another measure of how difficult the mapping problem is. The table also shows that in the presence of false negatives the solution of "greedy TSP" was fairly poor. It is fair to assume that results would be even worse if all kinds of errors were combined in the same instance. This motivates our next section, in which we present a heuristic that appears to be robust in the presence of false negatives.

Table 5.4 : Strong adjacency costs for two algorithms on matrices with different kinds of errors. Error rates are indicated in the heading of each column (only one type of error per column). Coverage in all cases is 10, where coverage is the ratio between the total length of all clones and target DNA length.

	C1P 0	Chimerism 0.5	False Positives 0.04	False Negatives 0.32
Greedy TSP	1.9	0.9	16.0	28.3
Random	86.4	89.7	94.4	94.9

MAPPING OF CHIMERIC CLONES

As the previous sections have shown, mapping is a difficult problem, and no general and good algorithms for it have been found. As a consequence, what we see in practice is that researchers resort to various heuristics to help them arrive at a solution. In this section we present two such heuristics that have yielded good results in hybridization mapping projects.

Screening Procedure

As noted above, chimeric clones occur with high frequency in clone libraries, and their presence brings serious problems to any mapping algorithm. In this section we present a simple heuristic that tries to split chimeric clones into fragments. Such a heuristic is very useful as a *screening procedure*, which can be used as a preprocessing step before employing more sophisticated techniques. The idea is very simple: if a clone is chimeric (and let us assume it is composed of only two fragments), the probes that hybridize to one of its fragments should not be related to the probes hybridizing to the other fragment. The key here is of

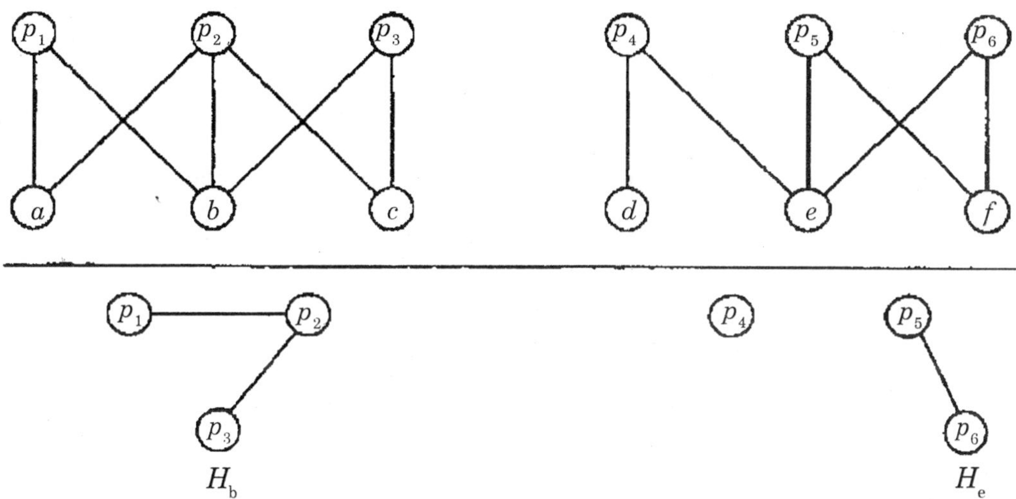

Fig. 5.12 : Above the line we show the hybridization graph H for some clones × probes matrix. Below the line at left is graph H_b for clone b. This clone is probably not chimeric, because H_b is connected. At right is graph H_e for clone e. This clone could be chimeric, because H_e is not connected.

course the concept of "relatedness". This is made concrete by looking at the following graph. Take clone i and the set of probes that hybridize to it, P_i. We create a graph $H_i = (P_i, E)$ for every clone i. We create an edge between two probes from P_i if they hybridize to a clone other than *i*. If the resulting graph is connected, we say that i is not chimeric. If it has more than one component, we say that *i* is chimeric and we replace i by new "artificial" clones, each new clone given by a connected component of H_i. This method can be refined by requiring that an edge exists between probes *p* and *q* in H_i only if *p* and *q* hybridize to at least *k* other clones, where k is a parameter depending on the particular problem at hand. Experience has shown that this heuristic behaves well. On the other hand it may consider a clone chimeric when in fact it is not. Therefore, another useful heuristic would be one that combines two clones that are actually one.

A Good Permutation

The heruristic presented here is more ambitious than the one we saw in the previous section. It aims at actually solving the problem, that is, obtaining a permutation of the probes. The idea is to estimate for every probe p the number of probes to its left and the number of probes to its right by looking at the hybridization graph *H*. We can then sort the probes using this estimate and thus obtain one "good" permutation. Given a probe we will be able to count the number of probes to its left and to its right if we can somehow split the other probes into *two* separate components: one left component and one right component. Probes near the ends of the target DNA may not split the others in two; therefore, the probes for which we can obtain two components will be called *splitters*.

The method for detecting splitters is described next. Given a probe p build a set of vertices S_p (whose elements are clones and probes) by including p in S_p and every other probe that shares a clone with p. Include in S_p all clones that are incident to any probe in S_p. Now remove all vertices in S_p from H. We will say that p is a splitter if the resulting graph has exactly two components. Using this method, the heuristic for obtaining a good probe ordering is described.

Algorithm *Probe Permutation Heuristic*

Input: clones × probes hybridization matrix

output: a "good" permutation for the probes

Identify splitters as described in the text

for every splitter i do Determine the components A_i and B_i

for each probe p do Initialize l_p and r_p with zero

Select a pair (A_k, B_k) arbitrarily

for each probe p do

  for each pair of components (A_i, B_i) do

  if $p \in A_i$ then

  if $|A_i \cap A_k| > |A_t \cap B_k|$ then

      Increment l_p

  else

    Increment r_p

  else if $p \in B_i$ **then**

    if $|B_i \cap A_k| > B_i \cap B_k|$ then

      Increment l_p

  else

    Increment r_p

Sort probes in decreasing order of $l_p - r_p$.

Fig. 5.13 : Heuristic to obtain a "good" probe permutation.

For each probe p, the algorithm keeps two counters, l_p and r_p, recording the number of left and right regions that contain p. These counters are initialized with zero and are incremented using the splitters. It is not obvious how to do this, because although we know the two components A_i and B_i of a splitter i, we cannot tell which one goes to the left or to the right. To solve this difficulty we rely on a fixed arbitrary splitter k that we use as a reference. We assume that A_k lies to the left of B_k. Then, given another splitter i, a component X of i (either A_i or B_i) is the left-most one when $|X \cap A_k| > |X \cap B_k|$. Once we have the final counts, we sort probes so that the ones with higher left bias $l_p - r_p$ come first.

A simple improvement to this heuristic is in the choice of splitter k. The more "central" this splitter, the better should the results be. As with the heuristic of the previous section, experience has shown that this one performs well. In particular, it is relatively robust with respect to false negatives. In addition, one can envisage this heuristic also used as delivering its result to other algorithms or heuristics that could try to improve the solution.

Chapter 6

RNA and Protein Threading

Up to now we have studied biomolecules considering only their basic, linear sequence. As discussed already that a biomolecule such as RNA or a protein is actually a three-dimensional object. Knowing a biomolecule's precise spatial structure is one of the foremost goals of molecular biology, for it is this structure that determines the molecule's function. However, determining the three-dimensional structures of proteins and RNA using techniques such as x-ray crystallography and nuclear magnetic resonance has proved to be difficult, costly, and not always feasible.

This presents a contrast to the ease with which it is possible to determine the primary sequence of these molecules (the bases in the case of the RNA and amino acids in the case of proteins). As a result, there currently exists a gap between the number of proteins for which the sequence is known (in the thousands) and the number for which the three-dimensional structure is known also in the hundreds); this gap has been widening every year. Such a situation has created an intense search for *structure prediction methods*: methods that can predict the three-dimensional structure of a molecule based on its primary sequence properties of its ranging from molecular dynamics simulation to neural net programs.

In this text, however, we are primarily interested in algorithmic techniques. Such techniques apparently are not yet applicable in general to molecule structure prediction, since they have to rely on reasonably accurate models of the molecule folding process; no such general models as yet exist. Nevertheless, for special cases an algorithmic formulation can be given, and its results have been useful in practice for a number of years. Such is the case of RNA structure prediction, which we address in the next section. We then discuss the protein folding problem, the solution of which is a sort of "holy grail" in molecular biology. We describe the problem and present one algorithmic approach that has shown great promise.

PREDICTION OF RNA SECONDARY STRUCTURE

RNA molecule is a single-stranded chain of the nucleotides A, C, G, U. The fact that RNA is single-stranded makes all the difference with respect of DNA: A nucleotide in one part of the molecule can base-pair with a complementary nucleotide in another part, and thus the molecule folds. Furthermore, the nucleotide sequence uniquely determines the way the molecule folds, and this is why we can attempt to predict RNA's structure by analyzing its sequence. However, the real three-dimensional structure is still too complicated for current prediction methods; instead, we concentrate on *secondary structure* prediction, which simply tries to determine which bases pair to which. We consider an RNA molecule as a string of n characters $R = R_1r_2 \ldots r_n$ such that $r_i \in \{A, C, G, U\}$. Typically n is in the hundreds, but could also be in the thousands.

The secondary structure of the molecule is a collection S of pairs (r_i, r_j) of bases such that $1 \leq i < j < n$. If $(r_i, r_j) \in S$, in principle we should require that r_i be a complement to r_j and that $j - i > t$, for t a certain threshold (because it is known that an RNA molecule does not fold too sharply on itself). However, the prediction method to be used is based on the computation of minimum free-energy configurations; thus we can avoid the situations above simply by imposing large free-energies on the corresponding configurations. There is one natural configuration that we shall exclude: the so-called *knots*. A knot exists when $(r_i, r_j) \in S$ and $(r_k, r_l) \in S$, and $i < k < j < l$. Without knots, S becomes a planar graph. Knots do occur in RNA molecules but their exclusion simplifies the problem.

A better reason for excluding knots is that secondary structure predictions are used to infer RNA three-dimensional structures, and knots can be inferred at that stage as well. Having laid the foundations, we now ask: What kind of algorithm could possibly be proposed to predict the secondary structure of an RNA molecule? Before we answer this question, let us examine the problem more closely. We have mentioned above that we will compute minimum free-energy structures. For this we need some way to attribute energy values to a given structure. We will simply assume that there exists one or more empirically derived functions that accomplish this task for us. Thus the simplest of all algorithms would be this: Enumerate all possible structures, and choose the one with lowest free energy.

Unfortunately, the number of possible structures is exponential in the number of bases, and such an algorithm would not be practical for large molecules (but it is worthwhile to note that some researchers have actually developed and used such enumerative algorithms). Therefore we need to look at other, more efficient ways of determining the lowest free-energy structure. We present in what follows two formulations that yield simple dynamic programming algorithms for attaining this objective.

Criteria for Pairing

The first formulation we describe for finding a structure of lowest free-energy makes the assumption that in any RNA structure the energy of a base pair is independent of all the others. This is a gross approximation, but it is useful and will help us to understand the foundations of more sophisticated algorithms. With such an assumption, the total free-energy E of a structure S is given by

$$E(S) = \sum_{(r_i, r_j) \in S} \alpha(r_i, r_j), \tag{6.1}$$

where α (r_i, r_j) gives the free energy of base pair (r_i, r_j). We adopt the convention that α $(r_i, r_j) < 0$, if $i \# j$, and α $(r_i, r_j) = 0$, if $i = j$.

The assumption that base-pair energies are independent provides the key to an efficient algorithm, since we can use solutions for smaller strings in determining the solutions for larger strings. Suppose for example that we want to determine a secondary structure $S_{i,j}$ of minimum energy of string $R_{i,j} = r_i r_i + 1 \ldots r_j$. Consider now what can happen to base r_j. Either it forms a pair with some base between i and j or it does not. If it does not, then $E(S_{i,j}) = E(S_{i,j-1})$. If r_j is base-paired with r_i, then we can say that $E(S_{i,j}) = \alpha(r_i, r_j) + E(S_{i+1, j-1})$. Finally, it could be that r_j is base-paired with some r_k, for $i < k < j$. In this case we can split the string in two: $R_{i.\ k-1}$ and $R_{k.\ f,}$ and say that $E(S_{i.\ j}) = E(S_{i.\ k-1}) + E(S_{k.j})$, for some k. (We can do tis because we have assumed there are no knots.

The problem here is that we do not know which k gives the lowest overall energy, so we should actually have $E(S_{i,\ j}) = \min\{E(S_{i,\ k-1}) + E(S_{k,\ j})\}$, where k should vary between i and j. What this all means is that we can compute the value of $E(S_{i,\ j})$ based on values for shorter strings. This situation is familiar to us: It means we can use a dynamic programming technique to solve the problem. In particular, we can write the above equations in the following way.

$$E(S_{i,\ j}) = min \begin{cases} E(S_{i+1,\ j-1}) + \alpha(r_i, r_j) \\ min\{E(S_{i,\ k-1}) + E(S_{k,\ j})\}\ for\ i < k \le j. \end{cases} \tag{6.2}$$

(The case in which j is not base-paired is subsumed by the second case above, because when $k = j$ we have $E(S_{i,\ j}) = 0$.) The algorithm to obtain $E(S_{i,\ n})$ is very similar and will not be repeated here. But notice that solving the recurrence provides us with a *value* for the lowest free-energy; the actual structure must be found by some post-processing on the dynamic programming matrix, just as we did to get an alignment from the sequence comparison matrix. Let us compute the running time of the above algorithm. In the basic sequence comparison algorithm we spend constant time per element of the matrix; hence the running times is $O(n^2)$, for an $n \times n$ matrix. Here things are a little different because we have to compute sums using a variable number of previous entries of the matrix for every pair i, j, as indicated in Expression (6.2). The exact overall number of sums is given by

$$\sum_{j=1}^{n} \sum_{i=1}^{j-1} (j - i - 1), \tag{6.3}$$

which is $O(n^3)$, and this is the algorithm's running time.

Clover Leaf

Now let us relax the independence assumption to include the fact that the free energy of a base pair depends on adjacent base pairs. This relaxation leads us to consider RNA structures called *loops.* Suppose (r_i, r_j) is a base pair in S, and consider positions u, v, and w such that $i < u < v < w < j$. We say that base r_v is *accessible* from (r_i, r_j) if (r_u, r_w) is not a base pair in S for any u and w. A loop is the set of all bases accessible from a base pair (r_i, r_j). The assumption that no knots occur implies that any structure S decomposes string R into disjoint loops. The size of a loop is the number of unpaired bases it contains.

We generalize the concept of loop to include base pairs as loops of size zero. When several base pairs occur next to each other, we get the so-called "stacked pairs" or "helical regions." Thus the number of loops in a structure is given by the number of base pairs it contains. An example of the dependence of free energy on the presence or absence of certain structures comes from the fact that helical regions have negative (stabilizing) free energy, whereas loops of size greater than zero have positive (destabilizing) energy.

This has the following consequence: When (r_i, r_j) is the base pair that closes a loop, structure $S_{i+1, j-1}$ may not be the minimum free energy for substring $R_{i+1, j-1}$. In other words, it could be that there exists structure $S'_{i+1, j-1}$ that has lower energy than $S_{i+1, j-1}$ but when we add the energy of pair (r_i, r_j) to the energy $S_{i+i, j-1}$, the stabilizing energy of the pair more than makes up for the difference. This means that we have to formulate new recurrence relations. These recurrence relations still contain some simplifications with respect to a fully general loop model.

These simplifications are necessary to make the dynamic programming algorithm more efficient. It is possible to show that a general model leads to an algorithm that runs in exponential time. The first simplification is that we will disregard the free energies of bases that do not belong to any loops. A second simplification is that in order to obtain the overall free energy of a structure we can simply add the energy of its components. Finally, we will assume we have to deal only with the kinds of loops. Let us see the details of this. We want to determine $S_{i, j}$ for string $R_{i, j}$. First suppose r_i is not base-paired. Then $E(S_{i, j}) = E(S_{i+1, j})$. Suppose now r_j is not base-paired. Similarly, $E(S_{i, j}) = E(S_{i, j-1})$. Now suppose r_i and r_j are base-paired, but not with each other. Then we can say (as we did for the situation without loops above) that $E(S_{i, j}) = \min E(S_{i, k}) + E(S_{k+1, j})\}$, for $i < k < j$, where r_k is any base between the pair of i and the pair of j. The final case happens when (r_i, r_j) is a base pair. This means that there may be one or more loops between i and j. Denote by $L_{i, j}$ the minimum free-energy structure in this case (to be detailed below). Writing up the full recurrence relation we get:

$$E(S_{i,j}) = min \begin{cases} E(S_{i+1,\, j}) \\ E(S_{i,\, j-1}) \\ min\ \{E(S_{i,\, k}) + E(S_{k+1,\, j})\}\ for\ i < k£\ j \\ E(L_{i,\, j}). \end{cases} \tag{6.4}$$

Now let us detail the case of $L_{i, j}$. We have to account for the various kinds of loops and for that we need to express free-energy values. As before, we will assume that experimental

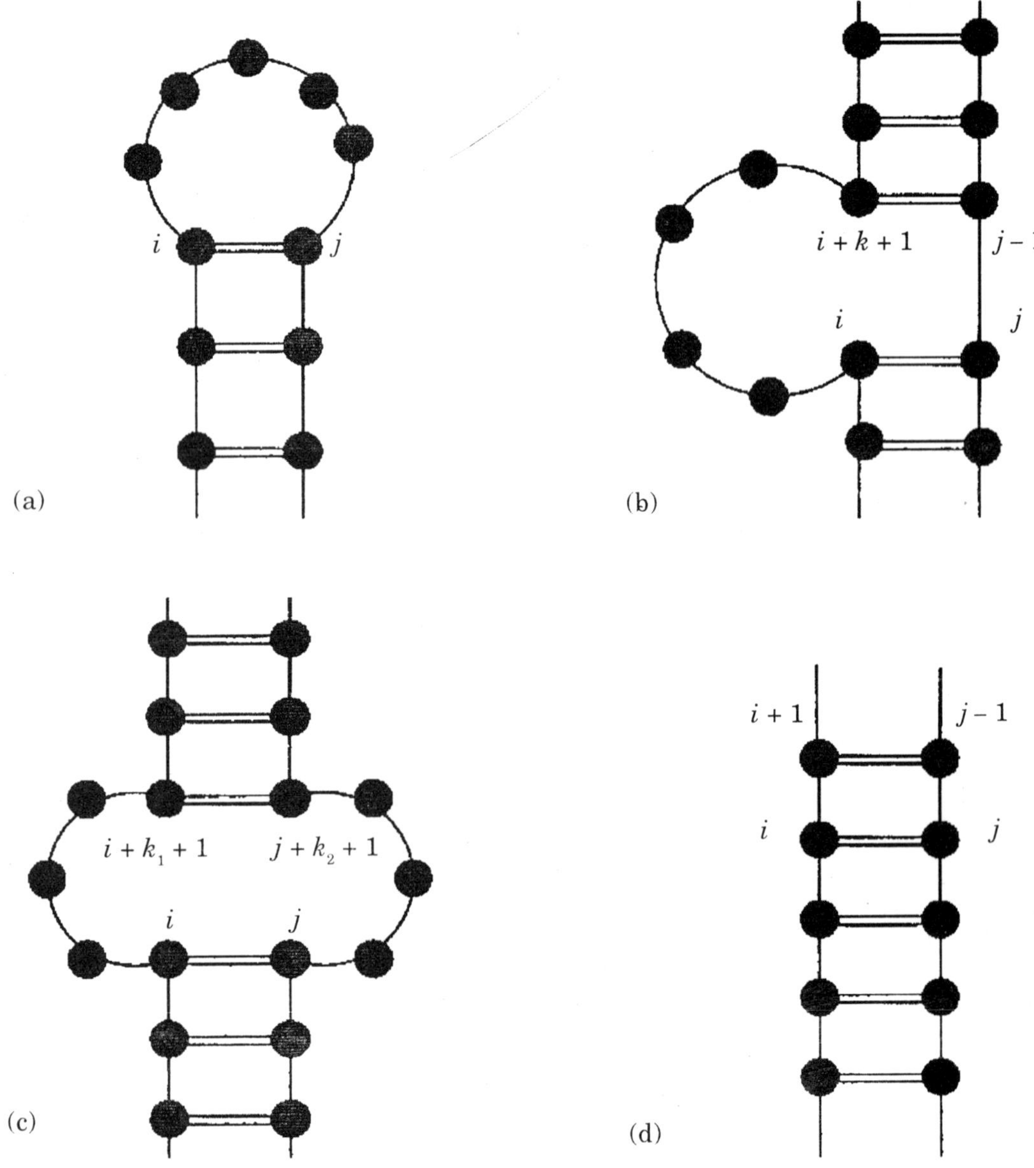

Fig. 6.1: RNA secondary structures without knots. The bullets are ribonucleotides and the horizontal double lines show the base pairs. (a) hairpin loop; b) bulge on i; (c) interior loop; (d) helical religion.

procedures have determined what these values are and that they are available to us in the form of functions. Again, $\alpha(r_i, r_j)$ gives the free energy of base pair (r_i, r_j). In addition, we have these other free-energy functions:

$\xi(k)$ = destabilizing free energy of a hairpin loop with size k

η = stabilizing free energy of adjacent base pairs

$\beta(k)$ = destabilizing free energy of a bulge of size k

$\gamma(k)$ = destabilizing free energy of an interior loop of size k

With these functions we can write the following recurrence relations for $E(L_{i,j})$:

$\alpha(r_i, r_j) + \xi(j - i - 1)$, if $L_{i,j}$ is a hairpin loop (6.5)

$\alpha(r_i, r_j) + \eta + E(S_{i+1, j-1})$, if $L_{i,j}$ is a helical region (6.6)

$\min_{k \geq 1}\{\alpha(r_i, r_j) + \beta(k) + E(S_{i+k+1, j-1})\}$, if $L_{i,j}$ is a bulge on i (6.7)

$\min_{k \geq 1}\{\alpha(r_i, r_j) + \beta(k) + E(S_{i+1, j-k-1})\}$, if $L_{i,j}$ is a bulge on j (6.8)

$\min_{k_1, k_2 \geq 1}\{\alpha(r_i, r_j) + \gamma(k_1 + k_2) + E(S_{i+1+k_1, j-1-k_2})\}$, if $L_{i,j}$ is an interior loop (6.9)

The value of $(L_{i,j})$ is found by taking the minimum over all values computed using the recurrences above. These equations, together with Equation (6.4) give us all the necessary information to set up a familiar matrix-based dynamic programming algorithm for predicting RNA secondary structure with loops. The schematic matrix layout shows which regions of the matrix must have their values known before we can compute $E(L_{i,j})$, for each of the recurrences involved. Because the most complicated part of the algorithm is the computation of the recurrence relations for $E(L_{i,j})$, let us now examine these relations in more detail.

The simplest cases happen when $L_{i,j}$ is a helical region or hairpin loop. In these cases we spend constant time per (i, j) pair, and thus the running time is $O(n^2)$. In the case of bulges, the parameter k ranges between 1 and j – i. Therefore this case is similar to the computation of minimum energies without loops, and the running time is $O(n^3)$. The final case is that of interior loops. Because we are dealing with the two parameters k_i and k_2, in this case we have to look at previous values of E which appear in the submatrix delimited by i and j. Intuitively, this means that for every pair i, j we have to look at a submatrix having as many as $O(n^2)$ elements, giving a total of $O(n^4)$. More formally, we have:

$$\sum_{l \leq i \leq j \leq n} \left(\sum_{i \leq i' \leq j' \leq j} c \right) = O(n^4),$$

for some constant c. The value of this constant is given by the time it takes to compute $\alpha(r_i, r_j) + \gamma(k_1 + k_2) + E(S_{i+1+k_1 j-1-k_2})$ for specific values of i, j, k_1, and k_2.

Computational Methods

It is thus clear that in the framework presented, the computation of interior loops is the most expensive. We now show how its running time can be lowered to $O(n^3)$. This can be done using the fact that along diagonals the size of a loop is constant. Let us give an example. The parameter for the γ function is the size of the interior loop, or $k_1 + k_2$, as above. The constraints on k_1 and k_2 are that $2 \le k_1 + k_2 < j - i - 2 - l_{min}$, where l_{min} is the minimum loop size. Suppose we want to compute $E(L_{4, 18})$ and that $l_{min} = 1$. Then $2 \le k_1 + k_2 \le 11$. We could have then at a certain point $k_1 = 2$ and $k_2 = 6$, meaning that we need to evaluate $\gamma(8)$ and $E(S_{10, 14})$, and at another point $k_1 = 5$ and $k_2 = 3$, meaning that we have to evaluate $\gamma(8)$ again and $E(S_{7, 11})$.

What is happening is that (10, 14) and (7, 11) lie along the same diagonal, and there loop sizes are constant. This suggests the following trick. We store for each pair (i, j) a vector $V_{i, j}$ indexed by all possible loop sizes. There are at most $2n - 1$ diagonals, so this vector has at most that many elements. In each element we store the minimum value found for each diagonal. Then, when we have to evaluate $E(L_{i, j})$ for an interior loop and given pair (i, j), we need only compute the minimum of $\{\alpha(r_i, r_j) + \gamma(l) + V_{i, j}[l]\}$, for all possible values of l, the loop size. This will take $O(n)$ per pair (i, j), giving a total of $O(n^3)$. For this trick to work we must have an auxiliary matrix to store all $V_{i, j}$ vectors, and we have to update the vectors as we scan the original matrix M.

Fortunately we can do this incrementally. For example, when we move from row i to row $i + 1$, to each element in row $i + 1$ there corresponds an element (a diagonal in M) from $V_{i, j}$. The elements of $V_{i + 1, j}$ can be obtained by comparing the elements in row $i + 1$ of M to elements in $V_{i, j}$, which also takes $O(n)$. Hence we also spend $O(n^3)$ overall updating the $V_{i, j}$ vectors. What this all means is that we can shave a factor of n from the running time for interior loop computation, but we double the storage space. This running time can be further improved if we look carefully at the kinds of free-energy functions we have, similarly to the improvements for gap-penalty functions. We will not describe these improvements, given that the resulting algorithms are much more complicated than those seen so far; some pointers are given in the bibliographic notes.

We remark that with *linear* destabilizing functions the time above can be improved to $O(n^2)$ (for interior loops and bulges). Experiments, however, have shown that destabilizing functions grow logarithmically. The algorithms that we have described for RNA structure prediction have one important flaw: They predict only one structure. In practice we would like to have a set of possible solutions, some of them possibly suboptimal in terms of free energy. The reason is the same as in other problems considered in this book: Given the abstractions in the model used to set up the algorithm, the optimal solution is not necessarily the true structure. For this reason it would be important to extend the above algorithms to find optimal as well as suboptimal solutions within a certain energy threshold.

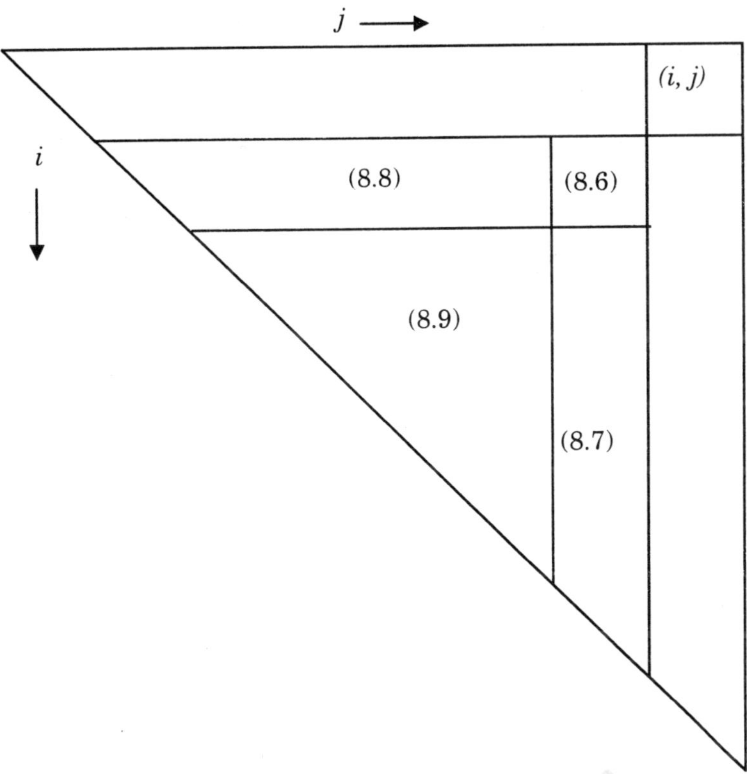

Fig. 6.2 : Matrix layout for computation of recurrence relations 8.6-8.9. The equation numbers label the matrix regions where previous values of $E(L_{i,j})$ must be known.

PROTEIN FOLDING

We recall that in the protein folding problem the main goal is to determine the three-dimensional structure of a protein based on its amino acid sequence. This problem has been intensely researched since the early 1950s, and as yet no completely satisfactory solution has been found. There is a vast body of work on this topic, and this section is intended to be a very brief introduction. In the next section we present one particular approach to the problem, among the many that exist. We believe that these two sections contain enough material to give the reader a good start on the subject. References for further reading are given at the end of the chapter. An important assumption of all protein folding prediction methods is that the amino acid sequence completely and uniquely determines the folding.

This assumption is supported by the following experimental evidence. If we unfold a protein *in vitro,* such that no other substances are present, and then release it, the protein immediately folds back to the same three-dimensional structure it had before. The folding process takes less than a second. Thus it appears that all information necessary for the protein to achieve its "native" structure is contained in its amino acid sequence. (This is not true of all proteins because there are some that need "*auxiliary molecules*" to fold.) Before discussing the difficulties of protein folding prediction, we briefly describe the main facts about protein three-dimensional structure.

Proteins have several levels of three-dimensional structure. Above the primary linear structure are the *secondary structures,* of which there are basically three kinds: the *α-helices*, the *β-sheets*, and structures that are neither helices nor sheets, called *loops.* An α-helix, is a simple helix having on average 10 residues, or three turns of the helix; some helices can have as many as 40 residues. Research has shown that some amino acids appear more frequently on helices than other amino acids, a fact that helps in secondary structure prediction but is not strong enough to allow accurate results. A β-sheet consists of the binding between several sections of the amino acid sequence. Each participating section is called a β-strand and has generally between 5 and 10 residues. The strands become adjacent to each other, forming a kind of twisted sheet.

Certain amino acids show a preference for being in β-strands, but again in this case these preferences are not so positive as to allow accurate prediction. Loops are the sections of the sequence that connect the other two kinds of secondary structure. In contrast to helices and sheets, loops are not regular structures, both in shape and in size. In general, loops are outside a folded protein, whereas the other structures form the protein core. Proteins that are evolutionary related usually have the same helices and sheets, but may have very different loop structures. In this case we say that core segments are *conserved.* The concept of a core is very important for the next section. Proteins have also higher-level structures. These are called *motifs* and *domains.*

A motif is a simple combination of a few secondary structures that appears in several different proteins. An example of a motif is the helix-loop-helix. Research has shown that this motif is used as a binding site for calcium atoms, so this structure has a clear function. Other motifs appear to play no role at all in protein function. A domain is a more complex combination of secondary structures that by definition has a very specific function, and therefore contains an *active site.* An active site is the section of the protein where some binding to an external molecule can take place. A protein may have only one domain, or may contain several. All of them taken together form the protein's *tertiary structure.* We can now state the folding problem in more detail. Given the amino acid sequence of a protein, we would like to process it and determine where exactly all of its α-helices, β-sheets, and loops are, and how they arrange themselves in motifs and domains.

Given enough and reliable chemical and physical information about the each component amino acid, in principle it should be possible to compute the free energy of foldings. A prediction method could simply enumerate all possible foldings, compute the free energy of each, and choose the one with minimum free energy (assuming that such a folding is the protein's native structure, which is not an established fact). Now let us have an idea of how many possible foldings a typical protein can achieve. The proteins can fold thanks in large part to the angles ϕ and ψ between the alpha carbon and neighboring atoms.

Experiments have shown that these angles can assume only a few values independently of each other; for each pair of values in each residue we would have one *configuration*. Let us suppose that each angle can assume only three values. Therefore each residue can be in only one of nine configurations. For a protein with 100 residues, which can be considered small, there are thus 9^{100} possible configurations, an astronomical number that rules out enumerative algorithms. To make matters worse, even if researchers agree that the native structure is the one with minimum free energy, there is no agreement on how to compute the energy of a configuration.

There are far too many factors to take into account, such as shape, size, and polarity of the various component molecules, the relative strength of interactions at the molecular level, and so on. From current knowledge it is possible only to formulate some general rules, such as the requirement that hydrophobic amino acids stay "*inside*" the protein and hydrophilic amino acids stay "*outside*" the protein, in the case of proteins in solution (water is the universal solvent found both inside and outside cells). Such a state of affairs has led researchers to concentrate on prediction methods for protein secondary structures.

The observation above about the difficulties in assigning free-energy values to tertiary structures applies to secondary structures as well. Hence, no general and reliable dynamic programming approach is known for the prediction of protein secondary structure. What we find in practice is the use of an assortment of methods, ranging from programs based on neural nets to pattern-recognition methods based on statistical properties of residues in proteins with known structure. However, these methods are still not as reliable as scientist want them to be. For this reason still other techniques have been development to help cope with the problem, and it is one such technique that is the topic of the next section.

PROTEIN THREADING

An early method used for secondary structure prediction was based on the idea that similar sequences should have similar structures. If we know the structure of protein A (from x-ray crystallography, say) and protein B is very similar to it at the sequence level, it seems reasonable to assume that B's structures will be the same or nearly the same as A's. Unfortunately this is not true in general. Similar proteins at the sequence level may have very different secondary structures. On the other hand it has been observed that certain proteins that are very different at the sequence level are structurally related in the following sense:

Although they have different kinds of loops, they have very similar cores. These observations have led researchers to propose the *inverse protein folding problem.* Instead of trying to predict the structure from the sequence, we try to fit a known structure to a sequence. To fit a structure to a sequence, we have to be more precise about what we mean by "structure." In this section, structure actually means *structural model,* and such a model can be obtained in the following way. Given a protein whose structure is known, we obtain a structural model by replacing amino acids by place-holders, but keeping associated with each place-holder some basic properties of the original amino acid. These properties can vary depending on the model adopted, but they should retain the fact that the original amino acid was in an α-helix or β-strand, or in a loop, and reflect spatial constraints of the structure such as distances to other amino acids, how much inside or outside the whole structure the place-holder is, and so on. Here we will not give details of such models but will assume instead that such a characterization of local structure is possible. With a given structural model we can now get a protein with unknown structure and try to align it to the model, taking into consideration the properties recorded for that particular model. This special kind of alignment is known as *protein threading.*

As we will see this technique results in a precise combinatorial optimization problem, since there can be many ways of threading one protein sequence into a model. As was the case with many other problems in this book, this one turns out to the NP-hard, too. However, the problem can be cast in such a way that a well-known technique for handling NP-hard problems can be used. The success obtained with this approach is our motivation for this section.

We assume that the input is given in the following way:

1. A protein sequence A with n amino acids q_i.
2. A core structural model C, with m core segments C_i. For this model we have the following information:
   a. The length c_i of each core segment.
   b. Core segments C_i and C_{i+1} are connected by loop region λ_i, and for each λ_i we know its maximum (l_i^{max}) and minimum (l_i^{min}) lengths.
   c. The local structural environment for each amino acid position in the model (chemical properties, spatial constraints, etc.).
3. A score function to evaluate a given threading.

The desired output is a set $T = \{t_1, t_2, \ldots, t_m\}$ of integers, such that the value of t_i indicates what amino acid from A occupies the first position from core segment i. It is thus clear that a threading is an alignment between the sequence and the core structural model. In this formulation the score function obviously plays a key role.

Given a threading and local structure environments, the score function should be able to evaluate sequence amino acids individually in their allotted positions in the model, as well as the resulting interactions between the various amino acids. Interactions can potentially be between pairs of amino acids, triples, quadruples, and so on. If we ignore interactions between amino acids, while allowing variable-length loop regions, the problem can be solve by standard dynamic programming techniques. If we allow pairwise interactions, the problem becomes NP-hard. Instead of oversimplifying in order to fall back on the (by now) familiar dynamic programming algorithms, here we will allow pairwise interactions and show how to solve the problem exactly with a standard technique for handling NP-hard problems. This technique is known as *branch-and-bound.* We now digress in order to explain this technique and then we will see how it can be applied to solve the threading problem with pairwise interactions.

Solution Space

In a combinatorial optimization problem we want to find an optimum solution among many possibilities in the *solution space.* The optimum refers to the maximum (or minimum) value of some function f that can be computed for each candidate solution. For most problems the solution space is exponentially large, although for problems in class P the optimum can be found in polynomial time. In the case of NP-hard problems, however, no algorithm is known that can search the solution space and find the optimum in polynomial time. Therefore the only alternative to obtain the optimum solution *for sure* is to enumerate and evaluate each possible solution one by one and choose the best.

However, we can try to speed up this process a bit by applying the following technique. First, it should be possible to separate the solution space into subspaces according to some constraints. For example, given solution space S, we could perhaps partition it into two subsets S_1 and S_2. S_1 could consist of all the solutions that have a certain property, and S_2 all of those that do not. This is the branching part of branch-and-bound. Second, it should be possible to obtain a bound on the value of the optimization function f for all candidate solutions in a set. This bound can be used to eliminate many solutions from consideration in the following way.

Suppose that we have a *minimization* problem and that we already know the value f(s) of some candidate solution s. Now consider a subset X of the solution space, and suppose that we can obtain a lower bound l on the value of *any* solution from X. Finally suppose that $l > f(s)$. This means that any solution from X cannot hope to improve on the solution we already have, and should therefore be discarded. If on the other hand $l < f(s)$, we should explore set X, because the optimum may lie there. Thus we see that the key ingredients in a branch-and-bound minimization algorithm are a rule to split the solution space in some efficient way and a lower bound on the value of the optimization function for candidate solutions.

The lower bound should be as close to the actual function value as possible, so that we can find the optimum faster than with a weaker lower bound; it should also be efficient to compute. However, we should keep in mind that the worst-case running time of branch-and-bound algorithms applied to NP-hard problems is always exponential. We will now apply these concepts to the protein threading problem. We first must define the possible solutions, and for this we must list the constraints faced by each possible threading. The spacing constraints can be given by the following inequalities:

$$1+\sum_{j<i}(c_j+l_j^{\min}) \le t_i \le n+1-\sum_{j\ge i}(c_j+l_j^{\min})$$

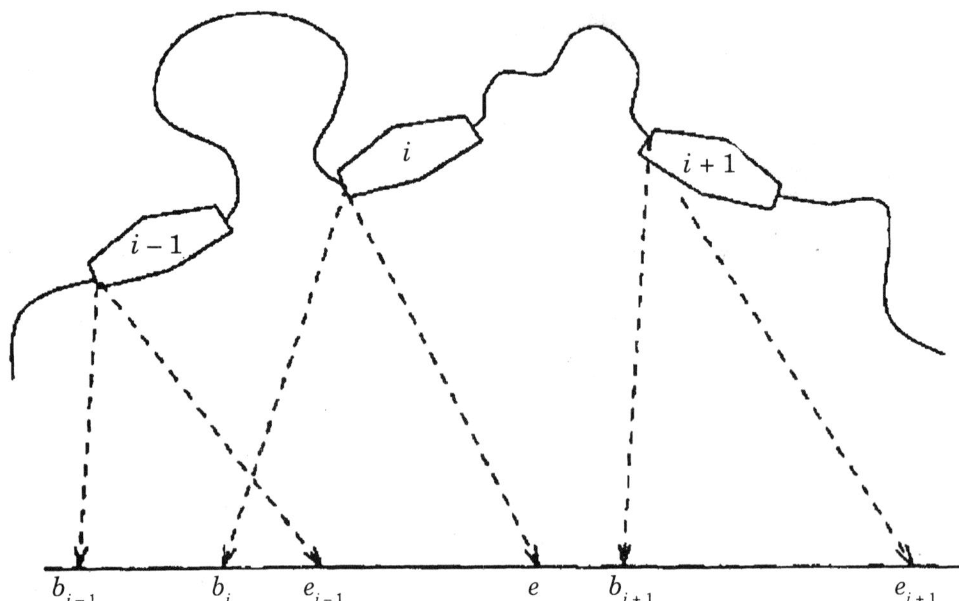

Fig. 6.3: Sets of threadings, Core segments are the polygons labelled by i - 1, i, and i + 1. Between them are loop regions. The line at the bottom is the sequence being threaded, and the dashed arrows show the intervals where the first position of each core segment can assume values in the sequence, hence depicting sets of possible threadings.

Another constraint comes from the fact that any threading should obey the order of core segments in the model. (This could be relaxed in a more general approach). This results in:

$$t_i + c_i + l_i^{\min} \leq t_{i+1} \leq t_i + c_i + l_i^{\max}.$$

These constraints imply that each t_i must lie in an interval given by values b_i and e_i. Thus a set of possible threadings can be expressed by a collection of intervals, one for each of the m t_i's. The branching part of the algorithm can now be explained. Choose one core segment, say i, and a position u_i inside the interval $[b_i, e_i]$. Split the solution set into three sets: one in which the new interval for t_i will be $[b_i, u_i - 1]$, another in which the new interval for t_i will be $\{u_i + 1, e_i\}$, and another in which t_i is forced to be equal to u_i. We have not specified which core segment to choose nor which position inside the segment's interval. These choices can be made based on considerations such as interval size.

We now turn to the lower bound. The score function can be given by the following equation, assuming T is some threading:

$$f(T) = \sum_i g_1(i, t_i) + \sum_i \sum_{j>i} g_2(i, t_i, t_j). \tag{6.10}$$

Note that the score function deals with core segments rather than individual amino acids. The reason is that the value of t_i determines the position of all amino acids from sequence A in core segment i, hence the pair of arguments (i, t_i) encapsulates all that is happening with segment i. The first term in the right-hand side above gives the contribution of each individual core segment to the total, whereas the second term gives the contribution of pairs of core segments. Function g_1 and g_2 are part of the input.

Given a set of threadings $\mathcal{T}$, a simple lower bound is given by:

$$\min_{T\in\mathfrak{I}} f(T) = \min_{T\in\mathfrak{I}} \sum_i \left[g_i(i, t_i) + \sum_{j>i} g_2(i, j, t_i, t_j) \right] \tag{6.11}$$

$$\geq \sum_i \left[\min_{b_i \le x \le e_i} g_1(i, x) + \sum_{j>i} \min_{\substack{b_i \le y \le e_i \\ b_j \le z \le e_j}} g_2(i, j, y, z) \right]. \tag{6.12}$$

This lower bound is obtained by summing up the lower bounds on each term separately. A stronger lower bound is possible. It is desirable to obtain a lower bound that corresponds to one particular threading, because this threading can then be used in the splitting process. Suppose threading T currently has the lowest value for the lower bound. We choose one core segment C_i from it and split it at point u_i given by the threading. The core segment chosen can be the one with the largest interval size (Fig. 6.4).

```
Algorithm Protein Threading
Input: protein sequence, core structural model, score functions,
    lower bound function f_lb
output: a threading T
T ← all possible threadings
lb ← f_lb (T)
|| We use a priority queue Q to keep threadings
|| In Q each element is a set of threading and its lower bound
insert (Q, T, lb)
while TRUE do
    ||We remove a candidate set of threadings from Q
    T_c ← RemoveMin (Q)
    if |T_c| = 1 then
        return T_c
        || The only remaining threading is the solution
    else
        Split (T_c)
        For each new subset T_i from T_c do
            lb ← f_lb (T_i)
            insert (Q, T, lb_i)
```

Fig. 6.4: The branch-and-bound protein threading algorithm.

This finishes the description of branch-and-bound applied to protein threading. We close this section with a few comments regarding the usage of such a technique. First, this is a recent development and must therefore undergo many tests. These tests can be roughly divided in three categories. In the first, after a structural model is obtained from a structure, we perform a *self-threading;* that is, we thread the very protein sequence that gave rise to the structural model.

Another test involves proteins that are genetically close (and hence have similar sequences) and share a similar structure (called *protein homologs*). A final test involves proteins that have different sequences but are structurally similar (called *structural analogs*). In these tests all structures are known beforehand, so we know exactly what the result should be and can detect anomalies. Encouraging results have been obtained, in that the structure obtained is usually very close to correct. When there are errors, it is relatively easy to insert additional constraints to correct them. This is another advantage of this technique, and it can thus use the extensive knowledge on the quirks of protein structure that has been accumulating over the years.

An application of this technique to the real world should work like this. A database of structural models is maintained. A researcher comes up with a new protein sequence and submits it to the database server. The server then threads the sequence on all models from the database, returning one or a few models where a good fit was found. Note that this would mean solving one protein threading problem for each structural model in the database, a task that requires a fast solution to each such problem.

CONCLUSION

In this chapter we presented a brief sampling of algorithmic results for molecular structure prediction. We studied two dynamic programming algorithmic for RNA secondary structure prediction, and showed how to improve the running time of the more complex algorithm from $O(n^4)$ to $O(n^3)$. We then briefly explained the protein folding problem, pointing out that extensive research has led to many approaches to the problem, of which we presented one, called protein threading. This approach allows an algorithmic treatment based on a familiar technique for handling NP-hard problems.

Chapter 7

Ribonuclease Protection Assay

The *ribonuclease protection assay* (RPA) is a sensitive technique for the analysis of total cellular RNA. It involves generating a specific antisense riboprobe, hybridizing the probe to total RNA, removing unprotected RNA by RNases, and finally isolating and analyzing the protected RNA on a denaturing gel. Although the RPA is somewhat more labour-intensive than Northern analysis, it has the advantage of being more sensitive (as little as 0.1 pg of target RNA can be detected with ideal hybridization conditions). RPAs are also more tolerant of partially degraded RNA (provided the area that is protected is intact). Although RPAs are not as sensitive as polymerase chain reaction (PCR)-based RNA analyses, the target RNA is analyzed directly; a reverse transcription step is not required. Finally, the RPA is quantitative as long as the probe is in excess. More important for the study of imprinted genes, the RPA can be designed to detect allele-specific expression of the target gene of interest.

DESIGN OF PROBES

There are several points to consider when designing a riboprobe for the typical RPA. The probe template is prepared by cloning the sequence of interest into a transcription vector such that the 3 end of the fragment is adjacent to bacteriophage promoter. The template can be cloned from cDNA or genomic DNA. The recommended probe size is 100-500 bases. In addition, the probe should be designed to be larger than the protected RNA fragment for ease of discriminating between RNA: RNA vs RNA: DNA hybridization products, the latter resulting from DNA contamination. Further useful considerations are presented by others. We subsequently focus on the design of probes for analysis of imprinted genes.

ALLELE-SPECIFIC ANALYSIS

The RPA is an excellent tool for the analysis of allele-specific expression of imprinted genes, since the methodology can be used to detect RNA polymorphisms between different (sub)species of mice and their Fl hybrid offspring. It was shown that RNase A is capable of cleaving base-pair mismatches. Consequently, the RPA can be used to detect single base changes between a probe from one species and RNA from a divergent species. Thus the products from each parental allele can be distinguished in an F I hybrid and can thereby be independently assayed. The RPA developed to detect allele-specific expression of the mouse *HI 9* gene was used to prove that *HI 9* is expressed exclusively from the maternal allele. To design a probe to detect allele-specific expression by an RPA, it is critical to identify polymorphisms within the transcribed region between the alleles of hybrid species.

For *H19,* the 5th exon contains the most divergent sequence between mouse subspecies. Figure 7.1 illustrates the template used to generated the allele-specific *H19* RPA probe and the position of polymorphisms between the *Mus musculus domesticus (M. domesticus)* and *Mus musculus castaneus (M. castaneus)* subspecies. A probe has also been designed to detect allele-specific expression of the imprinted mouse *Igf2* gene in Fl hybrids generated from the same inter-subspecific cross. Specifically, the *Igf2* probe expression vector contains a PCR generated fragment of 3 untranslated *Igf2sequence.*

We typically analyze *HI 9* as well as *Igf2* imprinted expression in Fl hybrids generated between C57BL/6 and B6(CAST-//YP), which has *M. castaneus* sequence at the distal portion of mouse chromosome 7, the region bearing *H19* and *Igf2.* Normally, *H19* is expressed exclusively from the maternal allele and *Igf2* is expressed primarily from the paternal allele in neonatal tissues.

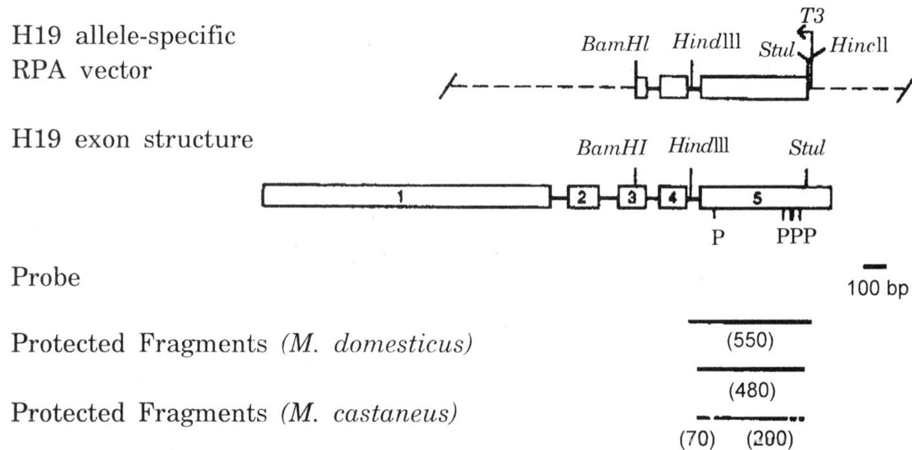

Fig. 7.1: Schematic of the allele-specific RPA for *HI9.* The top line depicts the RPA vector, which was constructed by cloning a 754-bp *BamHlStul M. domesticus* H19 genomic DNA fragment into the *BamHI* and *HincII* sites of Blucscript II KS plasmid (Stratagene). The T3 transcription start site used to generate the *HI9* antisense riboprobe is 3 of the *HincII* cloning site. The *H19* DNA fragment includes the exon 3 through exon 5 sequence (open boxes) as defined in the *H19* exon structure. We generally use the *HindIII* site in intron 4 to linearize the RPA vector. *In-vitro* transcription of *Hind*III-digested vector with T3 RNA polymerase results in a probe that is 550 nt long. Using this probe we detect one *M. domesticus* fragment (480 nt) and two *M. castamus* fragments (one ~290 nt and the other ~70 nt). Our assay conditions are such that we do not detect the shorter *castaneus-spectfic* protected fragments. Polymorphisms (P) have been mapped near the *Stul* site in exon 5.

Allele-specific RPAs of *HI 9* and *Igf2* have been instrumental in detecting the disruption of normal *HI 9* and/or *Igf2* imprinted expression in mice with targeted mutations at the *H19* locus. We generated a targeted deletion of important imprinting regulatory sequences 5 to *HI 9* that, when inherited from the paternal allele, results in biallelic expression of *H19* and when inherited from the maternal allele results in biallelic expression of *Igf2*. In our study, expression was assayed in Fl hybrids generated from reciprocal crosses of heterozygous mutants (maintained in a C57BL/6 background) with B6(CAST-H19). The targeted mutations were generated on a 129Sv/J allele, which results in the same *H19* and *Igf2* protected fragments as seen in the assay of C57BL/6 RNA.

QUANTITATIVE ANALYSIS

RNA may be assayed by the RPA with more than one probe simultaneously, thereby allowing for quantitation of relative RNA abundance among independent samples. When total RNA expression levels are to be assayed for imprinted genes, it may be useful to assay with probes that do not detect allele-specific expression *H19* and *Igf2* RPA probes simultaneously to determine the effect of the targeted deletion of the endodermal enhancer located 3 to the *H19* gene on levels of expression of both *H19* and *Igf2*. Although the probes used in this assay were not allele-specific, the unique size of the *H19* and *Igf2* protected fragments allowed for comparison of their relative RNA abundance.

To analyze the relative expression levels of *H19* and *Igf2* independently of each other, a riboprobe for the non-imprinted gene *rpl32* has been used in the RPAs of *H19* or *Igf2*. This allows for comparative quantification of *HI9* and *Igf2* levels in normal vs mutant mice. Having designed the probe(s) for allele-specific or quantitative analysis of the imprinted gene of interest, there are numerous variations to the RPA methods that one can use.

MATERIAL

In the preparation and handling of all materials required for this protocol, it is essential to avoid RNase contamination and degradation of samples. Work surfaces and pipetman must be cleaned before use and gloves changed often. Pipet tips and solutions are reserved for RNA use only. These precautions, together with the proper sterilization and storage of solutions, eliminates the need to treat solutions with the RNase inhibitor diethylpyrocarbonate (DEPC).

Preparation of Probe

In Vitro Transcription of Riboprobe

1. 5 transcription optimized buffer.
2. 0.1 M Dithiothreitol (DTT).
3. RNasin Ribonuclease inhibitor: 40 U/oL.

4. Nucleotide mixture: 2.5 m*M* ATP, 2.5 m*M* GTP, and 2.5 m*M* UTP (prepared from 10 m*M* nucleotide stocks and nuclease-free H_2O).
5. 0.1 m*M* CTP (prepared from 10 m*M* CTP and nuclease-free H_2O stocks).
6. Template DNA: linearized plasmid at 0.2-1.0 og/oL in H_2O.
7. [χ-^{32}P]CTP, 10 mCi/mL, 400-800 Ci/mmol.
8. Bacteriophage RNA polymerase (SP6, T3 or T7): 10-20 U/oL.
9. Nuclease-free H_2O.
10. RQ1 RNase-free DNase: 1 U/oL.

Purification of Probe by Gel Filtration

1. G-50 Sephadex columns for radiolabelled RNA purification (Boehringer Mannheim Biochemicals, Item 1 274 015).
2. Nuclease-free H_2O.
3. Hybridization buffer: 40 m*M* PIPES, pH 6.4, 400 m*M*NaCI, 1 m*M* ethylenediaminetetra acetic acid (EDTA), 80% deionized formamide freshly prepared from stocks.

Hybridization of Probe and RNA

1. 5 Hybridization buffer: 200 m*M* PIPES, pH 6.4, *2 M* NaCl, 5 mA/ EDTA (filter-sterilized, stored at 4°C).
2. Deionized formamide: prepared and stored in aliquots at -80°C.
3. Hybridization buffer: 1 hybridization buffer in 80% deionized formamide, freshly prepared.
4. RNA Probe.
5. Total cellular RNA.

Ribonuclease Digestion

1 Ribonuclease digestion buffer: 10 m*M*Tris-HCl, pH 7.5, 300 m*M*NaCI, 5 m*M* EDTA (sterilized by autoclaving).

2. RNase A: 4 mg/mL in 10 m*M*Tris-HCl, pH 7.5; 15 mA/NaCl: boiled 15 min to remove DNase; stored in aliquots at -20°C
3. RNaseT1: 100,000-500,000 U/mL, 4°C
4. 20% sodium dodecyl sulfate (SDS), filter sterilized.

5. Proteinase K: 20 mg/mL in 50 m*M* Tris-HCl, pH 8.0; 50% glycerol; stored at -20°C.
6. Phenol/chloroform/isoamyl alcohol, 25/24/1, equilibrated in 0.1 *M* Tris-HCl, pH 7.5.
7. Yeast tRNA: 10 mg/mL.
8. Ethanol.

Analysis of Ribonuclease Protected Fragments

1. Denaturing polyacrylamide-urea gel: 7% acrylamide, 7 *M* urea, 1 TBE.
2. 20 TBE: 1.78 *M*Tris-Borate, 0.08 *M*EDTA.
3. RNA loading buffer: 80% (vol/vol) deionized formamide; 1 m*M*EDTA, pH 8.0; 0.1% bromophenol blue; 0.1% xylene cyanol.
4. Radiolabelled marker.
5. Gel fixative: 10% methanol; 10% glacial acetic acid.

METHODS

As the protocol involves working with radioactive materials, appropriate laboratory safety procedures should be followed to prevent the exposure and spread of radioactivity.

Preparation of Probe

In Vitro Transcription of Riboprobe

1. Add the following ingredients in order to a microfuge tube at room temperature: 4 oL 5 transcription optimized buffer, 2 oL DTT, 0.8 oL RNasin ribonuclease inhibitor, 4.0 oL ribonucleotide mix, 2.4 oL CTP, 1.0 oL linearized template, 5.0 oL [χ-^{32}P]CTP, 0.5 oL RNA polymerase, 0,3 oL RNase-free water to a final volume of 20 oL.
2. Incubate reaction for 1 h at 37°C.
3. Add 1 oL of RQ1 RNase-free DNase and incubate the reaction for 15 min at 37°C
4. Place probe on ice.

Purification of Probe by Gel Filtration

1. Gravity drain the storage buffer from the G-50 Sephadex column into a microfuge tube, as per instructions of the kit manufacturer (Boehringer Mannheim Bio-chemicals). Place the column in a collection tube provided by the manufacturer. Insert this tube and column into a capped tube (e.g., Falcon 2059) (to prevent the spread of contamination) and spin at 1000*g* for 2 min in a centrifuge (e.g., Beckman GS-6R)

that is prechilled to 4°C. Discard collection tube with eluate and then place the column in the second collection tube provided by the manufacturer.

2. Add 50 oL RNase-free H_2O to probe, then load onto column. Again insert the collection tube and column in the capped tube to prevent the spread of contamination. Centrifuge at 1000g for 4 min at 4°C. Transfer purified probe from collection tube into a new microfuge tube containing 100 oL of hybridization buffer.
3. Measure the ^{32}P incorporation in a scintillation counter.
4. Store the probe at –20°C until ready for use.

Hybridization of Probe and RNA

1. Determine the amount of RNA to be assayed. For abundant RNA species, 1-3 og of total RNA is typically assayed. For rare RNA species, as much as 50 og of total RNA may be assayed. The maximum volume of RNA that can be incubated with the hybridization reaction is 7 oL. If necessary, ethanol precipitate the required amount of RNA.
2. Determine the total amount of probe that is needed for the hybridization of $n + 1$ samples. Each sample should be assayed with 50,000-500,000 cpm of probe (we typically use 50,000 cpm/sample). A negative control sample with 10 og tRNA should be included to verify that the ribonuclease digestion is complete. A second tRNA control sample may be included that will not be exposed to RNase, to analyze the integrity of the probe after hybridization. In addition, a range of RNA concentrations for a given sample should be assayed to verify that the probe is in excess. When analyzing allele-specific expression in RNA isolated from F1 hybrid tissue, it is important to include RNA samples isolated from the parental mouse strain, to ensure that allele-specific fragments will be detected.
3. Prepare the amount of hybridization buffer that is required for $n + 1$ samples (30 oL/sample). Add the appropriate amount of probe to the hybridization buffer and place on ice.
4. Add 30 oL of the hybridization buffer and probe mix to n number of microfuge tubes. Add the appropriate amount of RNA (not to exceed 7 oL) and mix by pipetting up and down. Alternatively, if a sample has been precipitated, add hybridization buffer and probe mix to the precipitated RNA. Pipet up and down to ensure that the pellet is dissolved.
5. To denature the RNA, incubate tubes in a hot block at 85°C for 10 min.
6. Rapidly transfer the tubes to a water bath set at the appropriate hybridization temperature for the specific RPA probe (30-65°C). Hybridize for 4 h to overnight, as required.

Ribonuclease Digestion

1. Prepare the required amount of ribonuclease digestion buffer (350 oL/reaction) containing 40 og/mL RNase A and 1 U/oL RNase Tl.
2. Remove the hybridization reactions from the water bath and spin briefly. Add 350 oL ribonuclease digestion buffer mix to each reaction.
3. Incubate at optimum temperature (room temperature, 37°C) for 60 min
4. To terminate the ribonuclease digestion, add 10 oL of 20% SDS and 2.5 oL of 20 mg/mL proteinase K.
5. Incubate at 37°C for 15 min.
6. Extract once with 400 oL of phenol/chloroform/isoamyl alcohol and remove the aqueous phase to a clean microfuge tube containing 10 og of yeast tRNA on ice. Vortex and then spin 5 min in microcentrifuge.
7. Add 1 mL of ethanol and precipitate on dry ice for 15 min or at -20°C for at least 2 h.

Analysis of Ribonuclease-Protected Fragments

Preparation of Denaturing Polyacrylamide-Urea Gel

1. Pour a 7% acrylamide/7 *M* urea/l TBE gel and allow to polymerize (30 min-1 h). (A lower per cent acrylamide may be used, particularly if analyzing long protected fragments.)
2. Assemble gel in gel apparatus, add 1 TBE buffer to upper and lower chambers, then carefully remove comb. Immediately flush the wells with buffer solution and, if necessary, adjust wells with syringe and needle.
3. Pre-run gel at least 30 min at ~250 V.

Analysis of RPA Products on Gel

1. Spin samples at the maximum speed for 15 min, preferably at 4°C.
2. Carefully remove supernatant with a pipetman and dry samples in speed vac.
3. Resuspend samples in 7 oL of loading buffer. In a separate tube, combine 0.5-2.0 oL (depending on specific activity) radiolabelled marker and 7 oL of loading buffer.
4. Immediately before loading, heat samples (including marker) at 85°C for 3 min and place on ice.
5. As samples are being heat-denatured, flush wells again from the prerun gel to remove urea that has leached out of the gel.

6. Load samples on gel, and run at ~ 250 V for appropriate length of time for the size of the protected RNA fragments.
7. Fix gel for 1 5 min, then dry before either exposing to X-ray film or a phosphorimage screen (Molecular Dynamics).
8. Quantitate relative RNA levels with densitometry of the X-ray film or by using Image Quant (Molecular Dynamics) to analyze the phosphorimage scan.

NOTES

1. All of the reagents required for the synthesis of the RNA probe, except for the template DNA and the [χ-^{32}P]CTP, are obtained from Promega Corporation and stored at –20°C. The reagents are available individually or as a kit (Item P1420 for SP6, P1430 for T3, P1440 for T7).
2. It is important to generate the probe template from purified plasmid devoid of nucleases. We use DNA isolated from a CsCl-ethidium bromide gradient. To prepare an RPA template, we linearize 20 og of plasmid with an appropriate restriction enzyme and verify that digestion is complete on an agarose gel. The DNA is subsequently phenol/chloroform-extracted, chloroform-extracted and ethanol-precipitated. The pellet is dissolved in RNase-free H_2O at a final concentration of 0.2-1 og/oL. The plasmid may be digested to generate multiple fragments. However, digests that generate a 3 overhang should be avoided, as 3 overhangs serve as templates for bacteriophage RNA polymerase.
3. We routinely use gel filtration to purify the probe by removing unincorporated nucleotides. Alternatively, the probe can be phenol/chloroform/isoamyl (25/24/1) extracted, ethanol-precipitated with 10 og tRNA, and resuspended in 100 oL of hybridization buffer. If partial length products are routinely synthesized by the *in vitro* transcription assay, then gel isolation of the full-length probe is recommended.
4. The RPA probes are labelled to high specific activity and are therefore prone to rapid radiolysis. Although it is best to use a freshly made probe, probes can be used for up to a week after synthesis. It is suggested that the integrity of the probe be analyzed on a gel prior to each use. If degraded, the probe should be discarded.
5. Total cellular RNA prepared from tissues by extraction with LiCl-urea is used for RPA analysis. The RNA is resuspended in TE (10 m*M*Tris-HCl, 1 *mM* EDTA), pH 7.5, and stored at –80°C.
6. Caution should be exercised when preparing the RNase A stock solution so that the surrounding work area is not contaminated. As the specific activity can vary from stock to stock, it is suggested that optimal RNase A concentration for the RPA be determined empirically each time a new stock is prepared.

7. The RNase Tl stock solution (1-2 mg/mL) that was previously provided by Sigma is no longer available. We currently use a Boehringer Mannheim Biochemicals stock solution that is provided at a concentration of 100,000 or 500,000 U/mL. We use 1 U/oL RNase Tl for the ribonuclease digestion. This is roughly equivalent to the recommended 2 og/mL of the Sigma solution, as determined from the specific activity.

8. Multiple probes may be used simultaneously in an RPA to detect more than one RNA product in a sample. However, if the relative abundance of the RNA products varies, it is advisable to alter the specific activities of the probes. For example, when using a probe to detect an abundant RNA product, the "cold" CTP concentration should be increased and the [χ-^{32}P] CTP concentration should be decreased; vice versa for the probe monitoring the rare RNA.

9. Removal of DNA template from the probe is very important. Contaminating DNA may be a source of background on a RPA gel that results from RNA: DNA hybrid formation. Contaminating DNA will also reduce the amount of probe available for RNA: RNA hybridization. If the protected RNA product is smaller than the probe, full-length DNA:RNA hybrids will be readily distinguishable from the protected RNA fragment.

10. If the probe activity is less than 100,000 cpm, incorporation of the radionucleotide may not have been sufficient. To verify the integrity of the probe before use, we suggest that 0.5 oL of the probe be electrophoresced on a poly aery lamide/urea gel. The wet gel covered with plastic wrap can then be exposed to a phosphorimage screen for 15 min or X-ray film for 1 h to analyze the integrity of the probe: a single discrete band should be observed.

11. The quantity of total RNA to be analyzed depends on the abundance of the target RNA in the tissue of interest. For *H19* and *Jgf2,* 1-3 og of total neonatal liver and skeletal muscle RNA is normally analyzed by RPA. In tissues where these mRNAs are less abundant, 10 og (neonatal kidney) to 50 og (neonatal brain) of the total RNA is analyzed.

12. It is important to eliminate secondary structure of the probe and RNA by denaturing the sample before the hybridization step. To avoid stable secondary structure, immediately transfer the denatured sample from the 85°C heating block to a water bath preset at the appropriate hybridization temperature.

13. Hybridization conditions are dependent on the melting temperature (T_m) of the probe and should be determined empirically. Specifically, for allele-specific analysis of *H19,* we hybridize samples at 65°C with a probe generated from the DNA template. For allele-specific analysis of *Igf2,* we hybridize the samples at 55°C with a probe generated from the DNA template. In general, longer hybridization times are required for less abundant targets. We typically perform alt RPA incubations overnight.

14. To detect a single-base-pair polymorphism between alleles of a gene it may be necessary to incubate the ribonuclease reaction at an elevated temperature. For example, we perform the ribonuclease digestion at 37°C for the allele-specific detection of *H19* RNA and we perform the digestion at 30°C for the allele-specific detection of *Igf2* RNA. Thus, the optimum ribonuclease digestion temperature and the hybridization temperature should be determined empirically. Alternatively, the nonspecific RNase I (cleaves the bond of single-stranded RNA 3 of any ribonucleotide) provided by Boehringer Mannheim Biochemicals may be utilized.

15. The gel running conditions are specific for 220 165 1 mm vertical gels.

Chapter 8

Recombinant DNA Constructs

DNA methylation at position 5 in the cytosine ring in the sequence CpG can be detrimental to the transcription of a variety of genes in higher eukaryotes. Although the significance of this transcriptional repression is currently under debate, there is little disagreement that it plays an important role in genomic imprinting and X-chromosome inactivation. To study the effects of DNA methylation on transcription in an experimental system, bacterial DNA methyltransferases have been used widely in order to mimic the DNA methylation pattern of eukaryotic genes. However, usually every target site in a given recombinant DNA molecule will be subject to DNA methylation by making use of those enzymes. This might result in an exaggeration of the effects of DNA methylation, as most recombinant DNA molecules contain a high degree of prokaryotic DNA, which is rich in CpGs. This methylated CpG-rich DNA can contribute to the effects of DNA methylation by formation of a repressive chromatin structure. In addition, selective DNA methylation is required to distinguish the effects of DNA methylation on transcription initiation and transcript elongation. Thus, there is a requirement for a method to generate recombinant DNA molecules that are methylated in a predetermined region.

The chapter following this one will describe a method that makes use of ligation of methylated DNA fragments into unmethylated vector DNA. This method relies on the availability of suitable restriction sites, which allow directional cloning of the fragment and, in addition, requires a highly efficient ligation reaction. The method described in this chapter generates CpG-methylated regions in recombinant plasmids by making use of targeted methylation on a double-stranded patch in a single-stranded DNA molecule.

PRINCIPLE OF THE METHOD

Any double-stranded DNA (dsDNA) fragment originating from the plasmid can be annealed to a single-stranded DNA (ssDNA) plasmid molecule to generate a single-stranded molecule with a dsDNA patch. A single-stranded DNA-binding protein is added in order to prevent the CpG methyltransferase M.*SssI* from methylating or binding to the ssDNA. Thereafter DNA methyltansferase M.*SssI* is used to methylate all CpGs in the dsDNA patch. Subsequently the ssDNA gap is filled in and a dsDNA plasmid is generated that contains a region of

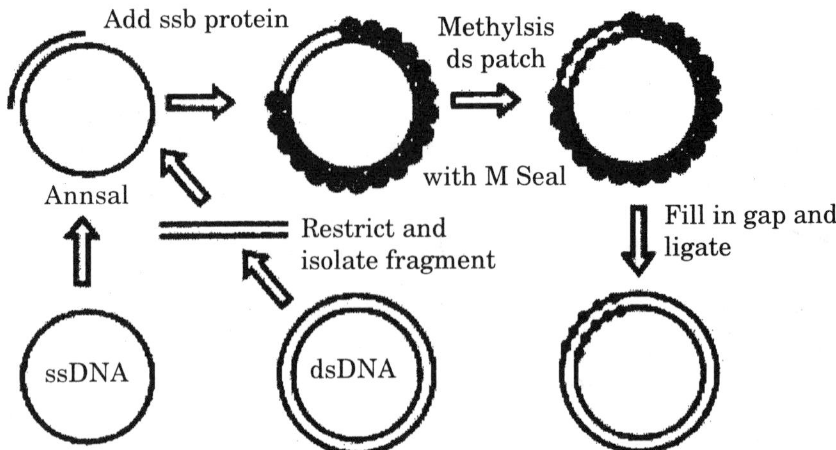

Fig. 8.1: Regional methylation of plasmid DNA. Single-stranded DNA (ssDNA) and a restriction fragment are isolated from the same plasmid and annealed. A single-stranded DA-binding (ssb) protein is added and the double-stranded DNA (dsDNA) patch methylated using MSssI DNA methyltransferase. The gap is then filled in ligated.

methylated DNA. The amounts and quantities specified in this method are sufficient to generate approximately 6-10 og of double-stranded plasmid DNA. This is sufficient to carry out at least duplicate transfections into tissue culture cells. We recommend carrying out mock methylation reactions which will serve as controls in the verification of DNA methylation as well as in transcription assays.

Materials

DNA

1. 5 og (approximately 3 pmol) of single-stranded (ss) plasmid DNA.
2. Approximately 9 pmol fragment DNA is required.

Annealing of dsDNA Fragment to ssDNA Plasmid

1. 99% ethanol.
2. 75% ethanol.
3. 3 *M* Na-acetate, pH 6.9.
4. Annealing buffer: 20 m*M* Tris-HCl, pH 7.5, 10 m*M* $MgCl_2$, 50 m*M* NaCl, 1 m*M* DTT.
5. 95°C water bath/heating block.
6. Annealing bath/incubator (from 70°C to 30°C in 1 h).

DNA Methylation Reaction

1. T4 gene 32 protein (10 og/oL; ssDNA-binding protein, Roche Molecular Biochemicals).
2. DNA methyltransferase *SssI* (2 units/oL).
3. 10 methylation buffer: 100 m*M* Tris-HCl, pH 7.9, 100 m*M* $MgCl_2$, 500 m*M* NaCl, 10 m*M* DTT).
4. S-adenosyl-L-methionine (32 m*M*).
5. Proteinase K (25 mg/mL).
6. Phenol, TE-equilibrated.
7. Chloroform.
8. Ethanol (99% and 75%).
9. 10% sodium dodecyl sulfate (SDS).
10. 0.5 *M* ethylenediaminetetraacetic acid.

Filling in and Ligation Reaction

1. Fill-in buffer; 100 o mol each of dATP, dCTP, dGTP, dTTP, 1 m*M* DTT, 50 m*M* Tris-HCl, pH 7.5, 10 m*M* $MgCl_2$).
2. Klenow fragment, 3 -5 exo- (5 units/oL).
3. T4 DNA ligase (400 units/oL).
4. ATP, 100 m*M*.
5. 37°C incubator/oven.
6. Nucleotide Removal Kit (Qiagen).

Agarose Gel Electrophoresis and Southern Blot Analysis

It is advised to verify the DNA methylation of the constructs using methylation-sensitive restriction enzymes with a subsequent agarose gel electrophoresis and Southern blot analysis. Standard protocols were used for these procedures.

Methods

Annealing of dsDNA Fragment to ssDNA Plasmid

1. Mix 3 pmol ssDNA plasmid with approximately 9 pmol of a restriction fragment derived from the same plasmid.

2. Add 1/10 volume of 3 M Na-acetate (pH 6.9), add 2.5 volume of 99% ethanol.
3. Mix, store for 10 min on ice, then centrifuge for 10 min at 17,900 relative centrifugal force (rcf) (note that all centrifugations are carried out in a microcentrifuge and 17,900 rcf corresponds to 13,000 rpm), room temperature.
4. Discard supernatant, add 500 oL ethanol (75%), vortex, store for 10 min on ice, then centrifuge as before.
5. Repeat **step 4**, then air-dry the DNA pellet for 10 min.
6. Dissolve the DNA pellet in 50 oL annealing buffer.
7. Heat for 5 min at 95°C.
8. Allow to anneal for 1 h from 70°C to 30°C.
9. Add 1/10 volume of 3 M Na-acetate (pH 6.9), and 2.5 volumes of 99% ethanol.
10. Mix, store for 10 min on ice, then centrifuge for 10 min, 15,000g, room temperature.
11. Discard the supernatant, add 500 oL ethanol (75%), vortex, store for 10 min on ice, then centrifuge as before.
12. Repeat **step 11**, then air-dry DNA pellet for 10 min.

DNA Methylation Reaction

1. Dissolve the DNA pellet in 119 oL H_2O.
2. Add 15 oL 10 methylation buffer.
3. Add 5 oL T4 gene 32 protein (10 og/oL).
4. Incubate 15 min, 37°C.
5. Add 10 oL DNA methyltransferase *Sss*I.
6. Add 0.75 oL S-adenosyl-methionine (32 m*M*; final concentration: 160 oM).
7. Incubate for 16 h at 37°C preferably in an oven, to avoid condensation).
8. Add 7.5 oL SDS 10%, 1.5 oL EDTA 0.5 M, 1.5 oL proteinase K (25 mg/mL); incubate for 30 min at 55°C.
9. Add 160 oL phenol and 160 oL chloroform, vortex, centrifuge 10 min at 17,900 rcf, room temperature.
10. Transfer aqueous phase to fresh tube, add 320 oL chloroform, vortex, and centrifuge as before.
11. Transfer aqueous phase to fresh tube, add 1/10 volume of 3 M Na-acetate (pH 6.9), and 2.5 volumes of 99% ethanol.

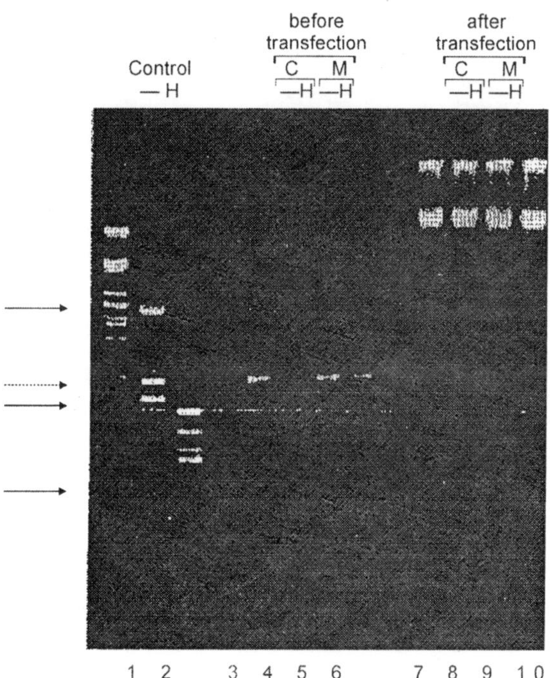

Fig. 8.2: Aragose gel analysis of regionally methylated plasmid DNA. A 1.1-kb restriction fragment was used to generate a methylated patch. The DNA was restricted with the appropriate restriction enzymes as indicated and resolved on a 1.2% gel. Lanes 1 and 2, control plasmid; lanes 3-6, mock-methylated (C) and methylated (M) constructs before transfection into cells; lanes 7-10, total DNA isolated from mammalian cells after transfection with constructs. All samples were digested to release the fragment used to generate the methylated patch (dashed arrow). H indicates digestion with HpaII.

12. Mix, store for 10 min on ice, then centrifuge for 10 min, 15,000g, room temperature.
13. Discard supermatant, add 500 oL ethanol (75%), vortex, store for 10 min on ice, then centrifuge as before.
14. Repeat **step 13**, then air-dry the DNA pellet for 10 min.

Filling in and Ligation Reaction

1. Dissolve the DNA pellet in 92 oL fill-in buffer.
2. Add 5 oL Klenow enzyme.
3. Incubate for 2 h at 37°C.
4. Add 2 oL T4 DNA ligase and incubate for 5 h at room temperature.
5. Purify the mixture using the Nucleotide Removal Kit (Qiagen).

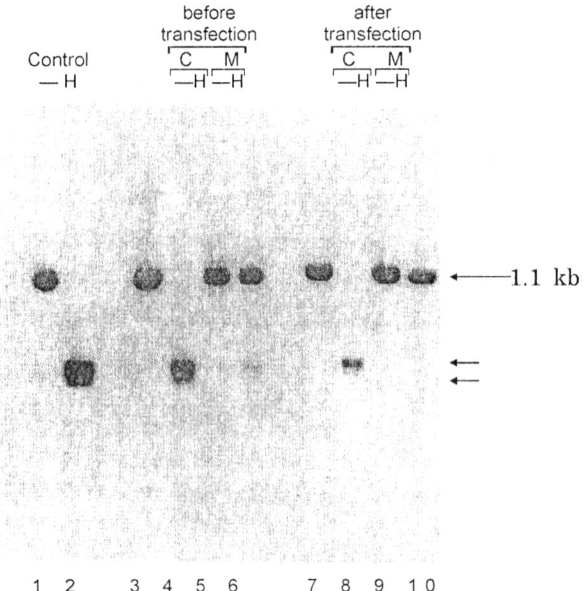

Fig. 8.3: Southern blot analysis of regionally methylated plasmid DNA. The gel from Fig. 8.2 was subjected to Southern blot analysis using the 1.1-kb fragment as a probe. Labelling is as for Fig. 8.2. Exposure of the X-ray film was longer for lanes 7-10 in order to obtain comparable band intensities.

Agarose Gel Electrophoresis and Southern Blot Analysis of Regionally Methylated Fragments

To test whether DNA methylation went to completion in the predetermined region, the DNA was digested prior or subsequent to transfection into mammalian tissue culture cells with the methylation-sensitive restriction enzyme *Hpa*II. A 1.1-kb DNA fragment was used in the methylation reaction, and it can be seen that this is resistant to digestion by *Hpa*II, when methylated, but not in the control (mock-methylated) sample. The gel was subjected to Southern blot analysis in order to reveal methylation patterns of DNA after transfection into mammalian cells. Upon *Hpa*II digestion, the methylated DNA remains largely undigested, whereas control plasmid and mock-methylated control reveal the expected digestion pattern.

Furthermore, after transfecting the plasmid DNA into cells, the same digestion pattern is observed, indicating that the preimposed DNA methylation on the 1.1-kb fragment was stable for at least 2 d after transfection.

Before transfection, small amounts of *Hpa*II digestion products are seen which are due to incomplete methylation of the threefold excess of fragment DNA. During transfection this fragment DNA is degraded and thus not detectable anymore.

Notes

1. Most plasmid vectors contain an f1 origin of replication, which allows the isolation of single-stranded DNA (ssDNA) from plasmid vectors. Blondel and Thillet describe an efficient method that we have used routinely.
2. Restriction fragments are annealed at a molar ratio of 3:1 (fragment DNA:ssDNA). Therefore, for each fragment to be methylated, approximately 9 pmol fragment DNA is required. High-quality fragment DNA can be isolated from agarose gel slices using the QIA quick gel extraction kit (Qiagen).

METHYLATION OF SPECIFIC REGIONS

The first imprinted genes were identified in the early 1990s and there are now over 40 mammalian genes known to be regulated by genomic imprinting. The details of the mechanism that discriminates between the active and silent alleles of these genes, based on their parent of origin, may differ from one imprinted gene to the next, but must include some form of epigenetic mark and distinguishes alleles that have passed through the male or female germline. The addition of methyl groups to cytosine residues of CpG dinucleotides might provide such a mark, since regions of differential methylation have been identified in the vicinity of many of the known imprinted genes.

Moreover, analysis of imprinted gene expression in a methyltransferase knockout (*dnmtl*$^{-/-}$) mouse has shown that the imprint is lost in a number of cases, resulting in either two silent alleles (Igf2, Igf2r and Kvlqt) or two expressed alleles (H19, p57^{kip2}, *Snrpn,* and *Xist*). Although there may be exceptions (for instance, imprinted expression of *Mash*2 is maintained in *Dmnt1*$^{-/-}$ embryos; differential methylation is likely to be an important aspect of the imprinting mechanism that is relevant to most of the imprinted genes in mammals.

There is accumulating evidence that methylation is also important for the imprinting of plant genes. In this section we describe a method that can be used to change the methylation status of defined regions in recombinant DNA so that effects on gene expression can be assessed following transfection of plasmid constructs into cultured cells. Using conventional molecular techniques, the region of DNA to be methylated is excised from the parent construct, purified, and then methylated in vitro using the CpG methylase SssI. The methylated region is then ligated back into the construct backbone, yielding a plasmid that is methylated only in the region of interest. Using this method we have been able to modify a variety of recombinant DNA constructs for testing in transfection assay.

Materials

Plasmid to Be Methylated

It is very important that the region to be excised and methylated is bounded by two different restriction sites that are each unique within the parent construct. This is to ensure that once the insert has been methylated, it can be ligated back into the construct backbone in the correct orientation. It also ensures that the vector cannot recircularise during ligation without taking up insert (although it does not exclude the possibility that two vectors might ligate to each other). A large amount of plasmid DNA starting material is needed (at least 20 og is recommended), especially if the region to be methylated is small in relation to the remainder of the construct and should be prepared using a method that yields plasmid of a quality suitable for mammalian cell transfection.

Enzymes

1. Restriction enzymes to excise the region of interest (used with buffers supplied by the manufacturer).
2. *Sss*I methylase (2000 units/mL, supplied with 10 buffer (500 m*M* NaCl, 100 m*M* $MgCl_2$, 10 m *M* DTT, 100 m*M* Tris-HCl, pH 7.9) and 32 m*M* S-adenosyl methionine (SAM) methyl donor cofactor.
3. Enzymes to check extent of methylation reaction. Widely available methyl-sensitive restriction enzymes are *Hha*I and *Hpa*II, recognizing the sequences GCGC and CCGG, respectively (used with buffers supplied by the manufacturer).
4. T4 DNA ligase. Use high-concentration preparations to maximize ligation efficiency (e.g., 20,000-30,000 Weiss units/mL enyme). Aliquot the 10 reaction buffer (100 mM $MgCl_2$, 10 mM DTT, 10 mM ATP, 500 mM Tris-HCl, pH 7.5, 250 og/mL bovine serum albumin) supplied with enzyme into single-use amounts, as it deteriorates with freeze-thawing.

DNA Purification

1. QIAquick gel extraction kit (QIAGEN).
2. QIAquick nucleotide removal kit (QIAGEN).

Electrophoresis

1. Agarose (ultrapure, electrophoresis grade; Gibco-BRL).
2. 1 TAE buffer: 0.04 M Tris-acetate, 0.001 M ethylenediaminetetraacetic acid (EDTA), pH 8.0. Note that a 50 TAE stock can be stored at room temperature and diluted as required.
3. Horizontal gel apparatus and power supply. Two sizes of gel combs are needed for moulding of either small wells, with approximately 20 to 30 oL volume capacity (for analysis of restriction reactions), or large wells, with approximately 200-oL volume capacity (for purification of relevant restriction fragments).

4. DNA markers of known size and concentration (e.g., phage v DNA digested with *Hind*III is widely available).
5. 10 gel loading buffer: 0.4% (w/v) bromophenol blue, 0.4% (w/v) xylene cyanol FF, 25% (v/v) ficoll (type 400, Pharmacia).
6. 10-mg/mL ethidium bromide stock solution for gel staining, and UV light source for visualization of stained DNA.

Miscellaneous

1. Absolute ethanol, for diluting DNA purification buffer solutions and for DNA precipitation steps.
2. 3 M sodium acetate, pH 5.5, for DNA precipitations.
3. Scalpel for removing bands from gels.

Method

1. Digest 20 og each of the insert-containing and vector-containing plasmid DNAs overnight with the chosen restriction enzymes. Use a reaction volume of 100 oL, containing 50 U of each restriction enzyme. If the enzymes require sequential digestion (i.e., because they have incompatible reaction conditions), carry out each reaction overnight, precipitating the DNA in between by adding 1/10 volume 3 M sodium acetate, pH 5.5, and 3 vol absolute ethanol. Mix gently, by inverting the tube several times, and place at –20°C for at least 1 h. Collect precipitated DNA by microcentrifugation at 10,000-14,000 rpm (9,000-17,000g) for 10 min, wash the pellet with 70% ethanol, spin again, dry, and resuspend in sterile distilled water.
2. After each phase of the digestion, remove 2-3 oL (around 500 ng) of each sample, add 1 oL 10 loading buffer, and make up the volume to 10 oL with water. Run this out on a gel to ensure that the digests are complete. Pour a 1% (w/v) agarose gel, first dissolving the agarose in 1 TAE buffer and then adding 0.5 og/oL ethidium bromide, using a comb with small wells (in adjacent wells run one sample of 500 ng undigested plasmid for comparison and one sample of DNA size makers). Run the gel, submerged in 1 TAE buffer, at 100-150 V for 90-120 min.
3. Having confirmed that the restriction digests have gone to completion, add 1/10th volume 10 gel loading buffer to the remainder of the digest and load this onto another 1% agarose (w/v) gel using a comb with large wells so that the entire sample can be separated.
4. Run the digest at 100 V until the required DNA fragment is well separated from any other fragments (progress can be monitored at intervals by viewing the gel with a UV light source, then returning it to the electrophoresis tank and continuing the run if necessary).

5. Once good separation has been achieved, excise the required DNA fragments (i.e., vector, or construct backbone, as well as the insert, or region to be methylated) from the gel using a scalpel. Trim the gel slice on all sides to remove as much as possible of the excess, unstained agarose. DNA can then be recovered from the gel slice using a QIA quick gel extraction kit (Qiagen) as detailed by the manufacturer, but note the following details: not more than 300 mg of agarose should be loaded per QIAquick spin column, and the final elution step is performed with 50 oL elution buffer (EB, 10 mM Tris HCl, pH 8.5) per column.
6. Run 1 oL of the vector and insert DNA on a 1% agarose gel to check its integrity. On the same gel, run DNA markers containing bands of known size and concentration to allow an estimation of the yield of the purified DNA fragment.
7. At this point, the vector (construct backbone) can be stored at -20°C until it is needed for the ligation.
8. Divide the insert (region to be methylated) DNA into two tubes. One will be methylated and the other mock-methylated, to serve as a control.
9. Set up the methylation reactions. Given here is a 50-oL speciment reaction 25 oL DNA, 5 oL 10 reaction buffer, 5 oL (2000 U/oL) *Sss*I methylase (replace this with water or enzyme storage buffer for the mock-methylated tube), 15 oL sterile distilled water, 1 oL SAM.
10. Incubate the methylation reactions for 4 h at 37°C. It is then vital that the SAM be replenished, as it is heat-labile and deteriorates quite quickly at this temperature. Add another 1 oL of the SAM and return the tubes to 37°C overnight.
11. Digest approximately 100 ng of the mock-methylated and methylated reaction with a methylation-sensitive restriction enzyme that will allow determination of whether the methylation reaction was successful and complete. Common enzymes to use are *Hba*I and *Hpa*II, as they each recognize a four-base motif that is frequently present in DNA sequences. Digest for 1-2h at 37°C using 5-10 U enzyme. Run the DNA on a 1% agarose gel; the mock-methylated DNA should digest while the methylated DNA remains intact. If the methylation is not complete, the overnight incubation can be repeated. Add 10,000 U more enzyme, 1 oL fresh SAM, and sufficient buffer and sterile distilled water to make the volume up to 100 oL.
12. Once the methylation is complete, the DNA can be purified away from enzymes and salts using a QIAquick nucleotide removal kit (Qiagen), following the manufacturer's instructions but with the final elution step performed using 50 oL elution buffer (EB, 10 mM Tris HCl, pH 8.5). Between 1 and 5 oL of the sample is then checked on a 1% agarose gel to confirm integrity and to estimate the yield by comparison with DNA markers of known concentration.
13. Ligate the insert (both methylated and mock-methylated fragments) with vector at a molar ratio of 3-5:1 insert: vector. Use all of the available insert and adjust the

amount of vector accordingly. Ligase buffer should be added to provide 1 final concentration, with 1 oL ligase (20-30 U/oL), with any extra reaction volume made up with sterile distilled water and the reactions incubated overnight at 4°C.

14. DNA can again be purified away from enzymes and salts using a QIAquick nucleotide removal kit (Qiagen).
15. It is necessary to check the progress of the ligation to see how much of the desired product is present. This is not always straightforward, as there are usually multiple possible ligation products. Digest the ligation mix with one of the enzymes used in the original insert/vector preparation. Cut an amount approximating 250 ng (base this calculation on the amount of DNA that went into the ligation reaction) for 2 h at 37°C, then run it on a 1% agarose gel at around 50V until the ligation products have separated as fully as possible. In addition to single-insert ligation products, you may also see on the gel bands that represent (a) single unligated insert fragments, (b) two insert fragments ligated together, or (c) two vector fragments ligated together. Typically, the yield of vector containing a single insert is around 30-40% of the ligation product (this is one reason that it is important to have a mock-methylated control sample). The described protocol has been successful for inserts of 0.8-2.8 kb, being ligated back into vector constructs of 6-14 kb.
16. The DNA constructs are not in a form that allows the effects of methylation of specific regions to be examined following their introduction into cells.

Notes

1. It is recommended that a large quantity of DNA be used from the outset, because there are several purification steps at which some loss of material is unavoidable. Beginning with 20-50 og DNA will usually result in around 2-5 og each of the methylated and mock-methylated construct samples. It is possible to prepare insert in batches that are then pooled in the ligation reaction.
2. S-adenosyl methionine (SAM) is unstable at 37°C and when subjected to repeated freeze-thaw cycles. Consequently, it is advisable to prepare several aliquots from the tube supplied with the enzyme on the first occasion that it is thawed and use a fresh aliquot each time these reactions are carried out.
3. A comprehensive database of restriction enzymes and their properties, including details of recognition sequences and methylation sensitivity, is maintained.
4. Sufficient DNA must be cut (a) to allow generation of a mock-methylated as well as a methylated plasmid (the mock-methylated plasmid will serve as a control, given the ligations will not be 100% efficient) and (b) to allow for the loss of some DNA during the purification steps. In some instances the insert-containing and vector-containing constructs will be the same plasmid, and it may be possible to purify both

fragments from the same digest. Exceptions to this are cases in which the two fragments are similar in size and cannot be adequately resolved on an agarose gel. This problem can be circumvented by carrying out two separate reactions, including in each an enzyme that cuts one of the fragments into two smaller pieces, allowing the other fragment to be cleanly excised from the gel.

5. For restriction digests that fail to reach completion, more enzyme can be added and the reaction continued for several more hours. However, the volume of enzyme in the reaction should never exceed 10% of the total reaction volume, because the glycerol used in enzyme storage buffers can interfere with enzyme activity at higher concentrations. Where the addition of extra enzyme would exceed this 10% limit, or for digests that fail completely, protein can be removed by phenol/chloroform extraction and the DNA recovered following precipitation. Repeat the digest, using a new batch of restriction enzyme if necessary.

6. Some care is needed to ensure that an optimal amount of digested DNA is loaded into the available wells of the agarose gel. A high concentration of DNA is desirable, as this will ensure a good yield of the purified fragment. However, if the concentration is too high, it will be impossible to achieve good separation of the required fragment from other products of the restriction digest. Aim for about 1 og of digested DNA for every 1 mm of well (measured across the gel), in the first instance, but note that quantities may need to be adjusted empirically for each particular digest (for example, more DNA can be loaded in cases where the size of the required fragment differs greatly from that of any additional fragments.

7. A number of kits are available for the purpose of DNA purification from agarose; the one recommended here works well. An effective alternative to commercial kits is the following, adapted from the method described by Heery: First, a spin column is made from one 1.7-mL and one 0.5-mL microfuge tube. Remove the lids from both tubes and pierce the bottom of the smaller tube using a narrow-gage syringe needle. Plug the small tube with a ball (approximately 5 mm diameter) of polyallomer wool (this is used as the filter material in tropical fish tanks and can be obtained from pet shops; it can be sterilized by autoclaving) and place the gel slice in the small tube, on top of this plug. Place the small tube inside the large tube and microcentrifuge at 10,000 rpm (9,000g) for 10 min. Buffer containing the DNA fragment will collect in the 1.5-mL microfuge tube, while the gel matrix will remain above the polyallomer wool plug (the eluted DNA can be visualized in solution under UV light). Next, the eluted DNA can be purified away from any contaminating small molecules by passing it through a column made from a 1-mL syringe barrel (again plugged with a ball of polyallomer wool, containing Sephadex-G50 (hydrate 10 g in 150 mL 1 TE buffer: 0.01 M Tris-HCl pH 7.6, 0.001 M EDTA pH 8.0, and sterilize by autoclaving). Place the filled syringe barrel in a 15-mL plastic disposable centrifuge tube and spin in a bench-top centrifuge at 250g (around 120 rpm) for 2 min to remove excess buffer. Transfer the column to a fresh 15-mL tube, add the solution eluted from the gel

slice to the top of the column, and recentrifuge at 250g for 2 min. A solution containing most of the purified DNA fragment should elute at this step (this can again be monitored under UV light, and more of the DNA fragment can usually be recovered by adding 100 oL 1 TE buffer and centrifuging once more.

8. The methylation reaction can be scaled up successfully to a volume of at least 150 oL.

9. The SssI enzyme mimics mammalian methyltransferase, in that it methylates cytosine residues in CpG dinucleotide pairs, but differs in that it will use both fully methylated and hemimethylated DNA as substrate. In the absence of Mg^{2+}, SssI is highly processive, tending to methylate cytosines at all CpGs along a DNA template molecule. It should be possible to produce random, distributed patterns of methylation by increasing the Mg^{2+} content of the reaction buffer, but we have not investigated this in our own experiments.

10. The methylase reaction shown will usually result in complete methylation of the DNA fragment. If this is not the case, then a fresh aliquot of SAM, and/or a new batch of enzyme, may be required.

11. Ligation reactions containing 20 og of either methylated or mock-methylated insert, together with 10 og vector work well. Ligations on this scale may require pooling of two or more batches of methylated DNA fragments, but allow efficient preparation of large quantities of the final product.

12. In our experiments the methylated constructs and controls have been assayed for gene expression following their introduction into various mammalian cell lines. A high transfection efficiency is desirable, since the yield of each construct preparation is typically very low (2-5 og of the correct ligation product). To achieve high transfection efficiencies with small amounts of plasmid DNA (0.25-1 og), we favor the use of FuGENE 6 transfection reagent (Roche). Gene expression from the transfected constructs can be measured in a transient assay or following stable maintenance of the exogenous DNA. In the latter case, the pattern of DNA methylation established in vitro can be maintained through DNA replication in vivo in at least some cell types.

Chapter 9

Promoter Recognition

In this chapter, we shall describe the problem of promoter recognition. We begin with a brief introduction to the biology of promoter structure and function. We then review some of the current computational approaches to the problem, with emphasis on basic concepts and methodologies in real applications.

There are two main classes of functional information encoded in the genomic DNA of every living organism. One class is the coding regions, which specify the structure and function of each gene product; the other class is the regulatory regions (occasionally, but very rarely, overlapping with a coding region), which control and regulate when, where, and how the genes are expressed. Promoter is the most important regulatory region that controls and regulates the very first step of gene expression: mRNA transcription. Promoter is commonly referred to as the DNA region that is required to control and regulate the transcriptional initiation of the immediately downstream gene.

For a typical eukaryotic (PolII or protein-coding) gene, it contains a core promoter of about 100 bp centred around the transcriptional start site (TSS), and a proximal promoter of about 500 bp immediately upstream of the core promoter. Often complex regulation in vivo can involve many more features, such as enhancers, locus control regions (LCRs), and/or scaffold/matrix attachment regions (S/MARs). Some people refer to enhancers as the distal promoter elements, which can be either upstream or downstream of the gene or within an intron and can be in any orientation. For our purpose, we use the region (–500,+100) with respect to a TSS as a specific definition. The main characteristic of a promoter is that it contains aggregates of transcription factor (TF) binding sites.

During the process of development, genes are turned on and off in a preprogrammed fashion, a process that eventually generates cell specificity. This developmental program is orchestrated by TFs, which bind to specific DNA sites in the promoters near genes they control. A single TF is not dedicated to each regulatory event. Instead, different combinations of ubiquitous and cell-specific regulatory factors are used to achieve a combinatorial control. Core promoter, approximately in (–50, +50), (1) binds to and controls assembly of the preinitiation complex (PIC) containing PolII, the general transcription factor (GTF), and coactivators; (2) positions the TSS and controls the direction of transcription; and (3) responds to nearby or distal activators (we use the same terms, "*activators*" or "*enhancers,*" to also imply "*repressors*" and "*silencers*," depending on the context, for simplicity) binding proximal promoter and enhancers. The PIC comprises the GFTs (PolII, TEIIA, TFIIB, TFIID, TFIIE, TFIIF, and TFIIH), and coactivators that mediate response to regulatory signals. A typical core promoter contains four DNA elements: TATA-box (binding site for the TBP subunit of TFIID.

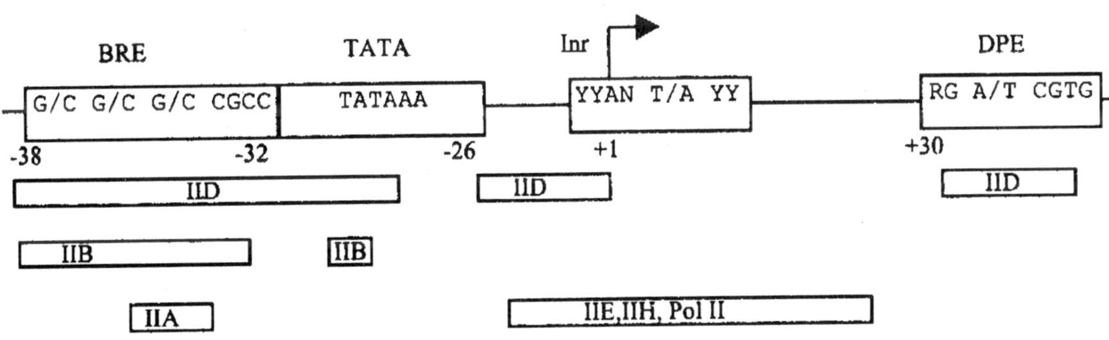

Fig. 9.1: Sequence elements and GTF footprints in a typical core promoter

Although recently some alternative TATA binding proteins have been discovered in a few specific types of cells, this review deals with only the major types of TBP: Inr (overlapping with TSS), DPE (downstream core promoter element), and BRE (TFIIB recognition element). Not every element occurs in a core promoter. People have classified core promoters according to the presence or the absence of TATA and/or Inr elements. Many "housekeeping gene" core promoters appear to lack both TATA and Inr elements but instead contain several TSSs, a high G+C content, and multiple binding sites for the ubiquitous TF Spl, which directs the formation of PIC to a region 40-100 bp downstream of its binding sites. Purified GTFs and PolII mediate basal (low-level) transcription on a core promoter in vitro but cannot support activated transcription in the absence of coactivators. More recent studies indicate that the functional form of PIC in vivo must also include coactivators/mediators.

The interaction of activators with any surface of this large GTF-containing complex (also called *holoenzyme),* allows recruitment of the complex to the core promoter and response of the polymerase to the regulatory signals. Transcriptional regulation is controlled by the binding of sequence-specific DNA-binding TFs to proximal promoters, approximately in (–500, –50) (also called regulatory promoters), and enhancers. It should be noticed that there is no real distinction between proximal and distal (enhancer) regulatory elements; they often involve the same set of TF binding sites. Some cooperative binding of activators to enhancers and proximal promoters can lead to the assembly of nucleoprotein structures termed "*enhanceosomes*". In a living eukaryotic cell, DNA is not naked; instead, it is wrapped into nucleosomes by histones.

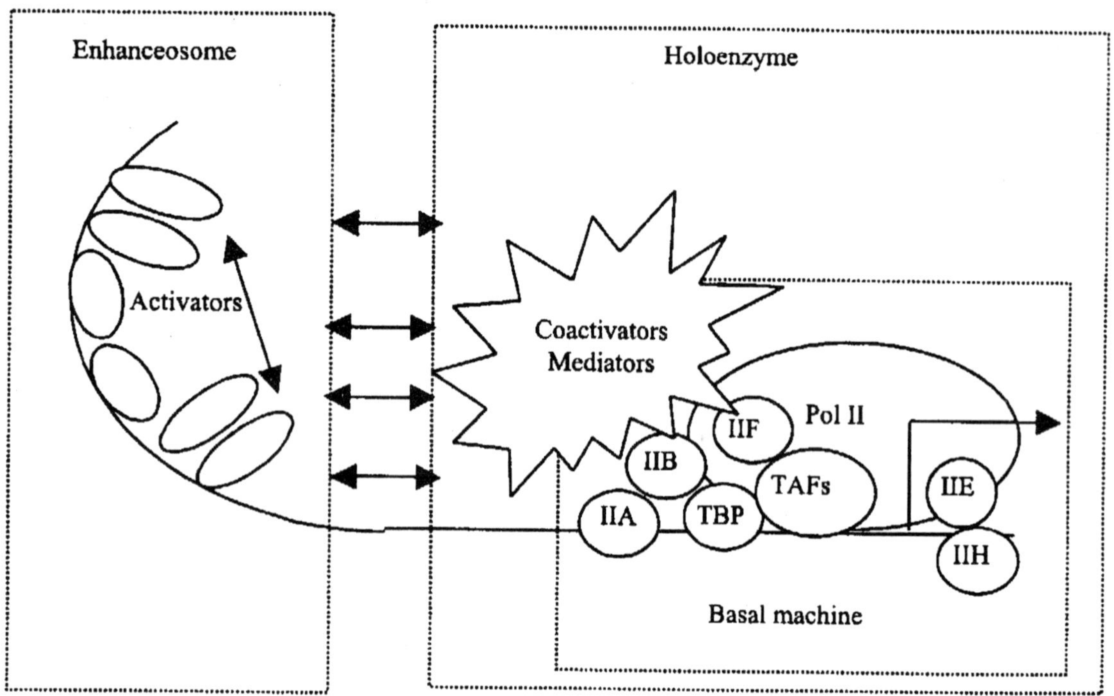

Fig. 9.2: Recruitment and activation of the PIC

With the help of many other non-histone proteins (NHP), nucleosomes are further condensed into chromatin filament. These higher order structures are believed to be necessary to keep most genes in a (*default*) repressed state. To activate a gene, the chromatin encompassing that gene and its control regions must be altered or "*remodelled*" to permit TFs to access their specific binding sites. Because of the complexity of such long-range interaction among many global regulators, chromatin remodeling is beyond the reach of current promoter recognition algorithms. Therefore, all existing computational methods implicitly assume all TF sites are accessible, which is the intrinsic source of a large number of false positives. In summary, promoter is the key DNA region that controls and regulates transcription.

Delineation of the promoter architecture is fundamental for understanding gene expression patterns, regulation networks, cell specificity, and development. It is also important for designing efficient expression vectors or to target specific delivery systems in gene therapy. In the large-scale genomic sequencing era, promoter prediction is also crucial for gene discovery and annotation. There have been many computational approaches to this extremely difficult problem. Depending on the goals, computational approaches can be divided into two classes: general promoter recognition methods and specific promoter recognition methods. The primary goal for the general methods is to identify TSS and/or core promoter elements for all genes in a genome; the specific methods focus on identifying specific regulatory elements (TF sites) that are shared by a particular set of transcriptionally related genes. Specific methods can have very high specificity when searching against the whole genome and can provide immediate functional clues to the downstream gene. But because of their broad coverage, the general methods are extremely useful for large-scale genome annotation.

FINDING TRANSCRIPTION FACTOR (TF)

As a specific promoter class is characterized by a specific set of TFs, finding TF binding sites is the most important step in promoter recognition. There are at least two classes of TFs (from now on, we use TF to refer to DNA binding transcription factors). One class is the general or ubiquitous TFs, such as TATA-box binding protein (TBP) or Sp1. There binding sites can be identified by simply collecting a large number of promoter (–500, +50) sequences. The other class is the specific TFs, which can only be identified by getting a specific set of promoters that share the same site (i.e., their target genes are co-regulated by the same TF). Experimentally, biologists are able to identify a TF site *de novo* with a single promoter sequence. They can characterize such a site by mutagensis and obtain a *consensus* description (such as the *E. coli* TATA-box TATAAT or allowing degeneracy TATRNT).

As more sites are known, one can get the same information by aligning the sites. Although it is easy to write a consensus pattern to represent aligned sites, it is difficult to find one that is optimal for predicting the occurrence of new sites (generalizability) or for discriminantly ranking the binding activities of the sites (differentiability). In most applications, a position *weight matrix* (PWM) is often superior.

Construction a Matrix Given the Alignment

To explain the high degeneracy found in many sites, Berg and von Hippel (1987) proposed a theory on the selection of DNA binding sites by regulatory proteins. They assumed that specific sites have been selected according to some functional constraint (e.g., the binding affinity or activity must be in some tolerable range), and all sequences that can fulfill this requirement are equally likely to occur. This theory, which has been checked by experiments, provided a link between a natural scoring function (minus binding energy) and observed base frequency at each position:

$$S_{bx} = \log(P_{bx}/P0_x) \tag{9.1}$$

where s_{bx} is the score (PWM element) for a base "b" at position "x", p_{bx} is the frequency of the base "b" found at "x" in the site, and "0" indicates the base (consensus base) that has the lowest energy. Often it is desirable to use a scoring function that can best discriminate the set of target sites from a set of control sites. In this situation, we replace p_{0x} by $p^0{}_b$, the base frequency evaluated in the control set (here, we assume they do not depend on the position). Choosing $p^0{}_b$ appropriately (representing the correct background contrast) can be very important for searching other such sites in a genome.

The total score of the site is the sum of individual position scores. Using a single base frequency implies the assumption of independence between any pair of bases, although this can be easily generalized to high-order Markov models. It should be mentioned that to construct a statistical measure, the quality of sequences is obviously extremely crucial. In addition to the integrity of the data, statistical independence of the sequences is essential so that the result will not be biased by the sample. There are many ways to reduce the redundancy; a simple criterion may be that no pair of sequences should have more than 90 per cent identity within the 100-bp surrounding region. There are many ways to introduce pseudo-counts in order to avoid a null frequency. This is equivalent to introducing a prior probability, which is necessary when an observed count is rare.

The Laplace plus-one method (i.e., add one to every base count at every position, and hence correspond to a uniform prior) is a simple and popular choice. How to determine the length of a motif can be a very subtle problem. Conventionally, one uses the relative information to measure the significance of each position:

$$I_x = \sum_b p_{bx} \log(p_{bx} / p_b^0) = \sum_b p_{bx} s_{bx} \quad (9.2)$$

which obviously has the meaning of the average binding energy of all the known sites at position "x". The optimal length of the motif can be obtained by an optimization procedure. One procedure, consistent with the Berg and von Hippel theory, would be to find an optimal window site so that the total information within the window (the area under the curve of I_x) minus the average of the total information in the two flanking windows become maximum. It can be further refined by the discriminant procedure described below.

Known Binding Site

Given a motif, either in the form of a consensus or a matrix, one must first assess the quality of the motif and determine a threshold value before one can use it to search for new members of the site. The way to do this is to perform a standard classification tset in which both the threshold score and the motif length may be optimized by minimizing the classification (Bayesian) error. Because a single TF site does not have enough specific information due to its short length (about 5 to 25 bp) and high degeneracy, any unconstrained genome search will almost certainly result in a lot of false positives. The specificity can only be achieved by combining interactive (correlated) sites into a promoter module (also called a *composite site*) and by higher order structure constraints (long-range control elements).

Unaligned Sequences

In order to discover novel motif sites, one has to use more sophisticated approaches. Given a set of related sequences, these methods must be able to find motif(s) shared by majority of the sequences and statistically significant, in a reasonable time. Although we are focusing on TF sites, all methods should be applicable to more general sequence motif discovery problems. There are existing numerous algorithms. Below is necessarily a personal selection that represents a short list of generic methodologies. HMM and neural network are not included because there are special books describing these machine learning approaches.

k-tuple or Exhaustive Pattern Search Methods

For finding short highly conserved motifs (such as many typical TF sites in yeast), *k*-tuple-based methods can be very effective. The basic idea is to detect over-represented (with respect to a control set or to a background set) *k*-tuples systematically.

Relative Information (RI): The simplest method is to calculate the frequencies for all *k*-tuples in both the target set and the control set. *k* is usually limited by the size of the sample data ($4^k \leq (L - k + 1)*N$, where *L* is the length of each sequence and *N* is the number of sequences). One can define a *RI* for every word w of length k as (also called LOG-ODD Ratio, or negative relative entropy):

$$RI(w) = \log(w)/f^0(w)) \tag{9.3}$$

Although the exact statistic is not known, one can use z = (RI – mean (RI)/std(RI) to estimate the significance. If f^0 represents a random background (usually a Markov approximation of order less than k, one could use chi-square test on = $\chi^2 = \Sigma w[O(w) - E(w)]^2/E(w)$, where $O(w)$ is the observed number of w and $E(w)$ is the expected number (calculated using f^0). This can been applied in yeast promoter analysis. This method can be easily generalized to allow limited degeneracy and/or iterative extension of the *k*-tuple motif.

WORDUP: This is a similar method but requiring the motif to be shared by majority of the sequences. The statistical significance of each *k*-tuple word w is determined by comparing, through a χ^2-test, the actual number of different sequences in which *w* is present with the expected occurrences. Expectations are calculated on the basis of two assumptions: (1) oligonucleotides are Poisson distributed, and (2) nucleotide sequences can be generated according to a first order Markov chain. Because the probability $p_i(w)$ that w is found at least once in the ith sequence is $pi(w)$=1-exp[$- \lambda_i(w)$] with $\chi_i(w) = p_i^0(w)\ (L_i - k + 1)$ and $p_i^0(w)$ is the Markov approximation of $f_i(w)$, namely $p_i^0(w) = f(w_{1,2})\, f(w_{2,3}) \ldots f(w_{k-1,k})/ f(w_2) f(w_3) \ldots f(w_{k-1})$. The expected number of sequences containing w is given by $E(w) = \Sigma_i p_i(w)$. If $O(w)$ is the observed number of sequences containing w, the standard χ^2-value given above can be used to rank significant *k*-tuples (the default threshold is 20) that form a vocabulary. An iterative procedure *k*-tuples (the default threshold is 20) that form a vocabulary. An iterative procedure was also used to construct a new vocabulary containing all significant words of length greater or equal to *k*.

A similar method was developed by van Helden *et al.* (1998) using slightly different statistical criterion. More sophisticated algorithms for detecting more complex patterns (with multiple sites) have also been developed recently.

Multiple Sequence Alignment Methods

For longer and more degenerate motifs, one has to use multiple sequence (local) alignment algorithms. Given N (number of sequences), L (length of each sequence) and k (length of the motif with indels), there are $(L - k + 1)^N$ possible alignments. Finding on optimal alignment that maximizes on objective function (say, $I = \Sigma_x I_x$) is a hard problem. Various heuristic approaches are available to attack this multi-dimensional optimization problem.

CONSENSUS

A greedy algorithm originally developed by Stormo and Hartzell (1989) and implemented in CONSENSUS this is a heuristic method, which is quite efficient and has been widely used in DNA motif discoveries.

The basic idea is illustrated in a toy example shown in Fig. 9.3. Given the three sequences (a) to be aligned, the algorithm starts by forming a frequency matrix for each of the k-tuples

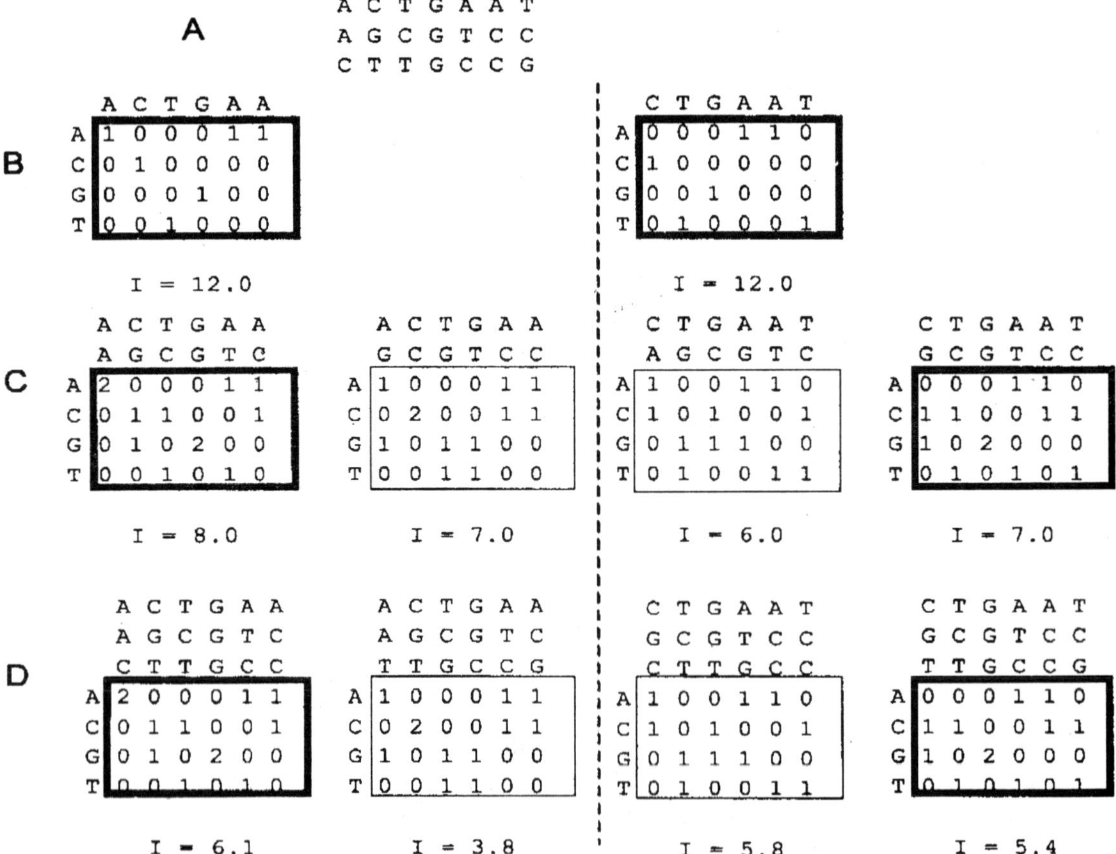

Fig. 9.3: A diagram of the algorithm in CONSENSUS.

in the first sequence (b). Each of these matrices is then combined with each k-tuple in the second sequence to form new matrices containing two k-tuples (c). However, for each k-tuple from the first sequence, the program only saves the "*best*" progeny matrix (measured by the information content). In the next cycle, each saved matrix is combined with each k-tuple in the third sequence to form new matrices, each containing three k-tuples (d). Again, the program only saves the best progeny of each matrix from the previous cycle. This cycle is repeated until the last sequence in the set has contributed a k-tuple to the saved matrices. Of the matrices saved after the last cycle, the one with the lowest probability of occurring by chance is considered to describe the consensus motif (the first in D). In practice, ties occur during the cycles, so that the number of matrices at the end is greater than the number of k-tuples in the first sequence.

The CONSENSUS program was first used to accurately identify the known consensus pattern for the *E. coli* CRP protein binding sites. It was then further improved and tested for robustness of the *E. coli* LexA protein binding sites. In both cases, the order in which the sequences are presented in not critical (the latest version of the program allows the user to set a parameter so that the result will not depend on the input order at all). The program is also robust enough to tolerate some sequences that do not contain binding sites. Thanks to the effort to Hertz, CONSENSUS has been constantly improved upon. Some of the important additions to the original algorithm are (1) independence of the input sequence order, (2) autodetecting motif length, (3) allowing limited insertions/deletions, and (4) more regorous statistical evaluation of the p-value.

EM AND MEME

EM (*expectation maximization*) is a standard technique widely used in maximum-likelihood estimations. Expectation maximization algorithms are named for their two iterative steps, the expectation (E) step and the maximization (M) step, which are alternately repeated until a convergence criterion is satisfied. Lawrence and Reilly (1990) first developed an EM algorithm and tested it on cyclic adenosine monophosphate receptor protein (CRP) binding sites. One starts with an initial guess on the base probability p_{bx} within the sites and the background p_b^0 for the nonsites, then the probability of the even B_{jy} that the site begins at position y in sequence j can be calculated by Bayes formula:

$$P(B_{jy} \mid \mathbf{p}, S) = P(S \mid B_{jy}, \mathbf{p}) / \Sigma_x P(S \mid B_{jx}, \mathbf{p}) \tag{9.4}$$

where S is the sequence data and the prior probability $P^0(B_{jy})$ is uniform, that is, $1/(L - k + 1)$ and $P(S \mid B_{jy}, \mathbf{p})$ = the product of probabilities for all bases. Using formula (9.4), one can complete the E-step by calculating the expected number n_{bx} of the base b at position x in the site and the expected number n_b^0 of the base b in the nonsites. The M-step is simply to replace $p = \{p_{bx}, p_b^0\}$ by $\{n_{bx}/N, n_b^0/N\}$ (the maximum likelihood estimators). One then iterates this to convergence when the parameter estimates no longer change. *MEME* added several extensions to EM to overcome some limitations. MEME chooses starting points systematically, based on all subsequences of the data. It eliminates the assumption of one motif per sequence and allows each sequence to contain zero, one, or several appearances of the shared motif.

Furthermore, *MEME* probabilistically *"erases"* the appearances of a site after is fond, and continues searching for other shared motifs in the dataset. The newer version has made MEME smarter and more robust as an unsupervised motif-discovering tool, as it will automatically determine the motif length and/or choose whether or not to enforce the palindrome constraint. Once a *MEME* motif is found, *MAST* can be used to search other sequences for new members.

Gibbs Sampler

As greedy or EM-based algorithms cannot guarantee to find the global maximum and may be prone to local optima, stochastic algorithms have been developed to overcome this problem. The Gibbs sampler, which consists of a site sampler and a motif sampler has been a very successful one. The site sampler assumes every sequence contains at least one site. The algorithm is initialized by choosing random starting positions within all the sequences. It then proceeds through two steps of Gibbs sampler iteratively. First, it builds a model step by constructing a model $p = \{p_{b,x}, p_b^0\}$, as in the EM case, using all the sequences with the selected sites except the first sequence. Then it samples a new site for the first sequence from every possible position according to a relative weight $p_{b,x}/p_b^0$. Then it repeats both steps for subsequent sequences.

A cycle is complete when the site for the last sequence is re-sampled. Theoretically, after an infinite number of cycles, the relative information $I = \Sigma_x I_x$ [formula (9.2)] will reach its maximum. In practice, the alignment often converges fairly fast. Sometimes the sampler can get stuck at sub-optima, which may require a simultaneous shift of all the aligned sites to the left or to the right by few bases. To speed up the convergence, the Gibbs sampler automatically samples a shift again according to the relative weight of the likelihood ratio after a certain specified number of cycles. This basic algorithm was also generalized to allow more than one type of motif per sequence.

In order to find sites that may have multiple copies in some sequences and zero copies in others, the motif (or Bernoulli) sampler was designed to concatenate all the sequences into a single one. It is initialized to an alignment of sites randomly spread throughout (no site can overlap another or across a sequence boundary); the rest are nonsites. The algorithm starts with picking the first possible site position out of either the aligned set or the non-site set, updates the model $p = \{p_{bx}, p_b^0\}$ (i.e., recalculates the base counts), then samples this position into the aligned set or the nosite set according to the odd ratio $[P(\text{site})*p_{bx}]/[P(\text{nonsite})*p_b^0]$, where the posterior (prior+pseudo-counts) P(site) and P(nonsite) may be specified by a user. Then the algorithm continues with picking the second site, and so on.

One cycle is complete when the last possible site position is sampled. Iterate enough cycles until convergence of the motif alignment (or the maximum number of cycles specified).

This has been generalized to handle more than one type of motif. Two other useful technical features are the column sampling (to allow automatically increasing the site length by sampling in more conserved flanking columns) and the near optimal sampling (to allow estimating relative probability, each site is sampled into the alignment). Two other modified versions of the Gibbs sampler for DNA sequence analysis have been reported in microarray data analysis applications.

STATISTICAL SIGNIFICANCE

It is also important to know how significant a particular alignment is with respect to a random model. Because all the alignments are ranked by the relative information I, it would be desirable to calculate the *p*-value, namely, the probability of finding an alignment with relative information greater than or equal to I. Assuming the null model for each alignment column is an independent multinomial model:

$$P_{matrix} = \sum_{x=1}^{k} \frac{N!}{\prod_{b=A}^{T} n_{bx}} \prod_{b=A}^{T} p_b^{n_{bx}} \tag{9.5}$$

If I is small and N is large, 2NI tends to a χ^2 distribution (df = 3k). Unfortunately, promoter analyses generally involve very large scores and frequently few sequences, and the limiting distribution tends to give poor probability estimates. Using large-deviation technique, the approximate mathematical formula for the *p*-value and the E-value (expected number of alignments with I or greater). They also implemented an efficient algorithm for calculating these values in CONSENSUS. Other methods have also been reported for estimating the statistical significance of a matrix search result.

Constructing Regulatory Modules

Because promoter is regulated by TF modules made of composite sites, simultaneous detecting of correlated sites is much more significant and hence provides better specificity. One of the earliest methods for detecting two sites in a fixed distance and orientation in the heat-shock promoters. Another interesting example was given by the identification of regulatory modules that confer muscle-specific gene expression, where a logistic regression analysis was used to combine matrix scores for multiple TF sites in each module.

This directly generalized the study of the two-site module. More recently, experimental analysis and computer prediction of CTF/NFI TF sites were reported where a generalized profile model for CTF-1 DNA binding specificity was proposed. This model consists of a conserved half-site (5 bp) + a spacer (5, 6, 7, or ∞) + a less conserved (palindromic) repeat. Detailed experimental analysis reveals the flexible and correlated nature of this protein binding site. With detailed modeling of TF modules, one will be able to recognize promoters of a specific class with extremely high specificity. Unfortunately, generation of these models requires high quality as well as systematic experimental data, which are still very rare. The development of composite site databases such as COMPEL and TRRD will greatly facilitate advances in this field.

Large-Scale Gene Expression

Recent advent of large-scale gene expression technologies is having a great impact on the understanding of gene regulation. By clustering gene expression profiles, different groups of co-regulated genes can be identified and their promoter elements may be detected by either *k*-tuple or multiple alignment methods once the upstream sequences become available for the transcriptionally co-regulated genes. Using large-scale expression data to detect novel promoters and to infer regulation networks will become the cutting-edge bioinformatics in the functional genomics era. Species-specific promoter databases such as SCPD shall become extremely useful resources for studying large-scale transcription data.

Phylogenetic Footprinting

As more genomes become available, comparative analysis of noncoding regions has also become an important approach for detecting promoters or regulatory regions in general. Phylogenetic footprinting is referred to as the identification of any functional region by comparison of orthogous genomic sequences between species. Although the orthologous coding regions (ORFs) are highly conserved, the conservation of regulatory regions varies widely with particular genes. To detect short TF sites, one would want to compare orthologous regulatory regions between species that are not too close (so that the sequences have enough time to diverge) and not too distant.

But we are currently limited to the few sequenced model systems. Several methods for detecting conserved blocks from a multiple alignment have been evaluated by Stojanovich *et al.* (1999). Programs designed for very long alignments of syntenic regions have also become available. Among many applications, PIPmaker was very successfully used for the identification of a coordinate regulator of interleukins 4, 13, and 5.

Predicting Transcriptional Start Site (TSS)

Because TF sites can occur anywhere, even with the location of a regulatory module in proximal promoter, identification of TSS is still not easy. General promoter prediction methods mainly focus on TSS site prediction in order to locate the beginning of a gene instead of seeking specific regulatory elements.

CpG Islands

Vertebrate genomic DNA is known to be generally depleted in the dinucleotide CpG. In the human genome, for example, the occurrence of CpG dinucleotides is five times less than statistically predicted from the nucleotide composition. CpG depletion is believed to result from methylation of Cs at 80 percent CpG dinucleotides, which leads to mutation of the methylated C to T, and thus conversion of the CpG dinucleotides of TpG. There are, however, genomic regions of high G + C content, termed CpG islands, where the level of methylation is significantly lower than the overall genome.

In these regions, the occurrence of CpGs is significantly higher, close to the expected frequency. CpG islands are greater than 200 bp in length, have more than 50 per cent of G + C content, and have a ratio of CpG frequency to the product of the C and G frequencies above 0.6. The CpG island is an important signature of the 5' region of many mammalian genes, often overlapping with, or within, a thousand bases down-stream of the promoter, where a promoter associate nucleosome is found. The identification of promoters by CpG islands with a resolution of 2 KB will be most useful for large-scale sequence annotation. Although visual inspection of CpG islands is often used for gene identification by many molecular biologists, the features that can best discriminate the promoter-associate CpG islands from the non-associated ones. This led to an effective algorithm (CpG-Promoter) for large-scale promoter mapping with 2 KB resolution.

Statistical tests showed that about 85 percent of CpG islands within an interval from –500 to +1500 around the TSS (transcriptional start site) were correctly identified, and that roughly 93 percent of the CpG island containing promoters were correctly mapped. The basic procedure is to use CpG Plot program of R.S. Lopez for mapping of potential CpG islands and then to use a quadratic Discriminant classifier for prediction of promoter associated islands. The EMBL CPGISLE database of human CpG islands was used for training the classifier on three discriminant features: length, G+C content, and the CpG ratio. It should be mentioned that, like CpG Plot, PIPmaker can also display CpG islands in a large genome.

TSS Prediction Based on TF Site Scan

As TF sites are over-represented in the promoter region, it is natural to seek a prediction program based on putative TF site density. PROMOTERSCAN is one such program developed by Prestidge (1995). It was based on the study of three datasets: a TF database, a promoter database and a nonpromoter set constructed from protein and RNA gene sequences. Density of all putative TF sites is calculated separately for promoter and non-promoter sequences within a 250-bp window upsteam of TSS for promoters) and use the radio $R=D_p/D_n$ of the two densities as the scoring function supplemented with a TATA-matrix score when scanning a test sequence using the same window.

TSS Prediction Based on *k*-tuples

Methods based on putative TF sites do have severe limitations: important context effect may be overlooked, the majority of putative sites are false positives, site matrices are scoring function/cut-off dependent, the database is biased by the limited number of known samples, and so on. A statistical learning approach without using any putative TF site information has become the attractive alternative. PromFind is based on the difference in 6-mer frequencies between promoters, coding regions and noncoding regions downstream of the first coding exon. Among all sites in an input sequence where the promoter versus coding region discriminant exceeds a certain threshold, the site where the promoter versus noncoding region discriminant reaches its maximum is taken as a promoter.

But this "content" approach would lose all the positional information. TSSG and TSW both use LDA (linear discriminant analysis) to combine (1) a TATA score, (2) triplet preferences around TSS, (3) 6-tuple score in three non-overlapping windows of 100 bp upstream TSS, and (4) putative binding site scores. TSSG is based on TFD; TSSW is based on TRANSFAC. Fickett and Hatzigeorgious (1997 had evaluated several earlier promoter prediction algorithms, including a Markov model-based algorithm. TSSW appeared to be ranked as one of the best, with a sensitivity of 42 per cent and specificity of 1 false positive per 800 bp. Using positional dependent 5-tuple measures, a QDA method for core promoter prediction was implemented in CorePromoter.

Statistical tests indicated that when given a 2 KB upstream region, CorePromoter was able to localize the TSS to a 100-bp interval approximately 60 percent of the time. The utility of CorePromoter and CpG_Promoter was recently demonstrated in the re-analysis of human chromosome 22 genes in conjunction with our internal exon finder MZEF and the terminal oxon finder JTEF. A recent algorithm Promoter Inspector is based on libraries of degenerate words extracted from training sequences by an unsupervised learning approach. It consists of three classifiers that discriminate promoter from intron, exon, and 3'-UTR separately and predicts a promoter when all three classifiers agree. Their test showed that 43 per cent of the predictions can be expected to be true positives, whereas 43 per cent of the annotated TSS were predicted correctly.

In principle, TSS may also be predicted by discriminant analysis of 5'UTRs. With more full-length cDNAs becoming available, 5'UTR features will become as important as promoter features in TSS prediction.

Chapter 10

Imprinted Genes

The advantages of using large genomic clones in the analysis of imprinted genes is described in Chapter 3 with particular reference to yeast artificial chromosomes (YACs). These contain on average 500-600 kb of DNA but can be much larger (>1 Mb). YACS are propagated in yeast and are therefore amenable to genetic modification by homologous recombination, and there are now many examples of their use to generate transgenic mice. This chapter describes a relatively new strategy for using large genomic clones that relies on *Escherichia coli*-based systems. Bacterial artificial chromosomes (BACs) are single-copy plasmids based on the *E. coli* fertility plasmid (F factor). P-1 derived artificial chromosomes (PACs) are based on the bacteriophage P1. Similar to YACs, these vectors are capable of holding large inserts (up to 300 kb) and are stably maintained as 1 or 2 copies. A number of genomic libraries are now available, both within the academic community and commercially. The main advantages of using these bacteria-based systems is the stability of the clones in culture and the ease of preparation of high-quality, intact DNA.

MODIFICATION OF BACs AND PACs

Until recently there were no methods for modifying BACs or PACs, which substantially limited their use. However, homologous recombination-based techniques for modifying *E. coli*-based clones have now been developed, two of which are summarized here. The protocols and plasmids can be obtained from the respective authors and are described there in detail. The first technique was used to modify a BAC and presumably will work as well with other *E. coli*-based systems. It takes advantage of a temperature-sensitive shuttle vector carrying an *E. coli RecA* gene. The required modification (marker insertion, deletion, or substitution) is constructed in a high-copy-number plasmid such as Bluescript with at least 500 bp of genomic DNA, which is homologous to the target locus, on either side of the modification.

This construct is in many ways similar to the type of construct used in homologous targeting of an endogenous locus in embryonic stem cells. This is cloned into the low-copy shuttle vector (pSV1.RecA), which is tetracycline resistant and can replicate only at 30°C. The shuttle construct is transformed into bacteria containing the target BAC (or PAC) and a series of selection procedures follows. In the first step, selection for tetracycline (shuttle vector) and chloramphenicol (BAC vector) at 30°C selects colonies in which both plasmids are present but replicating as two separate molecules. By raising the temperature to 43°C, only the replication origin of the BAC is active and colonies are selected in which the shuttle vector has recombined onto the BAC using the *RecA* pathway.

Once the integration event is identified (1/2 to 1/10 clones in our hands), the colonies are allowed to grow in the absence of tetracycline selection (shuttle vector), and some of these will lose the shuttle vector sequence and either the modification locus or the unmodified target locus. When the shuttle vector is lost, the *RecA* gene is also lost and the bacteria are no longer able to recombine homologous sequences, so the modified BAC is stably maintained. These tetracycline-sensitive colonies can be selected for on chloramphenicol/ fusaric acid plates at 37°C. The "resolved" BACs will either regenerate the original BAC or create the intended modification, which can be determined by analysis on an agarose gel. The second technique, known as "ET cloning," also relies on homologous recombination in *E. coli* and has been demonstrated to modify a number of targets including a P1 clone. This system allows recombination between a linear fragment with short (60-bp) homology arms and a circular target molecule.

The advantage of this system is in the construction of the modification since, in theory, the homology arms can be added by PCR amplification without the requirement for convenient restriction sites and the fragment can be used directly in a transformation without a cloning step. In addition, only a small amount of sequence information at the target site is required; 120 bp would allow the design of two 60-bp arms of homology for an insertional modification. However, in contrast to the first method, this protocol requires the final modification to carry a marker selectable in *E. coli.* This may be a major disadvantage if a subtle modification, such as a single-base-pair change, is planned. To remove the selectable marker necessitates one of two additional steps: (1) The marker may be flanked by either FRT or loxP sites and the modification exposed to the corresponding site-specific recombinase, which still leaves a small site-specific recombination target site at the point of DNA manipulation; (2) two rounds of ET cloning are required, combining positive and counter-selection steps. The first step in the ET protocol involves transformation of *E. coli* containing the target clone with the pBAD-ETι vector. This plasmid carries a truncated *Rec*E controlled by the arabinose-inducible promoter, a *recT* controlled by the EM7 promoter, and *red*i controlled by the EM7 promoter, and *red*i controlled by the Tn5 promoter.

The clones, which now carry both pBAD-ETι and the target clone, are exposed to arabinose, which induces expression of *RecE* and renders the cells capable of using a recombination pathway. These cells are then made competent and transformed with the linear molecule that carries the modification plus a selectable marker flanked by the short homology arms. The third selection step is for the marker within the modification, which is now integrated onto the P1 clone. Presence of the intended modification, which is now integrated onto the P1 clone. Presence of the intended modification can be determined by analysis on an agarose gel. We have used the first protocol successfully (1) to target two loci carried on BACs. In both cases the modifications were quite large, an addition of 7 kb spread over 10 kb The time required, after construction of the modification in pBluescript, was less than 3 wk, with a total of 10 h hands-on time.

The main rate-limiting step is in the final "pop-out" reaction, where there seemed to be as bias toward recovery of the original BAC, which was most likely due to the size of our insertion. In one case only 1/24 clones examined retained the required modification. With the availability of these techniques, it is now possible to use both YACs and BACs/P1 to generate transgenic animals. Two procedures can be used to get the transgene into mice: pro-nuclear injection and embryonic stem (ES) cell lines. Pro-nuclear injection is the more rapid way to generate a large number of unique transgenic lines and requires only small amounts of DNA (<1 og).

The ES cell method is more time-consuming and significantly more DNA is required, but full characterization of the integration events can be performed before chimeras are generated to identify single-copy integration events and confirm the intactness of the transgene. The preparation of BAC DNA for pro-nuclear injection is described in the first section of the protocol, and the preparation of DNA for the generation of embryonic stem (ES) cell lines is described in the second section. Similar protocols can be applied to any *E. coli*-based system with appropriate modification of the procedure (e.g., antibiotic selection, linearization agent).

Materials

Pro-Nuclear Injection

Preparation of BAC DNA

1. LB (1 L): 10 g bactotryptone, 5 g yeast extract, 10 g NaCl (autoclave).
2. 12.5 mg/mL chloramphenicol. Store at –20°C.
3. 14-mL polypropylene-round bottomed tube.
4. Solution I: 50 mM Tris-Cl pH 8.0, 10 mM ethylenediaminetetraacetic acid (EDTA) pH 8.0, RNAse A to 100 og/mL. Store at 4°C.
5. Solution II: 0.2 M NaOH, 1% sodium dodecyl sulfate (SDS). Prepare fresh.
6. Solution III: 3 M potassium acetate, pH 5.5. Store at 4°C.
7. 1 Phenol/1 (chloroform)/*iso*-amyl alcohol.
8. Chloroform/*iso*-amyl alcohol (v/v, 24/1).
9. Isopropyl alcohol.
10. 70% ethanol.
11. 10 mM Tris-HCl, pH 7.5.

Linearization of BAC DNA

1. *Srf*1 (Stratagene 501064) or lambda terminase.
2. Pulsed-field gel electrophoresis (PFGE) apparatus (Pharmacia Gene Navigator System).
3. Phenol.
4. Chloroform/*iso*-amyl alcohol.
5. 3 M sodium acetate, pH 5.2.
6. 100% ethanol.
7. 70% ethanol.
8. 10 m*M* Tris-HCl, pH 7.5.

Pro-Nuclear Injection

1. Injection buffer: 10 m*M* Tris-HCl pH 7.5, 0.1 m*M* EDTA pH 8.0, 10 m*M* NaCl, 30 o*M* spermine, 70 o*M* spermidine. Filter-sterilize and store at 4°C. Make fresh each week.

ES Cell Electroporation

Preparation and Linearization of BAC DNA

1. Approximately 30 og of DNA is required for one electroporation.
2. 50 mL polypropylene tubes (Falcon).

Electroporation of ES Cell Lines

1. 92-mm polylysine-coated plastic tissue culture plates.
2. 0.1% gelatin.
3. ES cell media (1 Dulbecco's modified Eagle's medium F-12/DMEM, 2 mM L-glutamine, 50 og/mL penicillin/streptomycin, 1 N nucleosides, 1000 units/mL LIF (leukemia inhibitory factor), 1 m*M* sodium pyruvate, 0.1 m*M* nonessential amino acids, 6% sodium bicarbonate, 15% fetal calf serum, 10^{-6} M δ-mercaptoethanol. Add 150-260 µg/mL G418 or 130 µg/mL hygromycin for selection.
4. Bio-Rad gene pulser.
5. 4 mm electroporation cuvette.
6. Trypsin/EDTA (10 solution supplied by Sigma T4174).
7. PBS without calcium and magnesium (1 L): 10 g NaCl, 0.25 g KCl, 1.44 g Na_2HPO_4, 0.25g KH_2PO_4, pH to 7.2 with HCl. Autoclave and store at room temperature.

8. Humidified CO_2 incubator at 38°C (5% CO_2 in air.
9. Hemocytometer.
10. 20-mL tubes.

Picking and Analyzing Colonies

1. 24-well tissue culture plates.
2. 90-mm bacteriological Petri dish.
3. 0.1% gelatin.
4. ES cell media.
5. Low-powdered microscope.
6. 1 trypsin/EDTA (10 solution supplied by Sigma T4174).
7. Dimethyl sulfoxide (DMSO).
8. Lysis buffer: 100 mM Tris-HCl pH 7.8, 50 mM EDTA, 0.2% SDs, 200 mM NaCl.
9. 10 mg/mL proteinase K.
10. 1 phenol/1 (Chloroform *iso*-amyl alcohol [1].
11. Chloroform/*iso*-amyl alcohol (24/1).
12. 3 M NaOAc, pH 6.0.
13. Isopropyl alcohol.
14. 70% ethanol.
15. TE pH 8.0: 10 mM Tris-HCl pH 8.0, 1 M EDTA pH 8.0.

Blocking Mouse Repeats for Southern Analysis of BAC Transgenic Clones

1. Church and Gilbert hybridization buffer (1 L): 623 g Na_2HPO_4, 23.4 g NaH_2PO_4, 70 g SDS, 2 mL EDTA pH 8.0.
2. G-50 sephadex spin column (Amersham).
3. TE, pH 8.0. 10 mM Tris-HCl pH 8.0, 1 mM EDTA pH 80.
4. 10% SDS.
5. Mouse genomic DNA, sheared by repeated passage through an 18-gage needle.
6. 2 SSC/0.5% SDS.
7. 0.1 SSC/0.5% SDS.

Generation of ES Cell Chimeras by Aggregation

1. 6- to 7-wk-old MF1 (Olac) female mice.

2. 6- to 7-wk-old F1 (C57BL6 CBA) mice in estrus, and vasectomized F1 males.
3. PMS, HCG, T6 + BSA, PB1 + BSA, Petri dishes, forceps, Pasteur pipets, light mineral oil, and CO_2 incubator.
4. Acid tyrode's (AT) 800 mg NaCl, 20 mg KCl, 20 mg $CaCl_2H_2O$, 10 mg $MgCl_2H_2O$, 5 mg $Na_2HPO_4I_2O$, 100 mg glucose. Make up to 100 mL with sterile water and pH to 2.5 with concentrated HCl. Filter-sterilize through 0.22-om filter. Store in 1.5-mL aliquots at 4°C.
5. 30-mm and 50-mm bacteriological Petri dishes.

Methods

Pro-Nuclear Injection

Preparation of BAC DNA

1. Pick a single colony into a 14-mL tube containing 5 ml Lb and 12.5 og/mL chloramphenicol. Incubate at 37°C with shaking for 8 h.
2. Centrifuge for 10 min at 1000g in a bench-top centrifuge.
3. Remove all supernatant and resuspend the pellet in 350 oL of solution I+100 og/ mL RNAse A.
4. Add 350 oL of solution II and mix gently by inverting the tube several times. Place on ice for 5 min.
5. Add 350 oL of solution III, mix gently but thoroughly, and leave on ice for 20 min.
6. Transfer to a 2-mL microfuge tube and spin for 10 min in a microfuge at maximum speed.
7. Transfer supernatant (1 mL) to a fresh microfuge tube, avoiding the white precipitate.
8. Add an equal volume of phenol/chloroform and mix gently.
9. Spin for 10 min in the microfuge.
10. Transfer the aqueous layer to a fresh microfuge tube, avoiding the interface.
11. Add an equal volume of chloroform and mix gently.
12. Centrifuge for 10 min at maximum speed in the microfuge.
13. Remove the aqueous layer to a fresh 2-mL tube.
14. Add 1 vol of isopropyl alcohol and mix gently.
15. Centrifuge immediately for 1 min at maximum speed in the microfuge.
16. Decant alcohol and wash with 500 oL of 70% ethanol.

17. Remove all traces of alcohol with a pipet tip and resuspend the pellet in 10 oL of Tris-HCl pH 7.5.

Linearization of BAC DNA—For Example, pBeloBAC

1. Digest 1 og of BAC DNA with either *Srf*1 (Stratagene) or lambda terminase (Epicentre) as recommended by the manufacturers in a 50-oL final volume.
2. Check for complete linearization by running 10 oL on a PEG.
3. To the remaining 40 oL, add an equal volume of phenol and mix gently.
4. Spin for 10 min in the microfuge.
5. Remove the aqueous layer to a fresh microfuge tube, avoiding the interface.
6. Add an equal volume of chloroform and mix gently.
7. Centrifuge for 10 min in the microfuge.
8. Remove the aqueous layer to a fresh microfuge tube.
9. Add 1/10th vol of 3 M sodium acetate pH 5.2, 2 vol of 100% ethanol, and mix gently.
10. Store the precipitating DNA at –20°C until the PEG result is known.
11. If the DNA is fully linearized and intact on the PEG, centrifuge the sample for 10 min in the microfuge.
12. Decant alcohol and wash with 500 ol of 70% ethanol.
13. Remove the alcohol with a pipet tip and resuspend the pellet in 10 oL of 10 mM Tris-HCl pH 7.5.
14. Check the DNA concentration by running 1/10 of the sample on a small agarose gel against samples of known concentration.
15. Store the DNA at –20°C. Avoid repeated freeze/thawing.

Preparation of DNA for Injection

One day of pro-nuclear injection requires 100 oL of a 1- to 3-ng/oL solution. Take 100-300 ng of the linearized BAC DNA and make up the total volume to 100 oL with injection buffer. The DNA is ready to inject.

ES Cell Electroporation

Preparation of BAC DNA for ES Cell Electroporation

1. Pick single colonies into ten 50-mL tubes containing 25 mL LB and 12.5 og/mL chloramphenicol (total volume 250 mL). Incubate at 37°C with shaking for 8 h.

2. Centrifuge the culture for 10 min at 1600g in a bench-top centrifuge.
3. Remove all the supernatant and resuspend each bacterial pellets in 1 mL of solution 1+100 og/mL RNAse A.
4. Add 1 mL of solution II and mix gently by inverting the tube several times. Place on ice for 5 min.
5. Add 1 mL of solution III, mix gently but thoroughly, and leave on ice for 20 min.
6. Combine the samples (20 mL total volume) in a 50-mL tube and spin for 10 min at 1600g.
7. Transfer the supernatant to a fresh 50-mL tube and spin again for 10 min at 1000g.
8. Transfer the supernatant to a fresh 50-mL tube, add an equal volume of pheol/ chloroform, and mix gently.
9. Spin for 10 min at 1000g.
10. Transfer the aqueous layer to a fresh 50-mL tube, avoiding the interface.
11. Add an equal volume of chloroform and mix gently.
12. Centrifuge for 10 min at 12,000g.
13. Divide the aqueous layer equally among four 14-mL tubes.
14. Add 1 vol of isopropyl alcohol to each tube and mix gently.
15. Centrifuge immediately for 10 min at 12,000g.
16. Decant alcohol and wash with 1 mL of 70% ethanol.
17. Remove the alcohol with a pipet tip and resuspend the pellets in a total volume of 100 oL 10 mM Tris-HCl pH 7.5.

Linearization of BAC

The BAC DNA can be linearised by scaling up the reactions 10-fold. Digest the DNA in a 500-oL reaction volume and resuspend in 40 oL of sterile water on the day that it is to be used.

Electroporation of ES Cell Lines

1. Grow ES cells under routine conditions to obtain at least 1 10^7 cells, which is sufficient for a single electroporation. This is approximately equivalent to three semiconfluent 92-mm plates.
2. Prepare six gelatinized 92-mm polylysine-coated plastic tissue culture plates. Add 5 mL of 0.1% gelatin to each plate and swirl. Aspirate the gelatin and leave for 5

min. Add 9 mL of ES cell medium to each plate.

3. Chill PBS on ice.
4. Set Bio-Rad gene pulser to 250 V, 500 oF.
5. Chill a 4-mm electroporation cuvet on ice.
6. Wash ES cells with 10 mL of ice-cold PBS.
7. Add 3 mL 1 trypsin/EDTA and swirl. Aspirate trypsin and incubate in CO^2 incubator for 3-5 min.
8. Add 2 mL of ES cell medium and pipet up down to break up the clumps, using a 1-mL Gilson pipetman.
9. Transfer the ES cell-containing medium to 20-mL tube and centrifuge for 500g for 5 min.
10. Aspirate medium and resuspend in 0.8 mL of cold PBS.
11. Count cells with hemocytometr. Adjust cell concentration to 1 10^7 in a final volume of 0.8 mL cold PBS. Keep on ice. Three semiconfluent 92 mm plates should be enough for one electroporation.
12. Mix 0.8 mL of cells with DNA in a maximum volume of 100 oL of H_2O on ice.
13. Transfer to the prechilled electroporation cuvet.
14. Deliver electric pulse (the time constant should be 8-14).
15. Quickly transfer the cells into 6 mL of ES cell medium + LIF in a 20-mL tube.
16. Divide the cell suspension among six 92-min gelatinized plates and grow without selection for 1 d.
17. Add ES cell medium containing selection agent and grow cells under selection for 7-10 d in a CO_2 incubator.

Picking and Analyzing Colonies

1. Prepare one 24-well tissue culture dish. Add 0.5 mL of 0.1% gelatin to each well. Aspirate. After 5 min add 1 mL of ES cell medium (without selection).
2. Set up tissue culture hood with a low-powered microscope.
3. Prepare a 92-min lid of a bacterial Petri dish next to the microscope stand with 30-oL drops of trypsin in a row.
4. On the plate from which ES cell colonies are to be picked, replace the medium with PBS.

5. Adjust a 200-oL Gilson pipet to 30 oL and with a yellow tip detach the first ES cell colony from the plate by gently scraping the top over the colony. Immediately suck up the colony into the yellow tip.
6. Expel the colony into the first drop of trypsin.
7. After 3-5 min, pipet up the drop containing the first colony and transfer it to the first well of the 24-well tissue culture plate. The trypsin and sucking force of the pipet should together cause the colony to disaggregate to a single-cell suspension.
8. Repeat until you have picked sufficient colonies.
9. Feed the cells with 1 mL of ES cell medium (+ LIF) for 2 d.
10. Prepare two fresh 24-well dishes with gelatin. Aspirate the gelatin and add 1 mL of ES cell medium to each well. Label plates A and B.
11. Trypsinize the cells and resuspend in 1 mL of ES cell medium.
12. Transfer 0.5 mL to the ES cell 24-well plate A. This well be the freezing plate.
13. Transfer 0.5 mL to the ES cell 24-well plate B. This will be the DNA plate.
14. Feed cells with 1 mL of ES cell medium for 2 d.
15. Trypsinize the cells in plate a and resuspend in 500 oL of ES cell medium. Add 50 oL of DMSO, wrap the plate in Parafilm, and freeze in a –80°C freezer.
16. When the cells in plate B are confluent, wash with PBS and add 0.75 mL of lysis buffer to each well.
17. Pipet up and down a few times to lyse the cells completely, then transfer each lysate to a 1.5-mL microfuge tube.
18. Add 7.5 oL of proteinase K.
19. Incubate at 55°C for 2 h or overnight at 37°C.
20. Add an equal volume of phenol/chloroform and mix gently for 5 min at room temperature.
21. Centrifuge for 10 min. Collect the upper phase and extract with chloroform.
22. Remove the aqueous phase and add 1/30th vol of 3 M sodium acetate pH 6.0 and 1 vol of isopropyl alcohol. Mix gently; the DNA will precipitate immediately. Hook out with a bent pasteur pipet sealed at the end. Rinse in 70% ethanol. Air dry for 10 min and resuspend in 5 oL of TE.

Blocking Mouse Repeats for Southern Analysis of BAC Transgenic Clones

1. Prehybridize Southern membrane in Church hybridization buffer at 65°C for a minimum of 5 min.

2. Radioactively label 0.5 og of BAC DNA.
3. Purify labelled DNA by passage through a G-50 Sephadex spin column.
4. Add 250 og of sheared mouse genomic DNA in a final volume of 250 oL T.E./0.5% SDS.
5. Boil for 10 min, then chill on ice for 10 s.
6. Add to 5 mL of hybridization buffer and incubate at 65°C for 1-2 h.
7. Add to membrane and hybridize for 14-16 at 65°C.
8. Wash in 2 SSC/0.5% SDS for 5 min, followed by 0.1 SSC/0.5% SDS twice for 30 min.
9. Expose to autoradiography film overnight at –70°C.

Generation of ES Cell Chimeras by Aggregation

1. Day 1: Superovulate MF1 (Olac) females by injecting 150 oL of PMS intraperitonealy using a 27-gage needle.
2. Day 3: After 48 h, inject 150 oL of HCG intraperitoneally using a 27-gage needle. Set up matings with stud MF1 (Olac) males.
3. Day 4: Check females for vaginal plugs and transfer to a separate cage.
4. Day 5: Set up recipients by mating 12 F1 females C57BL6 CBA) in estrus with 12 vasectomized F1 males.
5. Day 6: Check recipients for vaginal plugs and transfer to separate cages with the plug date recorded.
6. Place a 30-oL drop of T6+BSA in a 30-mm Petri dish. Cover with mineral oil. Incubate in CO_2 incubator until required.
7. Isolate morulae by standard procedures. Morulae should be at the 8-cell stage and beginning to compact. Using a mouth pipetting system, transfer the embryos to a drop of T6+BSA in a 50-mm bacterial Petri dish to wash.
8. Transfer the embryos into the 30-oL drop of T6+BSA in the 30-mm Petri dish. Incubate in the CO_2 incubator.

Preparation of ES Cells

1. Day 6: Trypsinize ES cells. One well of a 50-75% confluent 24-well plate is more than sufficient. Resuspend thoroughly in ES cell medium to give a single-cell suspension.
2. Count cells on hemocytometer. Approximately 1 10^6 cells/mL are required. Adjust volume with ES cell medium.

3. Place nine 15-oL drops of ES cells in a 50-mm bacterial Petri dish in 3 3 ros. Flood the dish with 6 mL of mineral oil and transfer to CO_2 incubator to allow the cells to settle to form an even lawn.

Preparation of Embryos

1. Place a 100-oL drop of PBI+BSA at the top of a 50-min bacterial Petri dish. Place two 30-oL drops of AT to one side and nine 30-oL drops of ES cell medium to the other side.
2. Transfer well-formed embryos from the T6+BSA drop to the drop of PBI+BSA.
3. Under a dissecting scope, take one embryo into the pipet in a minimal volume of PB1+BSA. Transfer to the top of the drop of AT. Quickly expel the remaining liquid from the pipet and then go back into the AT drop to catch the embryo before it hits the bottom of the dish. Transfer quickly to the first drop of ES cell medium.
4. Pick up the embryo and transfer it to the second drop of ES cell medium.
5. Repeat the process; 5-10 embryos can be taken through the AT drop at a time.
6. When 50-60 embryos have been processed through the AT, transfer all from the second ES cell medium drop through the fresh ES cell medium drops to the sixth drop to remove all traces of AT.
7. After 50-100 embryos have been treated, use a fresh drop at AT.
8. After all morulae have been processed, divide equally among *n* ES cell medium drops, where n=number of ES cell lines to be aggregated. About 20 embryos are sufficient for each line.
9. As soon as possible, transfer the embryos to the drops containing the ES cells.
10. Transfer to the CO_2 incubator to allow the ES cells to stick to morulae.
11. After 30 min, check the embryos. Ideally, about 10 or so ES cells should attach to each embryo. Transfer to a drop of T6+BSA to rinse away the ES cell medium and any loose ES cells.
12. Transfer the embryos to a 50 mm dish containing nine 30-oL drops of T6+BSA under mineral oil. Two to three embryos can be incubated in each drop, provided they are placed away from each other.
13. Culture for 2 d in the CO_2 incubator.

Transfer into Recipients

1. Day 8: Transfer the embryos into 30-oL drops of PB1+BSA under mineral oil (prewarmed in the CO_2 incubator). Allow 1 drop per recipient (10 embryos). The embryos will have expanded into blastocysts by this stage.

2. Using standard surgical procedure, prepare recipients to receive the embryos.
3. Transfer 10-20 embryos per recipient.
4. Embros are transferred into recipients at d 3 (d 1 is the day of vaginal plug), so the pups will be born 16-17 d later.

Table 10.1 : Agents for Linearization of Large Genomic Clones[a]

Site	No. of sites in pBeloBAC	No. of sites in pBACe3.6	No. of sites in pPAC4	No. of sites in pCYPAC2
AscI	0	2	2	0
FseI	0	0	0	0
NotI	2	2	2	2
PacI	0	1	2	1
PmeI	0	1	2	1
SgrAI	2	2	0	1
SrfI	1	0	0	0
cosN	1	0	1	0
loxP	1	1	1	1

a. All restriction enzymes recognize an 8-base sequence that is likely to occur infrequently within a BAC or PAC clone.

NOTES

1. One 5-mL culture should yield approximately 1-2 og DNA, which is sufficient for several days of pro-nuclear injection. The length of time in culture can affect the yield and quality of BAC DNA, and 8 h is optimal in our experience. A reliable alternative to the BAC DNA preparation method detailed here is provided by the Wizard PureFection kit.
2. The DNA used in generating BAC or PAC transgenic mice must first be linearized. All the commonly used vectors include a choice of unique restriction sites for this purpose. For example, the pBELOBAC vector contains an *Srf*I site that is usually not present in the insert (this can be assessed by digesting the BAC clone with SrfI and resolving the sample on pulse-field gel as described in the previous section). There are also two *Not*I sites that flank the cloning site, allowing for excision of the genomic sequence from the vector. Again, PFG electrophoresis is required to test for *Not*I sites within the genomic clone. If both *Srf*I and *Not*I sites occur within the genomic fragment, this vector also contains a *cos*N site, which can be opened using lambda terminase. Table 10.1 summarizes potential lineralization sites in some of the vectors commonly used to clone large genomic fragments.

3. It is possible to incorporate the modification of the BAC or PAC clone to contain a neomycin or hygromycin gene, either under a constitutive promoter or under the promoter of a gene on the BAC/PAC. For large constructs without a selectable marker (unmodified cosmids, BACs, etc.), a neomycin or hygromycin gene under a constitutive promoter can be co-electroporated with the BAC. For co-electroporation we use a ratio of 10/1 construct to neomycin cassette. If expression of the neomycin or hygromycin gene is driven by an untested promoter, the concentration of the selective agent may vary from the standard conditions.
4. Feeders: ES cells can be cultured on a layer of feeder cells in the absence of LIF (3). These are primary embryonic fibroblast cells obtained from 11.5-12.5 dpc embryos, where the nuclear material has been inactivated by mitomycin C. When under selection, feeder cells obtained from transgenic mice constitutively expressing neomycin or hygromycin-resistance can be used. We obtain our neomycin-resistant feeders from the ROSA-26 line. We use 2 10^6 feeder cells per 92-mm plate and 1 10^5 cells per well of a 24-wel tissue culture plate. Plate B, the DNA plate, should be grown without feeders.
5. Approximately 20 og of DNA should be recovered from one confluent well of a 24-well dish. 10 oL of this DNA (4 og) should be enough for digesting with appropriate restriction enzymes and analysis by Southern blotting.
6. It is sometimes possible to use the genomic BAC/PAC clone as a probe on a Southern blot to assess transgene integrity. This allows you to check for intactness of the integration event and approximate copy number in one hybridization step. We find that this protocol works well for constructs of up to 100 kb, but is less suitable for larger clones, where the complexity of a restriction digest is too great to allow all the individual fragments to be resolved on a standard agarose gel. In this case it is advisable to use smaller DNA fragments, such as vector sequences, probes from the ends of the genomic sequence or which detect the modification, which can be hybridized sequentially to ascertain integrity of the integration event.
7. There are two widely used techniques for generating ES cell chimeras, blastocyst injection and morula aggregation. We have used both techniques successfully. Overall, aggregation is the simpler technique and does not require any sophisticated equipment or specific training. It also has the advantage of allowing the aggregation of several lines simultaneously and therefore is the technique we recommend.
8. We use the albino MF1 (Olac) strain, as our ES cell lines are derived from mice with a pigmented coat colour that allows for simple identification of chimeric animals. Eight to 12 females should produce enough embryos for aggregation of several ES cell lines. When chimeras are ready to mate, they can be test-bred with MF1 (Olac) females. Pups inheriting the ES cells will be pigmented at birth. Earlier analysis can be performed at d 11.5 of embryogenisis, as the pigmented eye colour is easy to observe.

9. Avoid embryos in which the cells are uneven in size or in which some cells are brownish in colour, since these are not healthy. Make several glass drawn-out pipets to get the best size, which is not too much bigger than the morulae. Always draw some medium into the pipet before picking up the embryos. Never allow the embryos to go up as far as the widening part of the pipet, since either they will get stuck or they will make contact with air.

10. You should see the zona dissolving as the embryos sink in the AT, but it is important not to leave them here for more that 1-2s, as they will fall apart. With practice, multiple embryos can be treated with AT simultaneously.

11. The ES cells will have formed an even, single-cell layer at the bottom of the drop. Try to keep the embryos fairly separate, since if they touch they will fuse. Therefore transfer only 3-4 embryos per drop.

Table 10.2 : Transgenic Lines Generated Using This Protocol

ES cell line	Chimeric animals born	ES cell coat colour (%)	ES cell transmission (%)[a]
5D3	6	100 (n=6)	100
5A2	4	100 (n=3)	100
		30	n.t.
5A4	5	100 (n=4)	100
		60	100
6C1	1	100	100
5B6	2	50	100
		50	n.t.
5C2	1	80	12.5

a. n.t. = not tested.

12. The time of incubation with ES cells is crucial. Low-passage ES cells (6-10) need less time, as they seem to be more "sticky," so it may be necessary to experiment with different timing. If too many ES cells attach to the embryo, it will not survice; to few and there will be no or low-percentage chimeras.

13. Using this procedure we have generated chimeras for six ES cell over a 2-wk aggregation period. All of these lines resulted in germline transmission, with chimeras from four different lines transmitting the ES cells to 10% of their offspring.

PRONUCLEAR INJECTION

The production of transgenic mice using small DNA constructs has been widely used for many years to investigate the regulation of gene activity. Small plasmid-based constructs (less than 20 kb) have been favored for a number of reasons, particularly the ease with which they can be manipulated and purified in large quantities. While this approach is powerful, there are some problems associated with the size of these transgenes. In particular, many of these small transgenes do not reproduce accurately the expression seen from the endogenous gene. For some genes the regulatory elements that control activity are located at a distance from the promoter and can be omitted from the transgene. These may be enhancers, repressors, boundary elements, or even locus control regions (LCRs), which are responsible for maintaining te correct spatial and temporal expression patterns of a number of genes, such as the globin clusters in mouse and humans.

More important, small transgenes are susceptible to position effects from the chromatin environment in which they integrate, which often results in either ectopic expression (from trapping of nearby enhances for other genes) or suppression of gene activity. Finally, small transgenes usually integrate in a multicopy tandem arrangement that does not accurately reflect the situation seen at the endogenous locus. There is growing evidence from studies in mouse and humans that the regulatory elements for many imprinted genes may be widely dispersed within "imprinted domains," which may span hundreds of kilobases. Therefore, it is unlikely that analysis of small transgenes will provide much useful information concerning the expression or mechanism of imprinting for the majority of this unusual class of genes.

Availability of Larger Transgenes

In recent years technologies have been developed to make transgenic mice using large DNA constructs based on artificial chromosomes from a number of sources, including bacteria and yeast (YACs; for examples. These transgenes can first be manipulated in the host organism due to efficient mechanisms of homologous recombination, before purification and transfer into mammalian cells. Numerous constructs are available commercially for YACs, and methods of modification have been well documented that can be used for generating point mutations, deletions, replacements, and insertions. Therefore it is possible, for example, to introduce reporter genes into genes of interest, or to modify the size of the YAC construct under investigation, with relative ease.

The focus of this chapter is on the purification of YAC DNA and subsequent transfer into fertilized mouse eggs. As mentioned earlier, one of the major advantages that YAC-based transgenes have over smaller plasmid-based transgenes is that they are more likely to possess the full array of regulatory elements for the gene of interest, and thus are more likely to behave in a manner similar to that seen at the endogenous locus. There is also an increased likelihood of obtaining low-copy-number (or even single-copy) transgenic animals using YAC DNA, because fewer molecules are introduced, so these transgenes should provide a more accurate representation of both the patterns of expression and also the levels of expression of genes located therein.

Study of Imprinted Genes

To date we and others have investigated two regions containing imprinted genes using the YAC transgenic approach. These are a 130-kb region from mouse distal chromosome 7, which contains the reciprocally imprinted *Igf2* and H19 gens, and a 300-kb region from mouse chromosome 17, which contains the maternally expressed *Igf2r* gene. In both cases the YACs demonstrated appropriate imprinting of the genes at a number of ectopic genomic locations. For the 130-kb *Igf2/H19* YAC, little evidence was found for position effects such as the trapping of nearby enhancer elements. These analyses therefore demonstrate the power of using this technique as a first step toward defining the minimal critical regions for imprinting of particular genes, with the ultimate aim of identifying and characterizing the regulatory elements involved. The *Igf2r* YAC transgene was further modified by targeted deletion of a differentially methylated intronic CpG island, which led to loss of imprinting, thus defining a specific role for this region.

We have recently modified the 130-kb *Igf2/H19* YAC transgene by targeting *loxP* recombination sites around a putative control element of unknown function, which had allowed us to delete this element from the YAC after it had been integrated into the mouse genome. We were therefore able to compare directly the activity of the YAC transgene with and without the targeted element, at the same location in the mouse genome, providing rapid and conclusive data on the function of the deleted region. It is therefore evident that the use of large transgenes has many advantages over smaller, plasmid-based transgenes for initial investigations into dissecting imprinted domains, and is a powerful tool that is underexploited at the present time.

Materials

Preparation of High-Density Yeast Plugs

1. Synthetic minimal (SD) medium. 0.17% yeast nitrogen base, 0.5% ammonium sulfate, 2% dextrose. Add appropriate amino acids (nonselective). For plates, adjust pH to 5.8 with NaOH and add 2% bacto-agar.
2. Hemocytometer.
3. 50 mM ethylenediaminetetraaetic acid (EDTA), pH 8.0.
4. Yeast lytic enzyme (ICN Biochemicals), 50 mg/mL in water.
5. 1.5% low-melting-point (LMP) agarose (Seaplaque-FMC) in 125 mM EDTA, pH 8.0. Melt in microwave and equilibrate to 40°C.
6. Agarose plug molds (Pharmacia), sealed on one side with tape and cooled on ice.
7. LET buffer: 0.5 M EDTA, 10 mM Tris-HCl, pH 7.5. Just before use, add yeast lytic enzyme to a final concentration of 2 mg/mL.

8. Yeast lysis buffer (YLB): 100 mM EDTA, 10 mM Tris-HCl pH 8.0, 1% lithium dodecyl sulfate (LDS).
9. 100 mM EDTA, pH 8.0.

Purification of YAC DNA by Pulsed-Field Gel Electrophoresis

1. Pulsed-field gel electrophoresis (PFGE) apparatus.
2. Running buffer (0.5x TBE).
3. 1.2% LMP agarose (Seaplaque-FMC) in 0.5x TBE.
4. 0.5% LMP agarose (Seaplaque-FMC) in 0.5x TBE.
5. Edhidium bromide solution: 5 oL of 10 mg/mL per 100 mL 0.5x TBE.
6. Agarose digestion buffer: 10 mM Tris-HCl pH 7.5, 1.0 mM EDTA pH 8.0, 100 mM NaCl, 30 oM spermine, 70 oM spermidine.
7. δ-Agarase (NEB).
8. Injection buffer: 10 mM Tris-HCl pH 7.5, 0.1 mM EDTA pH 8.0, 100 mM NaCl, 30 oM spermine, 70 oM spermidine. Filter-sterilize and store at 4°C. Make fresh each week.
9. Dialysis membrane (Millipore 0.05 om, VMWP02500.

Preparation of Fertilized Mouse Eggs

1. F1 hybrid mice in which the maternal component is C57BL/6, paternak can be any vigorous inbred strain, e.g., CBA, C3H, DBA, 129/Sv. We generally use (C57BL/6 CBA) F1 mice; females for superovulation at 5-6 wk old, males mature and fertile.
2. Pregnant mare serum (PMS), "Folligon"; Intervet UK) 50 IU/mL sterile phosphate-buffered saline (PBS). Store at 4°C.
3. Human chorionic gonadotrophin (HCG), "Chorulon"; Intervet UK), 50 IU/mL sterile PBS. Store at 4°C.
4. 27G needles.
5. Phosphate-buffered medium (PB1): NaCl (5.97 g/L), KCl (0.2 g/L), KH_2PO_4 (0.19 g/L), $CaCl_2H_2O$ (0.14 g/L), $MgCl_2H_2O$ (0.1 g/L), glucose (1.0 g/L), sodium pyruvate (0.04 g/L), penicillin G (0.06 g/L), streptomycin sulfate (0.05 g/L), phenol red (0.01 g/L), Na_2HPO_4 (anhydrous, 1.14 g/L—dissolved seperately and added last). Filter-sterilize and store in aliquots at 4°C.
6. Bicarbonate-buffered medium (T6): NaCl (4.72 g/L), KCl (0.11 g/L), $NaH_2PO_4H_2O$ (0.06 g/L), $CaCl_2H_2O$ (0.26 g/L), $MgCl_2H_2O$ (0.1 g/L), glucose (1.0 g/L), sodium pyruvatae

(0.03 g/L), penicillin G (0.06 g/L), streptomycin sulfate (0.05 g/L), phenol red (0.01 g/L), 3.4 mL sodium lactate (60% syrup), $NaHCO_3$ (2.1 g/L—dissolved seperately and added last). Filter-sterilize and store in aliquots, with minimum air pocket above medium, at 4°C.

7. Bovine serum albumin (BSA), fraction V; Sigma). Add to PBI and T6 media at 4 mg/mL before use.
8. Hyaluronidase solution (6 mg/mL sterile water, stored at –20°C in 50-oL aliquots). Make up to 1 mL with PBI+BSA before use (final concentration 300 IU/mL).
9. Light mineral oil, embryo tested (Sigma, M8410).
10. Bacteriological Petri dishes (30 and 50 mm, Sterilin).
11. Dissection microscope, e.g., Wild M3C.
12. Fine forceps (e.g., Dumont no. 5).
13. Pasteur pipets, drawn to internal diameter of approximately 100 om, for egg transfer by mouth pipetting system.
14. Humidified CO_2 incubator at 38°C (5% CO_2 in air).

Injection of DNA into Pro-Nuclei

1. Standard micro-manipulation setup. We use a Leitz micro-manipulator and Zeiss Ergaval upright fixed-stage micro-scope equipped with Nomarski optics, and work in hanging drops in a Puliv chamber filled with heavy mineral oil (Sigma).
2. DNA delivery is by means of a 50-mL ground-glass syringe connected by an airline to the instrument holder.
3. Holding pipet, controlled by an oil-filled pressure device.
4. Equipment for glass instrument making: micro-forge, needle puller, Bunsen burner.
5. Thin-walled capillary with filament for injection pipets, prepared in house as required.
6. Thick-walled capillary without filament (Clark Electromedical Instruments, GC100-15) for holding pipets, prepared as required.
7. PBI+BSA (4 mg/mL), prepared.

Preparation of Recipients for Embryo Transfer

1. (C57BL/6 CBA) F1 females (6-8 wk old), mated naturally with vasectomized males of proven sterility. Day of vaginal plug counts as 1 of "*pregnancy*."

Screening for Transgenic Offspring

1. Tail lysis buffer (TLB): 50 mM EDTA, 10 mM Tris-HCl pH 8.0, 1% sodium dodecyl sulfate (SDS).
2. 1 Phenol/1 [chloroform *iso*-amyl alcohol].
3. 3 M Nacl.
4. 100% isopropanol.
5. TE: 10 m*M* Tris-HCl, 1 m*M* EDTA pH 8.0.

Methods

Prepare High-Density Yeast Plugs

1. Inoculate 100 mL of selection media (SD without selective amino acids) with a few colonies from a freshly grown selective SD plate. Grow with rapid shaking at 30°C for 16-24 h.
2. Count cell density using a hemocytometer, and determine the final volume that will give a density of 4 10^9 yeast/mL.
3. Spin cells at 1500g for 5 min and pour off media. Resuspend cells in an equivalent volume of 50 mM EDTA, spin, and repeat once. Briefly spin cells once more and remove residual liquid with pipet tip.
4. Using a wide-bore pipet tip, thoroughly resuspend cells in 50 mM EDTA to a concentration of 1.2 10^{10} yeast/mL.
5. Warm briefly to 37°C (30 s) before adding 5 oL yeast lytic enzyme/100 oL. Vortex briefly.
6. ADd 2 vol of 1.5% LMP agarose. Vortex briefly and quickly transfer 100 oL to each PFGE plug mold using a wide-bore pipet tip.
7. Leave 15 min to set and transfer plugs into LET buffer (20 mL/mL of plut), prewarmed to 37°C.
8. Incubate for 16-24 h at 37°C, with periodic shaking.
9. Replace LET buffer with equal volume of YLB, prewarmed to 50°C. Change for fresh YLB after 1 h and incubate at 50°C for 16-24 h.
10. Equilibrate the plugs in 100 mM EDTA at room temperature. Replace 3-4 times to remove residual traces of LDS before storing at room temperature or 4°C.

Purification of YAC DNA

PFGE Separation of YAC from Yeast Chromosomes

1. Equilibrate enough plugs for a single PFGE run (10-11) in 0.5 TBE for 30 min. Repeat two times.

2. Load plugs onto 1.2% LMP agarose gel. Marker lanes can be included at both ends of the gel. Seal the lanes with 0.5% LMP agarose.
3. Run gel under conditions appropriate for adequate separation of yeast chromosomes, depending on size of YAC of interest.
4. Cut off both edges of the gel including approximately one plug width at each side, and stain in ethidium bromide solution for 30 min. Store the central part of the gel in running buffer.
5. Rinse the stained gel edges and, under UV illumination, cut notches in the gel to mark the position of the YAC band.
6. Reassemble the gel using Saran wrap to keep the stained portions separated from the clean gel containing the YAC DNA to be purified. Slice out the unstained YAC band between the notches and transfer to 0.5x TBE at 4°C until ready to agarase treat.
7. Stain the remaining parts of the gel and reassemble under UV illumination to ensure adequate excision of the YAC band.

Agarase Treatment of Purified YAC Band

1. Equilibrate the gel slice in 20 mL of agarose digestion buffer at least three times for 30 min each.
2. Remove all the buffer with a pipet tip and transfer the gel slice to a 2-mL micro-fuge tube. Determine the weight of the gel by comparing to an empty tube. Use approximately 500 mg of gel/tube.
3. Melt the gel at 68°C for 3 min, spin briefly, then melt for a further 5 min.
4. Transfer to 42°C and equilibrate for 5 min.
5. Equilibrate 2 units of δ-agarase per 100 mg of gel in a 2-mL micro-fuge tube at 42°C for 30 s, then transfer the melted gel into the δ-agarase-containing tube with a pre-warmed wide-bore 1-mL tip, and mix very gently, but quickly, 3-4 times.
6. Incubate at 42°C for 3-4 h, then transfer to ice for 10 min.
7. Spin down any undigested agarose for 15 min at room temperature, and transfer the digested solution to a new tube.
8. Store at 4°C. Check intactness of the YAC DNA by running 15 oL on a pulsed-field gel, and determine the concentration by comparing serial dilutions to known concentrations of standard marker DNA. An approximate concentration of 0.5-1 ng/oL is ideal for injection to generate low-copy-number transgenic mice.

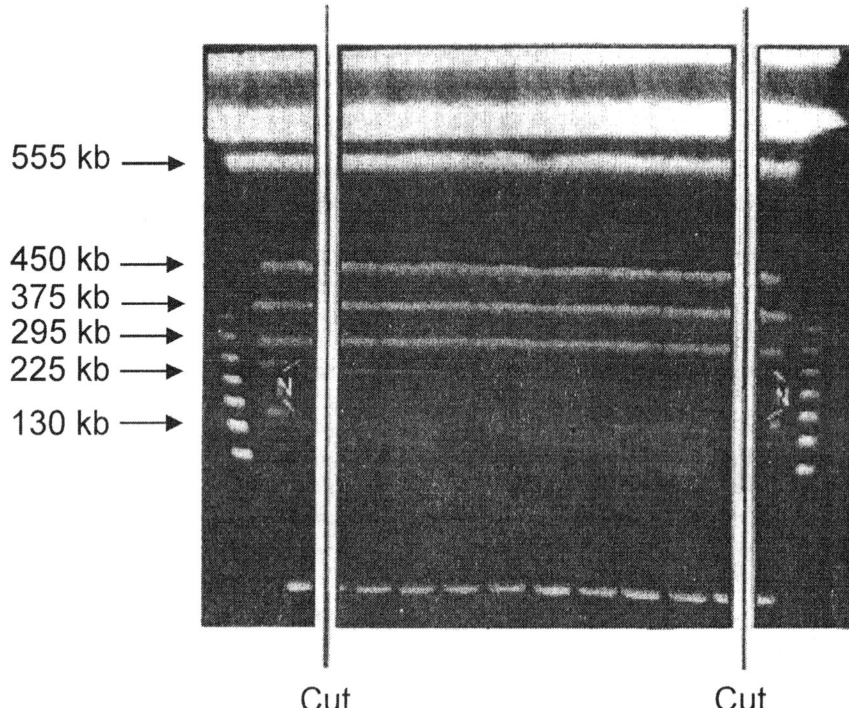

Fig.10.1: **Stained pulsed-field gel run for 24 h at 195 V with a pulse time of 40 s at 9°C, on a Pharmacia Gene Navigator system, containing yeast DNA plugs with a 130-kb YAC. Positions of where to cut off the edges of the gel for staining to reveal the location of the YAC band are indicated (cut). Notches cut into the stained gel edges at the positipn(s) of the band of interest are also shown (N). The gel is shown reassembled after slicing out the YAC band and the 225-kb yeast chromosome, demonstrating that the DNA has been excised efficiently before staining.**

Prepare YAC DNA for Injection

1. Dialyze 100 oL of YAC DNA solution against injection buffer *immediately* before injecting. Transfer the DNA, using a wide-bore tip, onto dialysis membrane floating on top of 20 mL of injection buffer in a Petri dish at 4°C for 1-2 h.
2. Transfer the dialyzed DNA into an microfuge tube and spin down any debris for 5 min before injecting. Keep on ice.

Preparation of Fertilized Mouse Eggs

1. Using a 27G needle, superovulate F1 females by intraperitoneal injection of 150 oL of PMS.
2. After 48 h, similarly inject 150 oL of HCG.

3. Immediately transfer each female to a cage containing a fertile male.
4. Check females for vaginal plugs the following morning.
5. At 20 h after HCG injection, sacrifice the females and excise the oviducts with care to avoid bursting the swollen ampullae. Transfer to a Petri dish.
6. Cover each pair of oviducts with a drop of hyaluronidase solution and open the ampullae with forceps under the dissecting scope.
7. A mass of eggs embedded in a cloud of cumulus cells should be released.
8. When most of the cumulus cells have detached (3-5 min), collect the eggs with a mouth-operated Pasteur pipet and wash through 3-4 large drops of T6+BSA.
9. Check that the eggs have been fertilized.
10. Incubate in a drop of T6+BSA covered with light mineral oil in the CO_2 incubator until ready to proceed with YAC DNA injection.

Injection of YAC DNA into Pro-Nuclei and Transfer into Recipients

Since the procedures for setting up the micromanipulator depend on the type of microscope and micromanipulator used, we will assume a working system is available to the user and provide only a brief outline of particular procedures we use in this section.

1. Prepare injection pipets immediately before use to avoid atmospheric contamination of the tip.
2. Transfer eggs in batches of 20-30 to drop(s) of PBI+BSA in the injection chamber.
3. Load the injection pipet from the rear by standing it in the DNA solution for a few minutes, tapping if necessary. Alternatively, place a drop of DNA solution within the pipet against the glass filament using a fine-drawn capillary.
4. Check that the piston of the syringe can move freely and without catching, then mount the injection pipet.
5. Open the tip of the injection pipet in the injection chamber by gently knocking it against the holding pipet until the outside diameter is approximately 0.75 om, preferably with a slight bevel.
6. Inject DNA into the pro-nucleus in a gentle and controlled manner to avoid shearing the DNA. Injection can be into either the male or female pro-nucleus, although the larger size of the male pro-nucleus makes the procedure easier. Inject until the pro-nucleus is observed to swell to about twice its normal volume. Withdraw the pipet gently.
7. If the tip is found to block after several injections, repeat the knocking procedure as many times as the needle shape will allow before the outside diameter gets to large (approximately 2 om).

8. After injection, transfer the eggs to a drop of T6+BSA, and remove any that have lysed. Incubate overnight in the CO_2 incubator.
9. Select eggs that have not blocked (now at the two-cell stage) to transfer to oviducts of d 1 pseudo-pregnant recipient females, following standard procedures.
10. Monitor the pregnancy by weighing at d 15 and d 18 after transfer (d 1). If the female has not given birth by mid-day of d 20, deliver the pups the Caesarian section and foster them to suitable foster mothers. This applies to any litter size, since birth at a later stage will result in increased perinatal lethality.

Screening for Transgenic Offspring

1. Take 0.5-cm tail snips from 10- to 14-d-old pups and incubate in 1 mL of TLB in a 2-mL microfuge tube at 37°C for at least 16 h.
2. Add 0.8 mL of phenol/chloroform and shake vigorously, repeated several times over a minimum of 15 min.
3. Spin at maximum speed in a microcentrifuge for 15 min at room temperature.
4. Transfer 0.8 mL of the supernatant into a 2-mL microfuge tube using a wide-bore pipet tip, taking care not to disturb the interphase.
5. Add 80 oL of 3 M NaCl.
6. Add 0.88 mL of isopropanol and mix gently until DNA strands form.
7. Leave at room temperature for 30 min for full precipitation, with occasional mixing.
8. Pellet the DNA in the micro-centrifuge at maximum speed for 15 min at room temperature.
9. Pour off the supernatant and rinse the DNA pellet once with 70% ethanol.
10. Briefly spin the tubes once more and remove residual ethanol with a pipet tip.
11. Resuspend the DNA pellet in 150 oL of TE. Vortex briefly, then stand at room temperature for at least 30 min. To ensure homogeneity of the DNA solution before use, it is recommended to pass the solution through a narrow-bore pipet tip at least 10 times.

NOTES

1. Longer incubation times of up to 48 h may be required for some strains.
2. δ-Mercaptoethanol can be added to a final concentration of 14 oM to increase the quality of the DNA preparation, although we have not generally found this step to be essential.
3. Ensure that the yeast cells are well mixed with the agarose solution, since homogeneity of the sample is very important for the preparation of good DNA plugs. We find

that brief vortexing at this step provides efficient mixing and is not derimental to the final DNA quality.

4. These incubation times can each be reduced to 4-6 h by changing the buffers regularly.
5. Plugs can be stored for many months in 100 mM EDTA at room temperature with little or no degradation. If storage at 4°C is preferred, it is essential that all traces of LDS have been removed, as this will precipitate as fine crystals.
6. We find conditions of 195V, 40-s pulse time, 9°C for 24 h is adequate to separate chromosomes up to 550 kb.
7. It is very important that no ethidium bromide comes into contact with the YAC DNA to be purified at any stage, as this will enhance degradation.
8. It is best to go on to the next step as soon as possible, although we have stored the gel slices at 4°C in 0.5xTE overnight with no detrimental results.
9. Immediate use for injection is preferable, but we have stored DNA in this way for up to 1 wk with little degradation.
10. In our hands the quality and quantity of DNA that can be purified by this method is suitable for the efficient generation of low-copy-number transgenic mice. It has been suggested, however, that more concentrated DNA samples can be advantageous, especially for very large transgenes (over 300 kb). Additional steps that can be taken to increase the final concentration of YAC DNA.
11. The aim is to produce about 200 fertilized eggs, from which 100-120 can be selected for injection. Yield varies from 20-40 eggs/female.
12. Eggs of this genotype enter the ideal stage for injection (enlarging pro-nuclei migrating toward the center of the egg) at around 23 h post-hCG injection, and are harvested 2-3 h before this. Harvesting earlier than this will increase the number of unfertilized eggs present.
13. For convenience, we generally produce a line of drops of PBI + BSA in the injection chamber and transfer 3-4 eggs into each one. Including a different number in the last drop provides an indicator of when the procedure is complete.
14. If more than 15% of the eggs have blocked, try using a more dilute DNA solution next time.
15. Do not overcrowd the uterus; a single healthy embryo can develop to term although it will probably require delivery by Caessarian section), whereas more than 8 surviving embryos will result in less vigorous pups at birth. We generally transfer up to 15 two-cell eggs into each recipient female, which results in a litter of between 1 and 6 pups on a good day.
16. DNA prepared in this way is suitable for use in both Southern analysis (use 10-15 oL per lane for identifying a single-copy fragment and PCR (use 0.25 oL in a 50-oL reaction).

Chapter 11

Allele-Specific Expression

One of the difficulties associated with the analysis of imprinted gene expression is the need to distinguish RNA synthesis occurring at the maternal vs the paternally inherited copy of the gene. Most of the techniques used to examine allele-specific expression exploit naturally occurring polymorphisms and measure steady-state levels of RNA isolated from a pool of cells. Hence, a restriction fragment length polymorphism (RFLP) an be exploited in a heterozygote, by a reverse transcriptase polymerase chain reaction (RT-PCR)-based procedure, to analyze maternal vs paternal gene expression. The human *IGF2R* gene was analyzed in this way. Smrzka *et al.* were thus able to show that the *IGF2R* gene possesses a hemimethylated, intronic CpG island analogous to the mouse imprinting box. However, *IGF2R* mRNA was detected that possessed the RFLP from both the maternal and paternal alleles in all but one of the 70 lymphoblastoid samples. (The one monoallelic sample reactivated its paternal allele with continued cell culturing.)

It was concluded that monoallelic expression of the human gene is a polymorphic trait occurring in a small minority of all tested samples. Although this is a sound conclusion, the question remains: Is the human *IGF2R* gene imprinted? The difficulty associated with this question reflects, to a large extent, the technical difficulties associated with the analysis of genes subject to epigenetic regulation. For example, useful parent-of-origin polymorphism in heterozygous states are not always readily available. Furthermore, procedures that measure steady-state levels of RNA can only determine the overall or average level of expression for all the cells within the examined tissue. This is fine for biallelically expressed genes that rely on cis-acting DNA regulatory sequences. However, imprinted expression states can be limited to a subpopulation of cells within a tissue, as observed for the mouse *Ifg2* gene. These subpopulations of monoallelic cells are hard to detect in homogenates of tissues. Another difficulty is provided by genes that randomly select one allele and express it exclusively. These monallelic genes can be difficult to detect in pooled RNA that possess transcripts that arose from either allele. Allelic exclusion has been reported for the mouse olfactory receptor cluster of genes and for the interleukin-2 locus in mature thymocytes and T cells. And then there are the genes that combine strategies. For instance, the *XIST* gene in females is imprinted in placenta (associated with inactivation of the paternal X chromosome) but is expressed from either allele in embryonic tissues, resulting in random X inactivation. To complicate matters further, the silent alleles of imprinted genes are known to "leak" reduced amount of RNA. This phenomenon has been observed at the mouse paternal *IGf2r* gene.

It is extremely difficult to determine the mechanism underscoring this leaky silencing. It may be due to a silencing mechanism that reduces transcription at the paternal allele, leaving a very low level of expression in all cells. Alternatively, a subpopulation of cells that fail to silence the paternal allele may be responsible for all of the leaky expression. Although the former possibility seemed the more likely, direct visualization of RNA synthesis by fluorescent *in situ* hybridization (RNA-FISH) provided evidence for the latter, at least for the cell lines examined (unpublished data). Although no paternal signal was observed in the vast majority of nuclei, a few cells were seen to express paternal *Igf2r* at levels similar to the maternal allele. This leaky expression seemed to occur in a nonclonal manner, suggesting that monoalleleic expression can be dynamically established in a stochastic manner as opposed to requiring a stably maintained, silenced state.

GENE EXPRESSION

The use of a fluorescent *in situ* hybridization (FISH) procedure that detects RNA in intact, interphase cells can circumvent some of the difficulties associated with the analysis of epigenetically regulated genes [*see*, for example, work on the human and mouse *Xist* gene]. The procedure involves hybridization of labelled probes to non-denatured cells that have been fixed on a microscope slide, followed by visualization using standard immunocytochemical methods. Expression of mRNA is observed at the single-nucleus level as discrete spots or "*transcription foci*" that localize, in cis, on the transcribed gene. Gene expression can therefore be observed at the single-cell level and within a proper cellular context, permitting the correlation of gene expression with other cellular events. Furthermore, the subcellular localization of imprinted, non-coding transcripts can be determined within both nuclear and cytoplasmic compartments, as was done when the noncoding *Xist* transcript was shown to coat the inactivated X chromosome. Although discrete spots have been observed by RNA-FISH in a variety of genes, the reason for the spatially restricted morphology of the fluorescent signal is unclear.

The spot could be due to the presence of multiple RNA polymerases on the gene, which produce a nested set of nascent transcripts that are targeted by the FISH probe. Alternatively, a rate-limiting step in the post-transcriptional processing of full-length transcripts (such as intron splicing) may act to retain RNA at the gene transiently. The question, therefore, arises as to which step in RNA synthesis is detected by RNA-FISH to produce the spots. The answer seems to vary depending on which gene is analyzed. Attempts to detect nascent transcripts specifically (by using probes that target the short-lived introns rather than the exons) have been made successfully (*Xist*, δ-globin). On the other hand, analysis of the inducible *hsp70* gene indicated that the predominant source of the signal was due to retention of full-length RNA at the gene. These differences should be kept in mind when setting up the procedure. It may be unwise to assume that the strength of the RNA-FISH signal is analogous to the transcription rate.

Post-transcriptional events affecting RNA (or its "processing fate") may also determine whether a spot is observed. Having said this, it should be clear that only a combination of RNA-FISH with nuclear run-on assays can provide unequivocal data about transcription rates. Using RNA-FISH to determine whether a particular cell line or tissue expresses a gene mono- or biallelically involves counting the number of fluorescent transcription foci within a nucleus and comparing it to the ploidy number for that cell. However, there may be some instances (possibly rare) when an average spot-per-nucleus number that is half the ploidy number is not an indication of imprinted expression. These would be instances where gene expression in cells that are at various stages of the cell cycle. However, expression is not continuous in cells.

For example, a general inhibition of transcription occurs during the mitotic stages and is reinitiated in interphase of the next cycle. If the gene to be analyzed is transcribed during a narrow window of the cell cycle and the two alleles express at different stages, then this allelic cell timing difference can result in nuclei that predominantly show a fluorescent spot at just one allele (despite the absence of imprinted expression). If there is an overlap in the timing of transcription between the two alleles, then a subset of nuclei will show two spots (the overall percentage will depend on the duration of the overlap). Cell timing phenomenon has been observed with δ-globin transgenes carrying an incomplete locus control region. Conversely, imprinted genes can be mistakenly classed as biallelic by RNA-FISH if the imprinting mechanism does not involve transcription silencing of one allele. This would occur if gene is transcribed biallelically but the transcripts from only one allele are stable enough to function. This has been observed for mouse *Xist* in differentiated embryonic stem cells, where use of different promoters at the two alleles produces transcripts with different half-lives.

Transcripts are detectable at both alleles by RNA-FISH with intronic probes, although only one allele transcribes the stable form of the transcript that coats the inactive X chromosome. With the appropriate controls, the RNA-FISH procedure can distinguish between the above possibilities and produce an increasingly more sophisticated picture of gene expression. Several variants of the RNA-FISH procedure have been developed. The primary difference between them is the way the fixed cells are permeabilized in order to permit the probe access to the nucleus.

Fixed cells can be treated with a detergent such as Triton X-100 or digested with a proteinase such as pepsin. Most cell types analyzed by RNA-FISH will produce results with either procedure, although the exact conditions may vary slightly for different cell types. The procedure outlined here is a simplified version of cell types, including cultured fibroblasts, embryonic stem cells, and cell suppensions from homogenized tissues such as fetal liver. Precautions required for preventing the degradation of RNA are discussed. The trickiest part of establishing RNA-FISH frequently involves the preparation of the haptenized probe. A novel, PCR-based procedure that reliably produces single-stranded probes is described, and troubleshooting hints are provided.

Materials

Probe Preparation

1. Taq polymerase (such as Gibco BRL).
2. 10 PCR buffer: 200 m*M* Tris-HCl, pH 8.4, 500 m*M* KCl.
3. 15 m*M* $MgCl_2$.
4. 10 nucleotide solution: 2 m*M* each of dATP, dCTP, and dGTP, 1.9 m*M* dTTP (lower concentrations of dTTP can be tried in order to increase the incorporation of the haptenized of dTTP can be tried in order to increase the incorporation of the haptenized or fluorochromized dUTP).
5. 1 m*M* digoxigenin-11-dUTP or biotin-16-dUTP (Boehringer Mannheim).
6. 3 *M* NaOAc, pH 5.5.
7. 70% and 100% ethanol.

Slide Preparation

1. Cell culture media and disposables.
2. Adherent cell lines can be grown on sterilized microscope objects slides soaked in 0.1% gelatin or poly-L-lysine (highest degree of polymerization). Alternatively, nondherent lines and cell suspensions can be spotted directly onto Poly-Prep slides (Sigma).
3. 10 PBS (pH 7.4): 80 g NaCl, 2 g KC, 14.4 g $Na_2HPO_4H_2O$, and 2.4 g KH_2PO_4. Adjust pH to 7.4 and the volume to 1 L Autoclave the solution and store at room temperature.
4. Fixatives: 4% formaldehyde, 5% acetic acid in phosphate-buffered saline (PBS); 1% formaldehyde in PBS.

RNA in situ Hybridization

1. 10% pepsin (Sigma) prepared in water and stored frozen.
2. Biotin- or digoxigenin-labelled DNA probes.
3. Formamide.
4. 30 SSC: 3 M NaCl, 0.3 M Na citrate, pH 7.0.
5. 2 M Tris-HCl, pH 7.5 (do not diethyl pyrocarbonate [DEPC]-treat).
6. 10 saline: 85 g NaCl dissolved in 1 L of distilled water. Autoclave the solution.
7. Tris-saline solution: 145 mM NaCl, 100 mM Tris-HCl, pH 7.5.

8. Tris-saline-Tween solution: 0.1 *M* Tris-HCl, pH 7.5, 0.15 *M* NaCl, 005% Tween-20.
9. 2 hybridization buffer: 4 SSC, 20% dextran sulfate, and 2% bovine serum albumin (BSA).
10. Cot-1 DNA.
11. Herring sperm or salmon sperm DNA (Sigma).
12. Yeast tRNA (Sigma).
13. Biotin and/or digoxigenin detection reagents:
    a. Avidin-FITC or TRITC for detection and biotinylated goat anti-avidin for amplification of signal (Vector).
    b. Sheep anti-digoxigenin-FITC or TRITC (Boehringer) for detection and rabbit anti-sheep-FITC or TRITC and goat anti-rabbit-FITC or TRITC for amplification.
14. 4, 6-Diamidino-2-phynylindole (DAPI).
15. Vectashield embedding medium (Vector).
16. Graded series of 70%, 90% and 100% ethanol.
17. Hybridization buffer: 50-65% deinoized formamide, 2 SSC, 50 mM sodium phosphate, 5% dextran sulfate, pH 7.4.
18. Blocking buffer: 4 SSC, 0.1% Tween-20, 4 mg/ml BSA.
19. Detection buffer: 4 SSC, 0.1% Tween-20, 1 mg/mL BSA.
20. Epifluorescence microscope equipped with appropriate filter sets and image-capturing devices (such as a CCD camera and imaging software).

Methods

Probe Preparation

Procedures for labelling DNA with haptenized (such as biotin or digoxigenin) or fluorochromized dUTP have been published elsewhere. Kits that incorporate these nucleotides by nick translation and PCR are available commercially (for example, see the Boehringer catalogue). When choosing a detection system, keep the following points in mind. Fluorochromized nucleotides (i.e., directly labelled probes) generally produce weaker signals compared to haptenized probes. For genes expressed at low levels it is preferable to use a detection system in which amplification of the signal is possible (i.e. digoxigenin and biotin-labelled dUTP). When using biotinylated probes, check the endogenous level of biotin in the fixed cell by hybridizing in buffer lacking probe and processing the slide through the detection system.

In mammalian cells, digoxigenin routinely produces lower background signals compared to biotin, especially if amplification is required. A convenient control for hybridization specificity is provided by co-hybridizing two different probes to the same target. Using different probe-detection systems (i.e., two different colours) will ensure that the probes can produce co-localized signals and that the signal is not due to spurious hybridization. Conversely, two probes to distinct targets can also be tested to ensure that distinct signals can be obtained. These controls necessitate using at least a two colour probe-detection system.

A third colour (usually DAPI) is used to counterstain the entire nucleus. Alternatively, a three-colour hybridization is possible by treating fluorescein as a hapten. The requirement for more than one detection system should be kept in mind when deciding what immunocytochemical method to adopt for detection of each probe, and therefore, what antibodies are purchased. The probes synthesized by the above procedures will produce double-stranded DNA that can simultaneously detect sense and antisense transcripts. However, the discovery of noncoding, antisense transcripts at an increasing number of imprinted loci produces the need for strand-specific probes.

The procedure listed below is a fast, PCR-based method first developed for the antisense transcript present at the mouse *Igf2r* locus (manuscript in preparation) and subsequently applied to study *Xist*. It relies on PCR amplification using oligos complementary to just one DNA strand (linear PCR).

1. Select the DNA template to be used for probe synthesis. Restriction fragments from genomic clones, PCR products, or cDNA clones can all be used, but gelpurify the required inserts beforehand. Design a primer complementary to the target strand (avoiding repetitive sequences). A cocktail of oligos distributed over several kilobases can also be used. Use approximately 10-50 ng of template DNA amplified with 1 oM final concentration of oligo(s).
2. Linear PCR reactions are performed in 50 oL in the presence of fluorochromized or haptenized dUTP as follows: 5 oL of 10 PCR buffer, 5 oL of 15 m*M* $MgCl_2$ (although a range of concentrations ought to be tested in order to optimize each PCR reaction), 5 oL of 10 nucleotide solution, 5 oL of 1 mM digoxigenin-11-dUTP or biotin-16-dUTP, 5 units Taq polymerase.
3. Labeled product is generated by 200 PCR cycles of 94°C for 20 s, the annealing temperature for the oligo for 30 s, and 72°C for 1-2 min. Check the linear PCR products by visualization on a 1% agarose gel. To test the strand specificity of the probe, Southern-blot the agarose gel and hybridise with radiolabeled sense and antisense oligos contained within the amplified sequence).
4. The hybridization mix is prepared by precipitating approximately 50 ng of labelled single-stranded DNA adding 0.05 vol 3 *M* NaOAc, 2.5 vol 100% ethaniol, and 10 og of carrier DNA (such as salmon sperm DNA) and 10 og of yeast tRA. Wash the precipitate several times in 70% ethanol and resuspend in 10 oL formamide (for a

50% final concentration). Increased hybridization specificity can be achieved by increasing the concentration of formamide up to 75%. Before hybridization, the mix is denatured at 75°C for 10 min, placed on ice, and diluted with 10 oL of 2 hybridization buffer. Cot-1 DNA (1 og) can also be added to the hybridization mix, in which case probe denaturation should be followed by a 15-min incubation at 37°C.

Cell Culture and Fixation

1. Culture subconfluent cells in the appropriate medium, CO_2 atmosphere, and temperature. Trypsinize adherent cells and seed onto gelatin or poly-L-lysine-coated glass object slides. Grow cells overnight to allow cells to attach. Non-adherent cells and cell suspensions from organ homogenates can be cytospun onto slides or spotted directly onto Poly-Prep slides (Sigma).
2. If the slides have been in cell culture medium, rinse the slides briefly in PBS.
3. Fix (4% formaldehyde, 5% acetic acid in PBS) for 20 min at room temperature. Prepare the fixative just before using.
4. Wash three times in PBS for 5 min. Slides can be stored at this stage in 70% ethanol at –20°C.

RNA in Situ Hybridization

1. Transfer slides from ethanol storage to Tris-saline solution for 5 min at room temperature.
2. Just before use, thaw and dilute the 10% pepsin stock (1/100 to 1/1000, depending on the cell type) in 0.01 M HCl prewarmed to 37°C. The time of protease treatment should be optimized for each cell type. Optimal incubation times commonly fall between 2 and 10 min.
3. Rinse slides in water for 1 min.
4. Incubate in 1% formaldehyde in PBS for 5 min at room temperature.
5. Rinse slides in PBS for 10 min.
6. Dehydrate the cells by immersinig successively in 70%, 90% and 100% ethanol for 3 min each. Air-dry the slides. Apply the denatured probe to the slide (use 15-20 oL with a 24 24-mm^2 coverslip). Hybridize overnight at 37°C in a humidified container. The hybridization temperature can be increased if more stringent conditions are required.

Immunocytochemical Detection

1. Prewarm a solution of 2 SSC at 37°C. Wash the slides in 2 SSC for 5 min and ensure that the coverslips are released. Wash the slides a further three times for 10 min.

2. Wash slides in Tris-saline-Tween solution for 5 min at room temperature.
3. Add 100 oL of blocking buffer to each slide and incubate under a coverslip for 30 min at room temperature.
4. Wash with Tris-saline-Tween for 5 min.
5. Dilute each antibody required for hapten detection just before use in detection buffer. Apply 100 oL under a coverslip and incubate for 30 min at room temperature.
6. Wash with Tris-saline-Tween two times for 5 min between each antibody application in the detection system.
7. Rinse slides in Tris-saline for 5 min.
8. Dehydrate the slides through 70%, 90%, and 100% ethanol. Air-dry the slides.
9. Mount in 60 oL of Vectashield containing 10 ng/oL DAPI and cover with 24 50 mm^2 overslip.
10. Examine the slides with a fluorescence microscope equipped with appropriate excitation and emission filters for FITC, TRITC, and DAPI fluorescence.

NOTES

1. To ensure the preservation of *in situ* RNA, a few simple precautions can be adopted. The first is to wear gloves throughout the procedure, including when preparing pipet tips, making solutions, and handling slides. The second is to set aside chemicals reserved for RNA use only. These need not be the entire set of solutions required for the procedure, but need only include those chemicals and solutions that are likely to be contaminated if in general use. The third is to heat-sterilize glassware and autoclave solutions. If feeling particularly cautions about the need to preserve RNA, then water and solutions (with the exception of Tris) can be DEPC-treated by adding 20 oL of DEPC (diethyl pyrocarbonate; Sigma) to 1 L of solution, mix, and leave overnight before autoclaving.
2. When setting up the technique it will be necessary to verify that the signal obtained is due to hybridization to cellular RNA. A useful set of controls involves RNase A-treating two slides. One is hybridized in its native form. The other is heat-denatured to make the DNA available for hybridization. This will produce a slide for RNA alone (no RNase treatment, nondenatured), a negative control (RNase treated, non-denatured), and a DNA control (RNase treated and denatured).

   The RNase treatment can be performed after pepsin treatment by incubating the slide in 100 oL of 0.1-mg/mL RNase A in 2 SSC. Apply a coverslip and incubate at 37°C in a moist chamber. Wash the slide three times in 2 SSC and continue as normal. Make sure the RNase A used in this way does not become a source of contamination for the RNA-FISH.

To heat-denature and make DNA available for hybridization, incubate the slide containing the hybridization mix on a hot plate set at 80°C for 5-10 min before incubating overnight at 37°C.

3. If multiple antibodies are used to amplify hapten detection, they should be applied sequentially and not combined except when an antibody and florochrome-conjugated avidin are involved.

4. Once fluorochrome-conjugated components of the hapten-detection system are applied to the slides, precautions should be taken to reduce the amount of exposure to light, to help prevent quenching.

5. Failure to obtain any signal may be due to inadequate permeabilization of the cells and, therefore, failure of the probe to access the nucleus. This may be due to overly long fixation times combined with inadequate pepsin treatment. However, as tempting as it may be to blame the histological side of the procedure for the failure to obtain a signal, it is generally more likely that the source of the problem is the probe. Possible problems include:

   a. Use of excessively long probes that fail to access nuclear RNA. Probes of a length of around 300-600 bp are optimal. Try decreasing the length of the PCR-amplified insert. However, note that, in a low percentage of nuclei, targeting just one short stretch of the cellular RNA can result in weak signal. If longer sequences need to be targeted, then label a large insert and digest the linear PCr product with a 4-bp restriction enzyme that possesses single-stranded DNA digestion activity (e.g., Mnl I).

   b. Excessive competition between the probe and the salmon sperm and COT-1 DNA. Your probe may become "competed out" by the carrier DNA. This will happen to any DNA fragment if the amount of competitor is high enough. Try reducing the amount of carrier DNA used. If the signal obtained this way is not specific for the targeted RNA, then consider labelling a different fragment with a high content of unique sequence.

   c. Inadequate incorporation of hapten into the probe. Try reducing the amount of dTTP without altering the amount of haptenized dUTP in the PCR reaction. This may only be necessary if the gene of interest is expressed in very low amounts.

   d. The simplest solution, of course, is to try and increase the amount of probe used in the hybridization and decrease hybridization stringency until some signal (even if just background) is obtained. It is easier to clean up a hybridization than to work with no signal at all.

6. Too much signal and background problems: Having gotten past the no-signal stage, you are likely to experience too much signal as the next problem. You now need to reverse most of the advice given above:

a. Increase the amount of competitor DNA.

b. Increase hybridization stringency, either by increasing the amount of formamide in the hybridization solution or increasing the hybridization temperature to 42°C.

c. Try including COT-1 DNA and decrease the amount of probe used.

Having sent you round in a circle, you may now discover that you are in a no-win situation: either you have no signal or you have persistent background problems. There are two situations when background signal will persist to the point that eliminating it will also eliminate that signal.

a. The first situation involves probes with high G+C content that can hybridize non-specificially to ribosomal RNA in the nucleolus. I know of no procedure that will circumvent this problem, other than to choose a different sequence, with a lower G + C content, for your probe.

b. The second situation involves persistent fluorescent spots on the surface of the nucleus that commonly appear as *"edge spots"* on the periphery of the nucleus. I suspect the cause of this problem is the presence of long strands of DNA that are unable to penetrate the nucleus but capture the probe, forming very intensely fluorescent spots. If the probe has been labelled by nick translation, then reduce the size of the template DNA either by increasing the concentration of DNase I or by using restriction enzyme digestion prior to nick translation. If the probe has been generated by the linear PCR procedure, then include a restriction enzyme digestion after amplification. Choose a combination of 4-bp cutters that will dramatically reduce the size of the unlabeled, double stranded template DNA and also digest single-stranded DNA. Heat-inactivate the reaction mixture before using.

7. Correcting problems of poor cellular morphology requires a balancing act between the fixation step and the pepsin treatment. You must balance the need to preserve cellular morphology and ensure probe accessibility to obtain an optimal signal intensity. Inadequately fixed samples that are overly pepsin digested can lead to loss of cellular structures, including the nucleus. Overly fixed cells with too little pepsin treatment may produce excellent morphology but too little or inconsistent signal. To correct tis problems you will need to determine optimal pepsin concentrations and digestion times empirically for each cell type that causes problems.

ALLELE-SPECIFIC REPLICATION

Chromosome replication banding studies show that homologous regions on a pair of autosomes generally replicate at the same time in S phase. Izumikawa *et al.* first observed that this was not the case for the imprinted chromosomal region 15q11-q13. This observation has been confirmed in other replication banding studies as well by the fluorescence *in situ* hybridization (FISH) replication assay. The latter technique has also been used to observe DNA replication synchrony in association with alelic inactivation of genes such as those encoding olfactory receptors and the cytokine, interleukin 2. The latter genes are not imprinted but display random silencing of an allele in individual cells. In imprinted regions, DNA replication was generally observed to occuı earlier on the homologue. The patterns of allele-specific replication in the cells of Prader-Willi (PWS) and Angelman syndrome (AS) patients, however, have generally allele-specific replication timing in the GABRB3/A5 cluster on 15q11-13 revealed that cells from PWS and AS have lost the strict replication timing observed on the parental chromosomes of normal cells. These results suggested the requirement of a biparental contribution for the regulation of replication asynchrony and lead to the hypothesis that allelic cross-talk, perhaps via pairing of homologous chromosomes, might play a role in the imprinting process. This possibility was tested directly using three-dimensional (3-D) FISH in spatially intact nuclei and confocal laser scanning microscopy to determine the distance between homologous chromosomes in cell cycle fractionated human lymphocytes. The results demonstrated that, in normal cells, the chromosome 15 homologs pair during the late S phase.

The most tightly paired region was observed to be 15q11-q13, consistent with the hypothesis that allelic cross-talk could be involved in the normal pattern of imprinting in chromosome 15q11-q13. In this chapter, we detail the protocols used to perform flow cytometry, to fractionate cells on the basis of DNA content, before FISh analyses of DNA replication and homologous chromosome pairing. Although FISH detection of allele-specific replication can be performed on unsorted cycling cells, comparisons between replication timing of loci in different cell lines (i.e., patient and normal) are imprecise due to differences in cell growth. The kinetic approach to investigating both allele-specific replication and homologous pairing outlined here was essential in differences in cell growth. The kinetic approach to investigating both allele-specific replication and homologous pairing outlined here was essential in determining the biparental requirement for both normal replication timing and homologous pairing that had been overlooked by previous studies of unsorted cells.

In addition, the protocol for culturing primary human T lymphocytes to obtain optimal numbers of cycling cells is included, as it varies from that of standard cytogenetic methods. Immortalized human cell lines such as lymphoblasts may be used instead of primary cultures, but epigenetic and cytogenetic alterations may occur as a result of long-term culture of these lines. FISH methods are given for both the standard 2-D preparation for allele-specific replication timing and the 3-D preparation for investigating homologous associations. Normally, well-spread 2-D nuclei are the best target for FISh replication studies because separated sister chromatids can be readily observed. Conversely, 3-D preparations in which the spatial organization of nuclei is maintained are ideally suited for investigations of nuclear chromosome organization. If a large number of cells in 2-D FISH preparations are scored, however, 3-D organization can be extrapolated. In the future, an automated approach to detecting homologous pairing on 2-D FISH preparations may be possible by the new technology of laser scanning cytometry.

Materials

Culture of Primary Human T Lymphocytes

1. 20-60 mL of human peripheral blood collected in heparinized tubes.
2. PBS: Phosphate-buffered saline without Ca^{2+} and Mg^{2+}, sterile.
3. Ficoll-Paque^{PLUS} (Pharmacia).
4. RPMI 1640 cell culture medium: Roswell Park Medical Institute media.
5. Human sera, pooled male AB (Sigma). Prescreen sera samples for T-cell growth and purchase enough sera to complete experiments.
6. Penicillin, steptamycin, and L-glutamine.
7. 24-well culture clusters (Costar).
8. PHA-P: Phytochemaggluthin P (HA17, Remmel).

Vital Staining for DNA Content

1. RPMI, prewarmed to 37°C in tissue culture incubator.
2. Hoechst 33342 (Molecular Probes), 1-mg/mL stock in water, store in dark at 4°C.
3. PBS, 4°C.

Two-Dimensional FISH for Allele-Specific Replication

1. 0.075 *M* KCl.
2. Fixative 3/1 methanol/glacial acetic acid, make fresh, 4°C.
3. 70%, 90%, 100% ethanol.
4. Clean glass microscope slides in 100% methanol, stored at –20°C.
5. Biotin- or digoxigenin-labelled DNA probes.
6. Human placental DNA (Sigma) or Cot-1 DNA (Gibco).
7. Deionized formamide. Store in aliquots at –20°C and use within 2 wk of thawing.
8. Hybridization solution (2): 40% dextran sulfate, 4 SSC. Make fresh, vortex to a uniform viscosity.
9. Denaturation solution: 70% deionized formamide, 2 SSC, make fresh.
10. Rubber cement.
11. Wash A: 50% formamide (inexpensive grade), 2 SSC.

12. Wash B: 0.1 SSC.
13. Wash C: 4 SSC, 0.5% Tween-20.
14. Blocking solution: 3% bovine serum albumin (BSA), 4 SSC, 0.1% Tween-20.
15. Detection solution: 1% BSA, 4 SSC, 0.1% Tween-20.
16. FITC-avidin Vector Laboratories) and/or Rhodamine-anti-digoxigenin (Roche Diagnostics).
17. Mounting media: Vectashield (Vector Laboratories) plus 1 og/mL DAPI (Molecular Probes.
18. Cover glasses, No. 1, 18 18 mm, 20 30 mm, and 20 50 mm.
19. Glass Coplin jars (at leat 4).

Three-Dimensional FISH for Homologous Association

Most solutions and materials are the same as in the above section, except the following list:

1. Poly-D-lysine, MW 30-70 kDa (Sigma), 100 og/mL in water, store aliquots at –20°C.
2. Microscope slides that have been wiped clean with 100% methanol.
3. Fixative: 4% paraformaldehyde/1% methanol in PBS. 10 stock, keep for 1 mo, make up as follows: to 4 g paraformaldehye (Kodak), add 6 mL water plus 2 drops 4 *N* NaOH, heat slowly while stirring and do not allow to boil. Heat until in solution. When solution is cool, add 1 mL methanol and bring volume to 10 mL if needed. Filter and store protected from light. Make fresh dilution in PBS before each use.
4. Postfixation wash and storage solution: 0.3 *M* glycine in PBS, dilute 1/10 in PBS from 3 M glycine stock, add 0.04% sodium azide from 100 stock. Make fresh each time.
5. Permeabilization solution: 0.5% Tween-20 in 0.2 N HCl. Make fresh each time.
6. Wash solution: 0.5% Tween-20 in PBS.
7. Nail polish.
8. Cytofuge and parts (StatSpin Technologies, Norwood, MA).

Methods

Culture of Human T Cells

1. Dilute peripheral blood 1:1 with sterile PBS in 50-mL conical tubes with a final volume of less than 30 mL per tube.

2. Underlay Ficoll (at least 1/3 blood volume) by placing a pipet at the bottom of the tube and slowly releasing the Ficoll until a lower gradient has formed. Handle tubes gently to maintain the gradient.
3. Centrifuge at 500g for 30 min with the rotor brake turned off.
4. Aspirate and discard the upper serum layer. The white blood cells are at the interface and should be carefully transferred with a pipet to a new 50-mL tube. Discard the Ficoll layer and red cell pellet.
5. Wash cells twice with 50 mL PBS. Centrifuge at 300g for 10 min for each wash. The centrifuge brake can be turned back on.
6. Prepare T cell growth media: 20% human sera in RPMI with 10-U/mL penicillin, streptomycin, and L-glutamine.
7. Resuspend the cell pellet in 10 mL T-cell growth medium and count a 1:10 dilution on a hemacytometer.
8. Add T-cell growth medium to bring cells to ~10^6/mL. Add 2 og/mL PHA-P.
9. Seed cultures in 24-well plates at 1-2 mL per well. Incubate for at least 72 h at 37°C in 5% CO_2. Cells should form large clumps from PHA-P stimulation after 24 h.

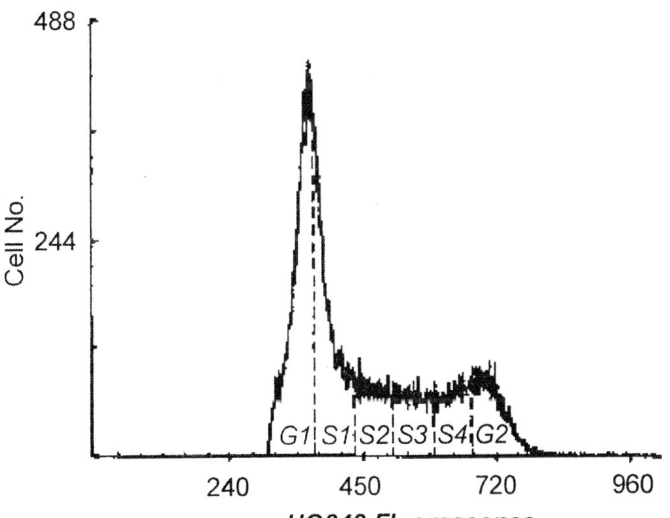

Fig. 11.1: **Flow sorting cells on the basis of DNA content. PHA-stimulated T lympho-cytes were stained with Hoechst 33342 (HO342) to allow flow cytometric analysis of DNA content in vitally stained cells. A representative histogram shows the distribution of DNA content (determined by HO342 fluorescence) in the cell population. Fluorescence-activated cell sorting gates were set at positions represented by the dotted lines. For kinetics of DNA replication, six fractions were collected and labelled G1, S1, S2, S3, S4, or G2 as shown. For detection of homologous pairing, four fractions were collected so that S1 and S2 were early S phase and S3 and S4 were late S phase.**

Vital Staining for DNA Content and Flow Sorting

1. Transfer cells from culture to a 50-mL tube, centrifuge at 300g for 10 min.
2. Make up staining buffer, use ~1 mL buffer per 10^6 cells, dilute Hoechst 33342 1/100 in prewarmed RPMI 1640 (10-o g/mL final concentration).
3. Resuspend cell pellet in Hoechst 33342 stain buffer, place in a tissue culture incubator for 30-45 min.
4. Centrifuge cells, resuspend in PBS at 10^7/mL and keep on ice prior to sort.
5. Sort cells into 4 or 6 fractions based on DNA content.

Two-Dimensional FISH for Allele-Specific Replication

Fixation of Cells

1. Centrifuge cells from cell cycle sort (0.1 to 1.0 10^6) at 300g.
2. Resuspend in 1 mL of 0.075 KCl.
3. Incubate at 37°C for 13 min. Centrifuge cells.
4. Aspirate, leaving some fluid, resuspend cell pellet in 1 mL fixative and incubate for 20 min on ice.
5. Centrifuge cells and resuspend in 20-100 oL fixative, depending on pellet size (roughly 10^7/mL). Store cells at –20°C.
6. Drop 10 oL of cell fixation solution on slides that have been immediately removed from methanol and wiped dry. The drop should spread and dry quickly (within 5-15s) for optimal spreading.
7. Air-dry slides overnight. Dehydrate in a 70%, 90%, 100% ethanol series for 10 min each. Store desiccated for up to 1 mo.

Hybridization and Washing

1. Prepare probe: combine 20-100 ng of labelled probe, 3 og of human competitor DNA, and 7 og of salmon sperm DNA per slide. Lyophilize DNA in a Speed-Vac. Resuspend DNA in 5 oL per slide in deionized formamide. Vortex for 5-10 min using a tube attachment head for the vortex machine. Add 5 oL per slide hybridization solution. Vortex for a further 5-10 min.
2. Prewarm slides for hybridization in a 55°C oven.
3. Prepare denaturation solution with sufficient volume to cover sample area in a Coplin jar. Cover jar, place in water bath, and heat bath to ~70°C.

4. Prepare postdenaturation ethanol series (70%, 90%, 100% ethanol in Coplin jars and chill on ice. Prepare an extra 70% ethanol jar for first quick wash.

5. Test temperature inside jar of denaturation solution. Once it has reached 70 ± 2°C, add 3-4 slides. Denature for 3 min, monitoring the time exactly. Transfer slides immediately to ice-cold 70% ethanol quick wash for 10 s. Repeat as needed for additional slides.

6. Wash slides for 5 min each in a cold ethanol series. Air-dry and prewarm slides in a 37°C oven.

7. Denature probe for 5 min at 72°C. Incubate at 37°C for 10 min to preanneal competitor DNA.

8. Add 10 oL denatured probe to slide. Spread the drop with a 18 18 mm coverslip, using care to exclude and remove bubbles.

9. Seal the coverslip with rubber cement. Place slides in a prewarmed moist chamber at 37°C. Incubate overnight.

10. Prewarm wash solutions A and C to 45°C and wash solution B to 60°C.

11. Carefully remove the rubber cement seal with forceps to avoid removing or dragging the coverslip. Place slides in Coplin jar with wash A at 45°C and agitate to remove coverslips for 5 min. Transfer slide to new wash A jar and wash twice for 5 min each at 45°C.

12. Wash 3 times in wash B for 5 min each at 60°C.

13. Blot excess liquid from the back and sides of slide. Add 200-oL blocking solution per slide. Spread solution with a 20 50 mm coverslip, being careful to exclude bubbles. The coverslip should float very loosely on slide. Do not allow slides to dry.

14. Incubate in a moist chamber for 30 m at 37°C.

15. Prepare fluorescent detection solution: dilute FITC-avidin 1/40 and Rhodamine-anti-digoxigenin 1/100 in detection solution. Centrifuge for 5 min to remove fluorescent debris. Protect from light.

16. Tilt slides to remove the coverslip and excess block solution. Pipette 200 oL of the fluorescent detection solution onto slide and spread with a 20 50mm coverslip.

17. Incubate in a moist chamber for 30 min at 37°C in the dark.

18. Tilt slides to remove coverslip and excess detection solution. Place slides in Coplin jar with 45°C wash C solution for 5 min. Wash twice more in wash C solution.

19. Blot excess liquid from the back and slides of slide. Pipette 30 oL of mounting medium per slide. Spread solution with a 20 3 mm coverslip, being careful to exclude bubbles.

20. Store sides upright in a covered slide box at 4°C.

Fluorescence Microscopy and Scoring of Replication Events

Examine slides on a fluorescent microscope with a mercury light source (100-W bulb) and a 100 oil objective with 1.4 numerical aperture. The microscope should be equipped with filter sets appropriate for DAPI, FITC, and Rhodamine/Texas red. A triple-bandpass filter is also useful for scoring replication events, but brightness of signals may be compromised by this approach. If FISH signals are dim and nuclear background is low, a cooled CCD-based imaging system may be useful for merging multiple colors.

A Photometrics Sensys CCD camera and IPLab Spectrum software were used to obtain the three-colour images. The FISH signals used to determine allele-specific replication events should be scored as singlets or doublets on each allele in interphase nuclei. Fine focus adjustment should be performed while scoring, to ensure that all FISH signals are counted. Doublets should be two closely spaced but discernable signals. For determining hybridization efficiency. A second FISH probe that detects a polymorphism can be used to distinguish parental alleles. In this case, the FISH replication signal closest to the FISH polymorphism signal should be scored as being on the same chromosome. The final data analysis of kinetic allele-specific replication is performed.

Three-Dimensional FISH for Homologous Association

Fixation of Cells

1. Centrifuge cells ~300g from the cell cycle sort (0.1-1.0 10^6 cells).
2. Resuspend in 1 mL fixative at 4°C and incubate for 20 min on ice.
3. Centrifuge cells and resuspend in 2-4 mL of 0.3 M glycine/PBS. Repeat once.
4. Resuspend cells at ~10^6/mL in 0.3 M glycine/PBS (with addition of sodium azide, cells can be stored for at least 2 mo at 4°C.

Slide Preparation

1. Polylysine coat slides: assemble cytospin gaskets and chambers. Pipette 20 oL of Poly-D-lysine per chamber, coat 5 m, wash twice with water and then air-dry.
2. Cytospin: pipet 100-200 oL of fixed cell suspension (~10^5 cells) into each chamber and centrifuge for 4 min at 55g. Aspirate solution from chamber. Without allowing slides to dry, disassemble chamber and place in a Coplin jar containing permeabilization solution for 10 min.
3. Wash twice for 5 in each in 2 SSC at 45°C. Slides can be held in the last wash prior to denaturation.

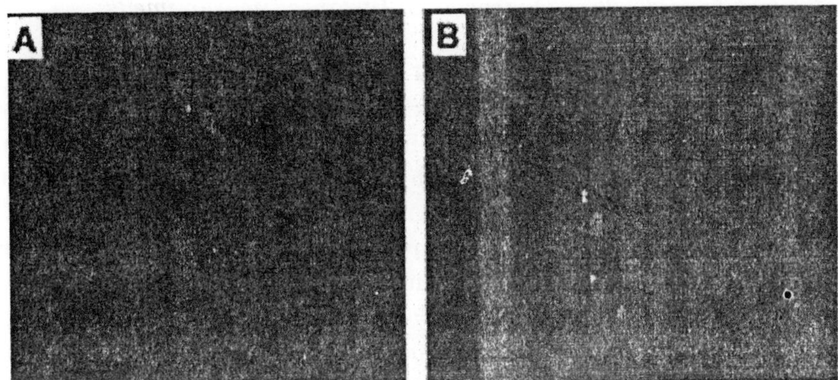

Fig. 11.2: Two-dimensional FISH detection of allele-specific replication. FISH detec-tion of allele-specific replication was performed on Inter-phase nuclei from a normal subject heterozygous for the chromosome IS satellite size polymorphism detected by digoxigenin-labelled D15Z1 (red fluorescence). The chromosome 15 with the larger D15Z1 signal is maternally derived and the shorter D15Z1 signal is paternal. Biotin-labelled plasmid DNA was simultaneously hybridized to the nuclei to detect replica-tion events (green fluorescence). Plasmid DNA signals appeared as either a singlet (unreplicated) or doublet (replicated) on each parental chromosome. Representative nuclei demonstrating the predominant parental allele specificity of asynchronous replication (singlet-doublet) are shown for two probes. (A) Probe p386 from the 5 end of GABRB3 displayed earlier replication on the paternal chromosome 15, as the doublet is associated with the shorter D15Z1. (B) In contrast, probe p438 from the GABRB3/ GABRA5 inter-genic domain detected earlier replication on the maternal chromosome 15 shown by the proximity of the doublet to the larger D15ZI signal.

Hybridization and Washing

1. Prepare probe.
2. Prepare denaturation solution.
3. Prepare postdenaturation washes (two of PBS/0.5% Tween, two of 2 SSC) in Coplin jars and chill on ice.
4. Denaturation: except first transfer slides to PBS/0.5% Tween and wash after denaturation.
5. Wash slides for 5 min each in a cold wash series. Do not air-dry. Slides may be held in the last wash.
6. Denature and preanneal probe.
7. Wipe excess liquid from around the cell sample, leaving 1-3 oL of fluid remaining over cells. Do not allow cells to dry.

8-19. Same as in Hybridization and working.

20. Seal coverslip with nail polish.
21. Store slides in covered slide box at 4°C.

Confocal Microscopy and Three-Dimensional Distance Measurements

Slides should be imaged with a confocal laser scanning microscope equipped with a 60 objective with 1.4 numerical aperture and appropriate excitation and emission to detect fluorophores. Clusters of 5-10 nuclei can be imaged in a single field if cytospin preparations are sufficiently dense [Fig. 11.3(A)]. Capture optical sections of 0.4 om z resolution through all focal planes containing FISH signals. Measure 3-D distances between FISH signals using appropriate software (such as Molecular Dynamics Image Space). For each nucleus in a data set of at least 50 nuclei, take six different measurements [Fig. 11.3(B)]. The two distances between homologous chromosomes (red to red and green to green) and four heterologous distances (red to green) are measured. Measurement data is graphed as described previously.

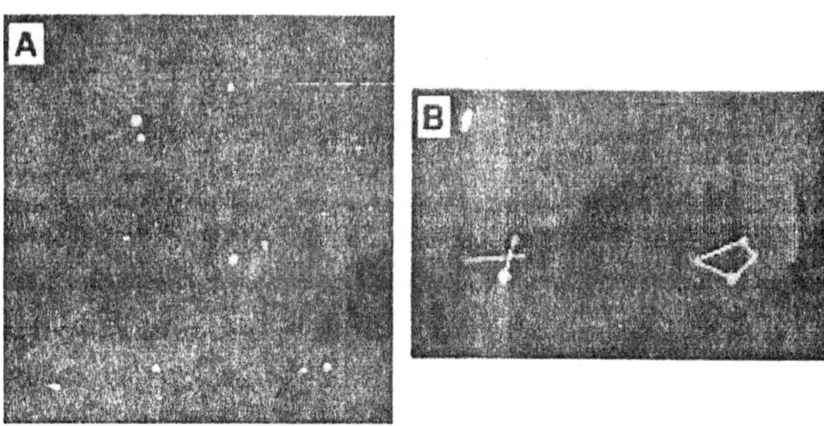

Fig. 11.3: Three-dimensional FISH detection of homologous association, (A) A field of densely distributed nuclei hybridized with D1SZ1 (green) and D12Z3 (red) was optically sectioned using a confocal laser scanning microscope. Images are three-dimensional projection images. (B) Three-dimensional distance measurements are determined using a mouse-driven cursor. Two homologous distances (red to red and green to green) and four heterologous distances are determined for each nucleus.

NOTES

1. If patient blood is limiting (less than 10 mL), the T-cell population in culture can be doubled by adding 5% Human T-stim (Collaborative Research) to the culture at 24-32 h and harvesting cells 4 d after PHA stimulation.

2. Fetal calf sera may be used in place of human sera, but the percentage of cells cycling will generally be lower.

3. Probes may be obtained commercially or labelled by nick translation or random priming methods. Probes may be labelled directly with fluorophores, but this is usually possible only for repetitive target sequence (centromeric satellite DNA). For replication studies,

target sequence should be single-copy, and the ideal probe size is generally 5-40 kb. Although larger probes are generally easier to detect, doublets may be difficult to distinguish with signals obtained from probes larger than 50 kb.

4. To control for pH variation, use MilliQ or double-distilled water for all solutions before and during hybridization. Posthybridization washes can be done in standard distilled water.

5. UV excitation with an argon-ion laser is required for cell sorting on the basis of Hoechst 33342 staining. For cell sorting we have used a Becton Dickonson FASC Caliber at 10 psi and a 70-om nozzle tip, collecting two fractions at a time. More recently, sorting times have been reduced by simultaneous four-fraction sorting on a Cytomation MoFlo run at 60 psi with a 75-om nozzle tip.

6. Slides should be examined under a phase-contrast light microscope following air drying of fixative. Interphase nuclei should be large and flat, with no cytoplasm on top. If cells are not well spread, try placing slides briefly on a warm surface for 1 min. If fixative dried too quickly and cells are not refractory to light, try breathing on the slides after dropping, to increase the humidity. These modifications are best practiced on more plentiful samples (unsorted lymphocytes), before dropping sorted fractions.

7. To avoid getting bubbles under the coverslip, hold the coverslip with forceps at a 45° angle adjacent to one side of the drop. Slowly lower the coverslip, allowing the drop's leading edge to exclude bubbls. If a few bubbles remain, they may be gently pushed to the edge of the coverslip with a pipet tip.

8. Hybridization efficiency of FISH probes should be tested on GI nuclei prior to analysis of replication events on at least 90% of the nuclei. If hybridization efficiency is low but background is also low, try increasing the amount of probe or decreasing wash stringency. If nuclear background is high, resulting in too many nonspecific signals, try reducing the amount of labelled probe, increasing wash stringency, or flattening nuclei further.

9. StatSpin Cytofuge 3-well cell concentrator gaskets and use one 4 mm 12 mm chamber closest to the middle of slide.

10. If probes labelled directly with fluorophores are used.

RT-PCR-BASED ANALYSIS

F1 hybrids resulting from intercrosses of inbred strains have provided an invaluable tool for the study of imprinting. The hybrids can be used to analyze parent-of-origin differences in expression of any gene, provided sequence differences exist between the two parental alleles. Methods used to detect allele-specific expression include ribonuclease protection assays and allele-specific RNA *in situ* hybridization, as well as a number of reverse transcriptase polymerase chain reaction (RT-PCR)-based assays. We describe here two such assays that are quantitative and require only single base differences between the two alleles.

Both assays rely on the amplification of the RNA of interest by RT-PCR using primer sets that flank the sequence polymorphism, a method shown previously to yield amplicons whose allelic ratio is proportional to the ratio in the starting material, regardless of the number of cycles of amplification. The first assay, single nucleotide primer extension (SNuPE), is adapted from a method first described for the detection of allelic variants in DNA. Briefly, it involves the following steps: (1) The gel-purified RT-PCR product is used as a template for the enzymatic extension by one radioactive nucleotide of a primer whose 3 terminus is adjacent to the position of the allelic difference. Each allele is probed separately, by use of the appropriate ^{32}P-labelled dNTP. (2) Following denaturing gel electrophoresis, the contribution of each allele is determined from the ratio of the two primer-extension products.

The method is highly reproducible, and allows the detection of an allelic variant in a 99-fold excess of the other. The second assay combines automated fluorescent-based sequencing of the gel-purified RT-PCR products with software allowing quantitative analysis of the signal from each allele. There have been previous reports of automated sequencing of allelic variants; the method as outlined here does not require specific tagged primers or a size difference between the two PCR products, and, rather than tracings, raw sequencing data are used to determine peak heights. PCR products differing as shown (two A/G mismatches and a G/T mismatch) were mixed in proportions ranging from 1:9 to 9:1, then subjected to automated sequencing with fluorescent-tagged dideoxynucleoside triphosphates.

The ratio of the peak heights varies with the input ratio; the relative proportion of a given allelic variant can thus be determined by comparison with a standard curve specific for the mismatch being measured. This rapid method has the following advantages: (1) As part of the automated sequencing reaction, one learns the identity and quality of the surrounding sequence, to verify the accuracy of measurement. (2) Where there are multiple mismatches, the allelic ratios obtained can be averaged. (3) The technique is rapid, nonradioactive, and amenable to the analysis of multiple samples. (4) The data appear top be as reproducible as the SNuPE assays, the major drawback being that, depending on the strength of the signal for a given base, ratios below 1:10 may be undetectable above background. Thus, in cases where detection of alleles expressed at relatively low levels is desired, the quantitative SNuPE assay is still preferable.

Analysis of Single Cells

Imprinting is not always an all-or-none process, in many cases varying with developmental state and tissue type. For this reason, it is sometimes desirable to analyze allele-specific expression in individual cells or cell types. While conditions for amplification of each gene must be determined empirically, we and others have found that even genes expressed at relatively low levels can be amplified from single cells, and describe here the method we used to amplify Ncam RNA from individual neurons.

Materials

Purification of RNA

1. Presiliconized RNase/DNase-free microcentrifuge tubes (0.65 mL).
2. RNAzol.
3. Mussel glycogen, 20 mg/mL in DEPC-treated water.
4. Chloroform.
5. 2-Propanol.
6. 75% ethanol.

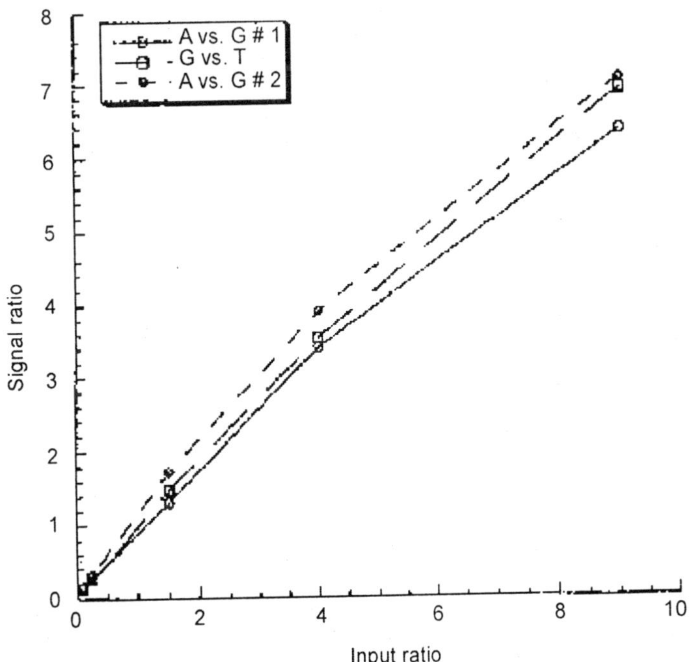

Fig. 11.4: Quantitative measurement of allele-specificity by direct sequencing. A region of the *Snrpn* gene containing two G/A mismatches and one G/T mismatch between *Mus spretus* and *Mus musculus* C57BL/6 DNA was amplified from DNA of the two strains. The PCR products were mixed in the proportions indicated (x axis), and the signal ratio determined by the combined use of automated sequencing and of a programme for peak analysis (y axis).

RT-PCR

1. DEPC-treated water.

2. 10 buffer: 100 mM Tris-HCl, pH 8.3, 500 mM KCl.
3. Magnesium chloride, 25 mM.
4. Upsteam and downstream gene-specific primers with T_m of ~60°C, 20 oM.
5. dNTPs (Boehringer Mannheim, Germany) diluted to 10 mM.
6. RNasin, 40 U/oL.
7. Moloney murine leukemia virus reverse transcriptase 200 U/oL.
8. AmpliTaq DNA Polymerase, 5 U/oL.
9. Mineral oil (not necessary if thermal cycler is equipped with a hot bonnet).

Gel Purification

1. 50 m*M* TBE buffer: Dilute from 1 M solution: 121.1 g Tris base, 51.4 g boric acid, 3.7 g disodium ethylenediaminetetraacetic acid (EDTA) in 1 L, pH 8.3.
2. Agarose, ultrapure.
3. 10 mg/mL ethidium bromide.
4. GeneClean.

SNuPE Assay

1. 10 buffer: 100 m*M* Tris-HCl, pH 8.3, 500 m*M* KCl, 20 m*M* $MgCl_2$, 0.01% gelatin (optimum pH and $MgCl_2$ concentration should be determined empirically for each primer).
2. SNuPE primer, 18-mer or longer, with 3 end just 5 to alelic base difference (20 o*M*).
3. [^{32}P]dNTPs corresponding to bases that differ, 3000 Ci/mmol, dilute to 1 oCi/oL in H_2O just before incubation.
4. 12% polyacrylamide-7 *M* urea gel in 100 m*M* TBE, pH 8.3.
5. Sequencing gel-loading buffer: 98% deionized formamide, 10 mM EDTA, pH 8.0, 0.025% xylene cyanol FF, 0.025% bromphenol blue.
6. AmpliTaq polymerase, 5 U/oL (Parkin Elmer).

Direct Sequencing

1. ABI Prism 377 DNA Sequencer (Perkin Elmer).
2. BigDye™ Terminator Cycle Sequencing Kits with AmpliTaq DNA Polymerase, FS, including dRhodamine Matrix Standards Kit and protocol # 4303237 (Perkin Elmer).
3. GeneScan 2.1.

Methods

Purification of RNA

1. Add 5 oL chloroform to each cell dissolved in 50 oL RNAzol in 0.65-mL microcentrifuge tube. Vortex vigorously. Keep on ice 5 min.
2. Centrifuge at 12,000g, 15 min, 4°C.
3. Transfer aqueous phase to fresh microcentrifuge tube.
4. Add 2 oL glycogen solution. Vortex.
5. Add an equal volume of 2-propanol.
6. Keep at -20°C overnight.
7. Centrifuge for 30 min at 12,000g 4°C.
8. Wash one time with 75% ethanol without mixing.
9. After decanting as much of ethanol as possible, air dry for 5 min.
10. Use sample immediately for RT-PCR.

RT-PCR

For each gene analyzed, a primer set should be selected that flanks the base(s) differing between alleles, and primers chosen that give a single strong band after amplification. The optimal concentration of $MgCl_2$ in both RT and PCR steps should be determined empirically for each primer set. In some cases the use of oligo($dT_{30}N_2$) or random primers in the RT step in place of the downstream primer improves the RT-PCR signal. Following is the procedure that we have used to amplify Ncam RNA from single mouse hippocampal neurons.

1. Prepare pre-RT master mix: Mix together (per sample) 2 oL 10 buffer, 4 oL $MgCl_2$, 2.5 oL downstream primer, and 6.5 oL DEPC-treated water. Suspend each sample in 15 oL of the mix and add mineral oil, 50 oL. Subsequent incubations are all done using a thermal cycler.
2. Incubate at 64°C for 10 min, then cool slowly at ~5 s/°C.
3. Prepare RT master mix: Mix together (per sample) 2 oL dNTPs, 0.25 oL MMLV-RT, 0.5 oL RNasin, and 2.25 oL DEPC-treated H_2O. Add 5 oL to each sample, and incubate at 42°C for 1 h.
4. Inactivate the MMLV RT at 99°C for 10 min, then keep samples at 55°C while preparing PCR master mix below (maximum time of incubation, 30 min).
5. Prepare PCR mix: Mix together (per sample) 8 µL 10×buffer, 0.5 µL Amplitaq polymerase, 2.5 µL upstream primer, and 69 µL H_2O. Add 80 µL to each sample.
6. Amplify using PTC-100 Thermal Cycler (MJ Research) programmed as follows: 95°C,

2 min, 1 cycle; 95°C, 1 min, N°C, 1 min, 72°C, 1.5 min, 2 cycles, where N is successively 60, 57, 54, 51, 49, and 47°C; 95°C, 1 min, 45°C, 1 min, 72°C, 1.5 min, 40 cycles; 95°C, 2 min, 72°C 5 min.

Gel Purification

1. Electrophorese each sample (up to 100 oL) on 1.5% agarose gel in 50 mM TBE.
2. Excise band from agarose gel using GeneClean, following manufacturer's instructions.
3. Elute DNA with water.
4. To determine the concentration of each sample, run serial twofold dilutions on an agarose gel stained with ethidium bromide (use 10 oL of 10-mg/mL ethidium bromide per 100 mL of agarose solution), and compare with known quantities of DNA.

SNuPE Assay

1. Dilute each sample to 2 ng/oL and distribute 5 oL (10 ng) to each of two microcentrifuge tubes.
2. Prepare master mix: Mix (per tube) 1 oL 10 buffer, 0.5 oL SNuPE primer, 0.15 oL AmpliTaq Polymerase, and 1.35 oL H_2O. Add 3 oL of master mix to each tube.
3. For each sample, add 2 oL (2 o Ci) of one allele-specific [^{32}P]dNTP into one tube, and 2 oL of the second [^{32}P]dNTP into another.
4. Incubate the samples in a thermal cycler for one round of denaturation, annealing, and synthesis: 95°C, 1 min, 42°c, 2 min, 72°C, 1 min, Place on ice.
5. As controls for background and maximum incorporation, incubate 10 ng of amplified product from each allele with the "incorrect" and appropriate [^{32}P]dNTP, respectively. Initially, controls should also include various ratios of the two alleles, mixed before amplification, to establish a standard curve.
6. After addition of sequencing gel-loading buffer (10 oL), incubate each sample at 90°C for 1-2 min, and place on ice. Load 10 oL of each sample on a 1-mm-thick 12% polyacrylamide-urea gel (18.5 cm long 16.5 cm wide) and electrophorese at 30 mA/ gel until the bromophenol blue reaches two-thirds of the way to the bottom of the gel (~40 min).
7. Cut out the portion of the gel between the bromphenol blue and xylene cyanol markers. Wrap the gel slice in Saran wrap, and determine the amount of radioactivity above lane background in each (n + 1) band (n = length of SNuPE primer) by use of a radioisotope scanning system, such as the PhosphorImager (Molecular Dynamics, Sunnyvale, CA).
8. For each sample, calculate the ratio of cpm incorporated with each of the two [^{32}P]dNTPs. Correct for the following: (1) background incorporation of the incorrect

[^{32}P]dNTP, calculated from the ratio of the signals seen when the control templates for each allele are incubated with the incorrect vs correct [^{32}P]dNTP—the ratio should be 1%; (2) different efficiency of incorporation with each [^{32}P]dNTP, determined from the ratio seen when equal amounts of each control template are incubated with the corresponding [^{32}P]dNTP.

Direct Sequencing

1. Perform automated DNA sequencing using 10 ng of PCR product for each 100 bp in length, and the protocol supplied with the BigDyeTM Terminator Cycle Sequencing Kit. Use dRhodamine Matrix Standards and filter E when performing acrylamide gel electrophoresis.
2. For each sample, open the corresponding sequencing file in the GeneScan program.
3. In the Settings menu, under Preferences, vary (1) Results Dye Scales, to optimize ability to select minor peaks, and (2) Dye Indicators, so that the colour correspondence of the bases on the screen is the same as that of the sequencing program. This window also provides a reference for the default colour correspondence used by GeneScan in tabulating peak heights of the bases. Then, under Auto-Analysis Defaults, select GS2500.1 as the size standard, and R as the dye.
4. In the sample menu, under Install New Matrix, open the dRhodamine matrix (this must be done each time a sample is analyzed). Then select Analyze Sample.
5. Once the results of quantitative analysis appear in the window, use the magnification tool to locate the base mismatch(es) of interest. Select the bases, then in the View menu, select View Only Selected Rows.
6. Copy resulting table to Microsoft Excel or other spreadsheet.
7. To determine allelic ratio, compare ratio of peak heights to standard curve obtained.

NOTES

1. While overnight precipitation in 2-propanol has given in most consistent results, in some cases a 2-h incubation at –20°C is sufficient.
2. Use of variants of AmpliTaq Polymerase such as AmpliTaq Gold (Perkin Elmer) and HotStarTaq together with Q-Solution (Qiagen) often result in increased RT-PCR signals. While we have not yet confirmed quantitative allele-specific amplification from single cells using these polymerases, they may ultimately prove to be superior to AmpliTaq.
3. It is best to use [^{32}P]dNTPs within several days of arrival, ad to store templates and SNuPE primers at 4°C or frozen in aliquots, to avoid repeated freezing and thawing.

4. The sensitivity of the SNuPE assay depends on the particular mismatch. While C/A can be measured down to the 0.001% level, G/C, C/G, and G/A mismatches have a higher limit of ~1%. C/T, T/C, and A/G mismatches give backgrounds similar to C/A.

5. When the base 3 to the incorporated [^{32}P]dNTP is identical to it, (n+2) bands of varying intensity will be obtained, where n is the length of the SNuPE primer. These products should be included in the calculation along with the (n+1) products.

6. Occasional minor bands of incorrect size may be seen. These can be ignored as long as control experiments show that the expected standard curve is obtained by calculations using only the SNuPE product of the expected size(s).

Chapter 12

Computational Genome

When comparing entire genomes across species, we want to compute a form of *distance* between them, but we want to use a set of basic operations. We are no longer interested in point mutations such as substitutions, deletions, and insertions, but rather in larger mutations. Pieces of chromosomes can be exchanged in various ways, affecting sections consisting of a significant percentage of the chromosome. A piece can be moved or copied to another location, or it can leave one chromosome and land in another. Such movements are called **genome rearrangements**. In some cases, for instance, when studying chloroplast or mithchondrial genomes, a single type of rearrangement event—the *reversal*—appears to be almost the only way to effect evolutionary change.

It makes sense in these cases to measure the evolutionary distance between two genome organizations by the number of reversals needed to transform one of them into the other. As an example, consider the chloroplast genomes of alfafa and pea, two related plant species. There each labelled arrow denotes a *block.* A block here is a section of the genome possibly containing more than one gene, which is transcribed as a unit. The arrow denotes the fact that blocks have orientation. The reason that blocks are oriented has to do with the particular strand from where transcription takes place.

Two blocks in different genomes have the same numerical label if they are *homologous,* that is, if they contain the same genes. Notice finally that the distances between genes are

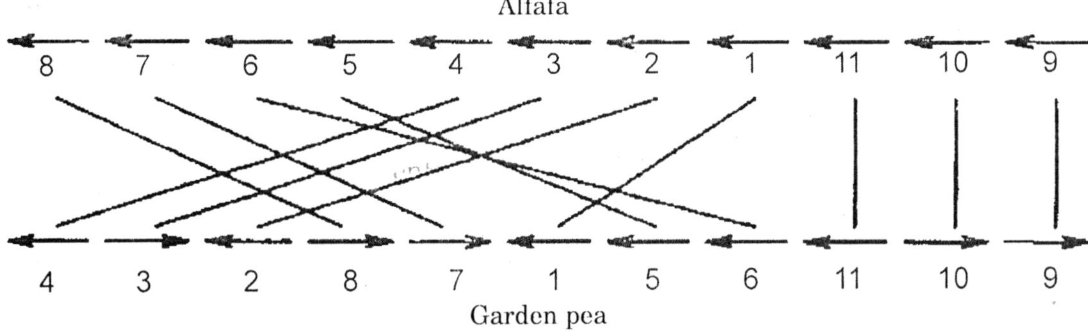

Fig. 12.1: Relation between the chloroplast genomes of alfafa and garden pea.

not always the same in both genomes. The reversal operation for oriented blocks can be defined as follows. It operates in a contiguous segment of blocks. It inverts the order of the affected blocks and also flips their arrows. Given this definition, we are interested in the following question: What is the minimum number of reversals that leads from one genome to the other? The first line represents the alfafa genome, the last line represents the garden pea genome, and each line except the first is obtained from the previous one by a reversal involving the underlined blocks.

But, how do we know that no shorter series of reversals exists to accomplish the required transformation? We would like to have a polynomial-time algorithm to give us such an answer, and that is precisely the topic of the this chapter. The reason that we want a shortest possible series of reversals is the same *parsimony assumption* given in other sections of this book that deal with minimization problems. We assume that Nature always finds paths that require a minimum of change, and therefore, if we want to know how alfafa could have changed into pea or *vice versa*, we should try to find a minimum series of reversals that accomplishes this task. A similar rearrangement problem without the orientation can be conceived.

Sometimes the information that we have about genes is not accurate enough to ascertain their relative orientation, but it does give out their relative order. In this case, we would look for a shortest series of reversals that transforms one block order into the other, without regard to orientation. For instance, suppose that we did not know the orientations of the blocks. In that case a series of three reversals would be enough to solve the problem. The next sections explore both versions of the problem, which is known as sorting by reversals.

$$\overleftarrow{8}\ \overleftarrow{7}\ \overleftarrow{6}\ \overleftarrow{5}\ \overleftarrow{4}\ \underline{\overleftarrow{3}}\ \overleftarrow{2}\ \overleftarrow{1}\ \overleftarrow{11}\ \overleftarrow{10}\ \overleftarrow{9}$$

$$\overleftarrow{8}\ \underline{\overleftarrow{7}\ \overleftarrow{6}\ \overleftarrow{5}\ \overleftarrow{4}\ \overrightarrow{3}\ \overleftarrow{2}}\ \overleftarrow{1}\ \overleftarrow{11}\ \overleftarrow{10}\ \overleftarrow{9}$$

$$\overleftarrow{8}\ \overrightarrow{2}\ \overleftarrow{3}\ \overrightarrow{4}\ \overrightarrow{5}\ \underline{\overrightarrow{6}\ \overrightarrow{7}\ \overleftarrow{1}}\ \overleftarrow{11}\ \overleftarrow{10}\ \overleftarrow{9}$$

$$\underline{\overleftarrow{8}\ \overrightarrow{2}\ \overleftarrow{3}\ \overrightarrow{4}}\ \overrightarrow{5}\ \overrightarrow{1}\ \overleftarrow{7}\ \overleftarrow{6}\ \overleftarrow{11}\ \overleftarrow{10}\ \overleftarrow{9}$$

$$\overleftarrow{4}\ \overrightarrow{3}\ \overleftarrow{2}\ \overrightarrow{8}\ \underline{\overrightarrow{5}\ \overrightarrow{1}\ \overleftarrow{7}}\ \overleftarrow{6}\ \overleftarrow{11}\ \overleftarrow{10}\ \overleftarrow{9}$$

$$\overleftarrow{4}\ \overrightarrow{3}\ \overleftarrow{2}\ \overrightarrow{8}\ \overrightarrow{7}\ \overleftarrow{1}\ \overleftarrow{5}\ \overleftarrow{6}\ \overleftarrow{11}\ \underline{\overleftarrow{10}}\ \overleftarrow{9}$$

$$\overleftarrow{4}\ \overrightarrow{3}\ \overleftarrow{2}\ \overrightarrow{8}\ \overrightarrow{7}\ \overleftarrow{1}\ \overleftarrow{5}\ \overleftarrow{6}\ \overleftarrow{11}\ \overrightarrow{10}\ \underline{\overleftarrow{9}}$$

$$\overleftarrow{4}\ \overrightarrow{3}\ \overleftarrow{2}\ \overrightarrow{8}\ \overrightarrow{7}\ \overleftarrow{1}\ \overleftarrow{5}\ \overleftarrow{6}\ \overleftarrow{11}\ \overrightarrow{10}\ \overrightarrow{9}$$

Fig. 12.2: A solution to the problem of Figure 12.1.

8	7	6	5	4	3	2	1	11	10	9
8	7	1	2	3	4	5	6	11	10	9
4	3	2	1	7	8	5	6	11	10	9
4	3	2	8	7	1	5	6	11	10	9

Fig. 12.3: A solution to the same problem, but with unoriented blocks.

It is interesting to note that while there is a polynomial-time algorithm to find a minimum series of reversals in the oriented case, the unoriented version has been shown to be NP-hard.

BREAK POINTS

We start the study of the oriented version of the problem with an example. Suppose we have two genomes divided in homologous blocks. Each of these blocks represents a contiguous segment that is well conserved in both genomes. In this example, a series of three reversals does the job and is the shortest possible as we show below. There are many other series that work, but all of them have three or more reversals. Figure 12.5 displays one possible solution, with labels on each block showing the action of reversals. The arrows above the labels indicate the orientation of homologous blocks. Because the labels are arbitrary, we use the numerals from 1 to n (n=5 in this example) and assign them in the natural order to the upper genome. With this, the labels in the lower genome are uniquely determined.

Each block gets the same number as its homologous block, and the arrows reflect the actual orientations. In the figure, reversals are indicated by a line running under the affected blocks. The initial permutation is changed step by step to the final arrangement. How do we know that no shorter series of reversals exists that accomplishes the required transformation? In the case of this particular setting there is an easy argument that confirms that less than three reversals are not enough. The argument relies on the concept of a

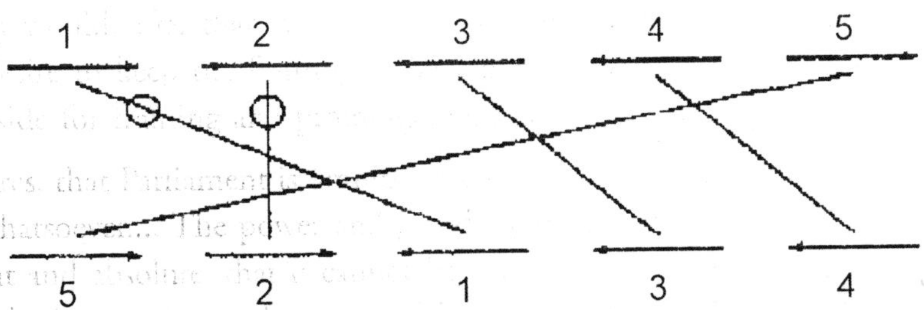

Fig. 12.4: Two genomes with homologous blocks. Arrows represent the blocks in each genome. Homologous blocks are linked by lines. A small circle around a line indicates that the block has different orientations in the two genomes.

$\vec{1}$ $\overleftarrow{2}$ $\overleftarrow{3}$ $\overleftarrow{4}$ $\vec{5}$

$\vec{1}$ $\overleftarrow{2}$ $\overleftarrow{3}$ $\vec{4}$ $\vec{3}$

$\vec{1}$ $\overleftarrow{2}$ $\overleftarrow{3}$ $\overleftarrow{3}$ $\overleftarrow{4}$

$\vec{5}$ $\vec{2}$ $\overleftarrow{1}$ $\overleftarrow{3}$ $\overleftarrow{4}$

Fig. 12.5: A solution to the problem of the previous figure. Each line is obtained from the previous one by a reversal involving the underlined elements.

• $\vec{1}$ $\overleftarrow{2}$ • $\overleftarrow{3}$ $\overleftarrow{4}$ • $\vec{5}$ •

Fig. 12.6: Each • is a breakpoint.

breakpoint. A breakpoint is a point between consecutive labels in the initial permutation that must necessarily be separated by at least one reversal in order to reach the target permutation.

In other words, these two consecutive labels are not consecutive in the target, or, even if they are, the orientations are not the same in a relative sense. In addition, a breakpoint exists before the first label if it is not the first label—with the same orientation—in the target. Figure 12.6 shows the breakpoints for our initial permutation relative to the target in Fig. 12.4. Each breakpoint is indicated by a bullet (•). For instance, the location between labels $\overleftarrow{2}$ and $\overleftarrow{3}$ is a breakpoint because in the target blocks $\overleftarrow{2}$ and $\overleftarrow{3}$ are not consecutive. If no reversal separates these labels, they will remain together until the end and will be then in the wrong order. In contrast, $\vec{1}$, $\overleftarrow{2}$ is not a breakpoint because this is exactly by way these two blocks appear in the target. Our initial permutation has four breakpoints altogether. A reversal can remove at most two breakpoints, because it cuts the permutation in exactly two locations.

The target permutation has zero breakpoints and it is the only permutation with this property. Hence we need at least two reversals to fix the four break-points in our example. But a more careful inspection reversals to fix the four break-points in our example. But a more careful inspection reveals that no single reversal can remove two breakpoints from our initial permutation. Indeed, to remove the first break-point, the one before 1, we would have to move $\vec{5}$ to the first position, and even this is not enough. The only reversal that can bring $\vec{1}$ there also flips its arrow, and the resulting permutation still has a breakpoint before the first position. The label is in its correct position but the orientation is wrong. A similar argument convinces us that there is no way of removing the last breakpoint with one reversal.

Our only hope of removing two breakpoints relies on the two central ones. However, the only reversal affecting these two locations does not remove any breakpoint! So, we know now that the first reversal will decrease the number of break-points by at most 1. Then we will have at least three breakpoints left, and another two reversals are needed, for a total of at least three reversals. This shows that the solution presented earlier is optimal.

Permutation

Genome rearrangements is a topic that requires us to be very careful in our definitions and to employ a coherent notation. Let us first define what an oriented permutation is. Given a finite set L of *labels*, of oriented labels from *L*, where

$$\mathcal{L}^0 = \bigcup_{a \in \mathcal{L}} \{\overleftarrow{a}, \overrightarrow{a}\}.$$

$2\,\overrightarrow{6}\,\overleftarrow{1}$

The set $\mathcal{L}^0$ has twice as many elements as $\mathcal{L}$. For x ∈ $\mathcal{L}^0$, denote by $|x|$ the element of $\mathcal{L}$ obtained by removing x's arrow. Thus, $|\overleftarrow{a}|$ $|\overrightarrow{a}|$ = a for any a ∈ $\mathcal{L}$. Let n be the number of elements of $\mathcal{L}$. An *oriented permutation* over $\mathcal{L}$ is a mapping α: [1 . . . n] → $\mathcal{L}^0$ such that for any label $a \in \mathcal{L}$ there is exactly one $i \in$ [1 . . . n] with |α(i)| = a. Informally, a permutation "picks" for each index i an oriented label $\overleftarrow{a}$ or $\overrightarrow{a}$ from $\mathcal{L}^0$, but if $\overleftarrow{a}$ is picked for some index, $\overrightarrow{a}$ cannot be picked by any other index and *vice versa*. There being n = $|\mathcal{L}|$ indices guarantees that for each a ∈ $\mathcal{L}$ either a or a will be picked exactly once. We represent a permutation a as a list of the elements α(1), α(2), . . . , α(n). Usually we just list the elements separated by blanks as in the previous section, but sometimes we use commas and parentheses to avoid ambiguities. In most of cases our set $\mathcal{L}$ will be just the first n positive integers: 1, 2, . . ., n. Thus, for instance, for n = 6, the following is an example of a permutation over $\mathcal{L}$ = {1, 2, . . ., 6}:

$$\alpha = (\overleftarrow{2}, \overleftarrow{3}, \overrightarrow{1}, \overrightarrow{6}, \overleftarrow{3}, \overleftarrow{4}). \qquad (12.1)$$

So, in the above example, α(1) = 2, α(2) = 3, and so on. The *identity* permutation is the permutation I such that I (i) = *i* for al *i* between 1 and n.

A reversal is an operation that transforms one permutation into another by reversing the order of a contiguous portion of it, and at the same time flipping the signs of the elements involved. Let i and j be two indices with 1 ≤ i, j ≤ n. A reversal affecting elements α(*i*) through α(*j*) is indicated by [*i, j*]. The reversal [*i, j*] is actually a mapping from {1, 2, . . ., *n*} into itself, and it can be *composed* with an oriented permutation α to give another mapping from {1, 2, . . ., n} to $\mathcal{L}^0$, that is, another permutation. The *composition* of mappings α and ρ, denoted by αρ, is the mapping defined by

$$(\alpha\rho)\ (k) = \alpha(\rho(k)).$$

For instance, applying the reversal [2, 4] to the permutation α of Equation (12.1) results in

$$\alpha[2, 4] = (\overleftarrow{2}, \overleftarrow{6}, \overleftarrow{1}, \overrightarrow{3}, \overleftarrow{5}, \overleftarrow{4}).$$

In general, a reversal ρ = [*i, j*] changes a permutation α into αρ defined by

$$\alpha[i, j](k) = \begin{cases} \overline{a(i + j - k)} \\ \alpha(k) \end{cases} \quad \textit{if } i \le k \le j \textit{ otherwise},$$

where *l* is label *l* with its arrow flipped. There are *n*(*n* + 1)/2 different reversals in a setting involving n blocks. This includes the unitary reversals [*i, i*] that just flip the arrow of α(*i*).

Our main interest in this section is in the problem of *sorting by reversals.* Given two oriented permutations *α* and *β* over the same label set $\mathcal{L}$, we seek the minimum number of reversals that will transform α into β. In other words, using the notation just introduced, we seek a series of reversals $\rho_1, \rho_2, \ldots, \rho_t$, with as few reversals as possible, such that

$$\alpha\rho_1\rho_2 \ldots \rho_t = \beta.$$

The number *t* is called the *reversal distance* of α with respect to β and is denoted by dβ(α). Oriented permutation α is called the *initial* permutation and β is the *target.* This process is called *sorting* α *with respect to* β.

Notice that while t = dβ(α) is fixed for a given pair α, β, the actual series of reversals that sort α is generally not unique. Usually we will be interested in finding a reversal ρ that "makes progress toward β. Such a reversal is called a sorting reversal of α with respect to β; it is characterized by the property

$$d_\beta(\alpha\rho) < d_\beta(\alpha),$$

which is equivalent to

$$d_\beta(\alpha\rho) = d_\beta(\alpha) - 1$$

because the distances of α and αρ cannot differ by more than one.

It is important to observe that the reversal distance is symmetric, that is, dβ(α) = dα(β). This stems from the fact that any reversal has no net effect if applied twice. In other words, for every reversal ρ, $\rho\rho$ has no effect. Therefore, if

$$\alpha\rho_1\rho_2 \ldots \rho_1 = \beta,$$

then, composing both sides of this equation successively with ρ_1, r_{t-1}, . . ., ρ_1, we get

$$\alpha = \beta\rho_t\rho_{t-1} \ldots \rho_1.$$

We conclude that the same reversals transform β into α, but they have to be applied in reverse order. Thus, the sorting reversals of β with respect to α are not necessarily the same as the sorting reversals of α with respect to β.

We will abbreviate dβ(α) to just d(α) when the target β is clear from the context. Likewise, if this is the case, we will omit β when talking about sorting reversals, break-points, and so on.

Breakpoints

Let us now formalize the concept of a breakpoint. Given a permutation α, the first thing we have to do in order to define breakpoints is to consider the *extended* version of α, which is obtained from α by adding a new, artificial label $\mathcal{L}$ (for left) before the first real label α (1) and also another new label R (for *right*) after the last label α(n). Thus, for the premutation α of Equation (12.1), the extended version is

$$(L, \overleftarrow{2}, \overleftarrow{3}, \overrightarrow{1}, \overrightarrow{6}, \overleftarrow{5}, \overleftarrow{4}, R).$$

A *breakpoint* of α with respect to β is a pair x, y of elements of $\mathcal{L}^0$ such that xy appears in the extended version of α but neither xy nor the reverse pair $\overline{x}\overline{y}$ appear in extended β. For instance, if α is the permutation in Eq. (12.2) and β is the identity, then break-points are L $\overleftarrow{2}$, $\overleftarrow{2}$, $\overrightarrow{3}$, $\overleftarrow{3}\ \overrightarrow{1}$, $\overrightarrow{1}\ \overrightarrow{6}$, $\overrightarrow{6}\ \overleftarrow{5}$, and $\overleftarrow{4}$ R. The pair $\overleftarrow{5}\ \overleftarrow{4}$ is *not* a breakpoint, because its reverse pair $\overrightarrow{4}\overrightarrow{5}$ appears in β.

The number of breakpoints of a permutation α is denoted by $b_\beta(\alpha)$ or just by $b(\alpha)$ if β is clear from the context. In our example, $b(\alpha) = 6$.

Notice that artificial labels L and R never more. So, L is involved in a breakpoint only if the first labels of α and β are different, or are equal but with different orientations. An analogous observation holds for *R*.

A Lower Bound

Let β be fixed. As we mentioned earlier, a reversal can remove at most two breakpoints from a permutation. In our notation,

$$b(\alpha) - b(\alpha\pi) \leq 2, \tag{12.3}$$

for any permutation α and any reversal ρ. This observation leads immediately to a lower bound on the reversal distance as follows. Let $\rho_1, \rho_2, \ldots, \rho_t$ be a series of reversals (not necessarily optimal) that converts α into β. Then

$$\alpha\rho_1\rho_2 \ldots \rho_t = \beta,$$

so

$$b(\alpha\rho_1\rho_2 \ldots \rho_t) = b(\beta) = 0. \tag{12.4}$$

Notice that β has zero breakpoint with respect to itself. But, by Equation (12.3)

$$\begin{array}{lcll}
b(\alpha) & - & b(\alpha\rho_1) & \leq 2, \\
b(\alpha\rho_1) & - & b(\alpha\rho_1\rho_2) & \leq 2, \\
 & & & \vdots \\
b(\alpha\rho_1 \ldots \rho_{t-1}) & - & b(\alpha\rho_1 \ldots \rho_t) & \leq 2.
\end{array}$$

Adding these inequalities, and taking Equation (12.4) into account, we get

$$b(\alpha) < 2t.$$

This is true for any sorting of α. Now if we apply this result to a shortest series, then $t = d(\alpha)$ and we have

$$\frac{b(\alpha)}{2} \leq d(\alpha).$$

This is a lower bound for d(α), that is, a quantity that specifies a minimum value under which we can be sure d(α) will not lie. If we ever find a series of reversals with the size of a lower bound for a particular permutation α, we know we have found a solution in this case. However, this lower bound is not always achievable. For instance, we saw at the beginning of Break Point section described above. A pair of permutations α and β with $b_\beta(\alpha) = 4$, hence the lower bound is 2, but $d_\beta(\alpha) = 3$.

A surprising fact in this context is that there are reversals that remove two break-points but are not sorting reversals, that is, they lead us in the wrong way as far as sorting α is concerned.

Reality and Desire

The lower bound $d(\alpha) \geq b(\alpha)/2$ is not very tight. For a significant portion of the permutations it is far from being an equality. In this section we derive a better lower bound, based on a structure called the reality-desire diagram of a permutation with respect to another, which will be defined shortly.

To acquire some intuition about this structure, let us try to determine what makes a reversal remove breakpoints from α. A general situation is depicted below, where the reversal is indicated by the two vertical lines:

. . . x | y . . . z | w . . ..

After the indicated reversal is performed, we end up with the sequence

. . . x $\bar{z}$ . . . $\bar{y}$ w . . .,

where, recalling from above, $\bar{l}$ means the label l with flipped arrow. Assuming that the breakpoint xy was removed, we have that either $x\bar{z}$ is in β or $z\bar{x}$ is in β. However, xy was a breakpoint, so xy and $\overline{yx}$ do not appear in β, although xy appears in α.

When we have a breakpoint, it is caused by two labels that are adjacent in α but that do not "want" to be adjacent. They want to be adjacent to other labels. Thus, we observe a deviation between the *reality,* that is, the neighbour a certain label has in α, and its *desire*, that is, the neighbour this same label has in β. Actually labels are oriented entities, so they have a specific desire for each one of their ends. An oriented label can be viewed as a battery with charged terminals indicated by positive and negative signs (Fig. 12.7).

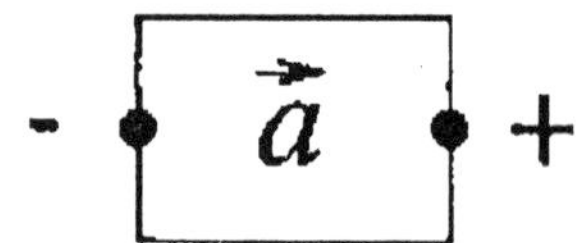

Fig. 12.7: Oriented label as battery. The positive terminal is always at the tip of the arrow, while the negative terminal is at the tail.

Let us then analyze the particular reality and desire situation in the situation above. We draw each label as a box. In addition, we draw "reality" lines linking label ends that are in contact in the present permutation, and we draw "desire" lines linking label ends that will be in contact in the target permutation. Notice that the removal of a breakpoint resulted in a desire line linking the same terminals: that is, desire became reality after this reversal. In this particular case we do not know which terminals are positive or negative, but this is not the most relevant issue. The above considerations motivate us to construct the **reality** and **desire diagram** of a permutation α with respect to β, denoted $RD_\beta(\alpha)$, as follows.

We want to represent each label by two nodes, so that we can keep track of the orientation of the labels. For each label $l. \in \mathcal{L}$ we include two noes $\overleftarrow{l}$ and $\overrightarrow{l}$ in $RD(\alpha)$. The note $\overleftarrow{l}$ represents the tail of the arrow and $\overrightarrow{l}$ is the head. We also include two extra nodes L and R to be able to record the reality and desire of end labels. This is the same for all permutations.

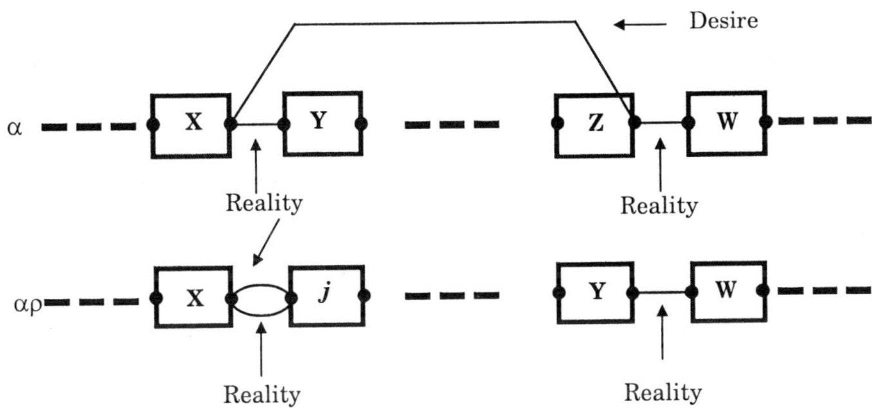

Fig. 12.8: Reality and desire lines in breakpoint removal.

The diagram $RD(\alpha)$ has two types of lines or *edges:* the reality edges and the desire edges. Reality edges connect terminals of consecutive labels in permutation α. Also, L is connected to the first terminal and the last terminal is linked to R. Desire edges connect terminals according to permutation β. Finally we arrange all terminal nodes around a circle, with L and R occupying the top two positions, L to the left of R, and the other nodes following α counterclockwise. Figure 12.9 displays a step-by-step construction of the reality and desire diagram of the permutation $\vec{3}\,\overleftarrow{2}\,\overleftarrow{1}\,\vec{4}\,\overleftarrow{5}$ with respect to the identity.

Figure 12.10 shows the diagram arranged in a circle. Reality edges go along the circumference. Desire edges are chords. Notice that sometimes there are two parallel edges, one of type reality and one of type desire, between the same pair of vertices. This occurs exactly in the points that are *not* breakpoints. Reality equals desire there. On the other hand, there are no self-loops. The reality and desire diagram is sometimes called *graph of reality and desire.* We prefer to use the term *diagram* to stress the fact that we draw it in a particular way.

A graph can be drawn in any way as long as the vertices are correctly labelled and the edges make the correct connections. But our diagram RD(α) must be drawn in such a way as to follow the order of labels in the permutation α. This will be particularly important when we talk about crossing desire edges in some of the later sections. In spite of that, we will use the standard graph terminology when referring to the diagram, with the obvious meanings. An important property for us is that every node has degree two in RD(α).

This is because both A, the set of reality edges, and B, the set of desire edges, are *matchings* in RD(α). This means that each node is incident to exactly one edge from A and exactly one from B. The union A $\cup$ B forms a number of connected components, each one being a cycle with edges alternating between A and B. Furthermore, each such cycle has an even number of edges, half of them reality edges and half of them desire edges. The number of cycles of RD(α) is denoted by $c_\beta(\alpha)$, or just $c(\alpha)$ if β is implicit. Then note that $c_\beta(\beta)$ = n+1, because β has no breakpoints, so all the cycles are composed by two parallel edges between the same pair of nodes. Because there are 2n+2 nodes, we have n+1 cycles. Furthermore this is the only permutation for which $c_\beta(\alpha)$=n+1.

So, in a way, we can see the process of sorting a permutation as a process of transforming RD(α) into a graph with as many cycles as possible. That does not necessarily mean that every reversal that increases the number of cycles is good—see Exercise 6. However, a very natural question in this context is, How does a reversal affect the cycles in *RD*(α)? This is what the following result tells us. Observe first that a reversal is characterized by the two points where it "cuts" the current permutation, which correspond each to a reality edge.

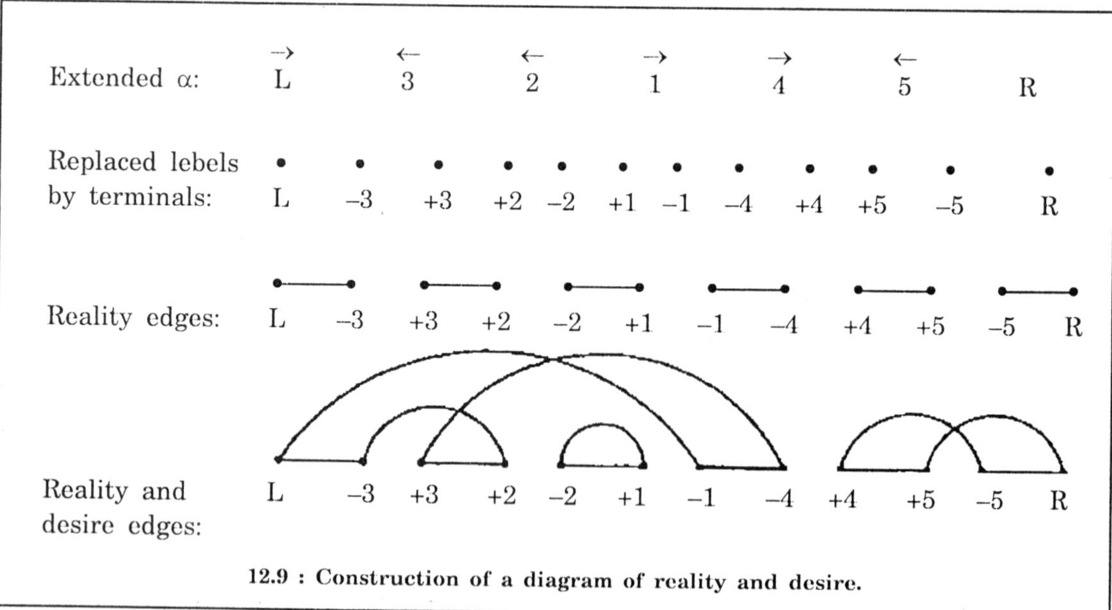

12.9 : Construction of a diagram of reality and desire.

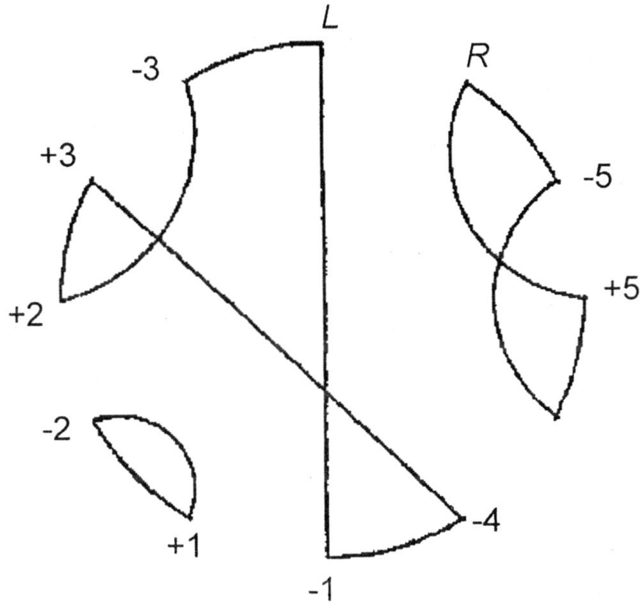

Fig. 12.10: Diagram of reality and desire.

Theorem 12.1: Let (s, *t*) and (*u*, *v*) be two reality edges characterizing a reversal ρ, with (*s*, *t*) preceding (*u*, *v*) in the permutation α. Then *RD*(αρ) differs from *RD*(α) as follows.

1. Reality edges (*s*, *t*) and (*u*, *v*) are replaced by (*s*, *u*) and (*t*, *v*).
2. Desire edges remain unchanged.
3. The section of the circle going from note t to node u, including these extremities, in counterclockwise direction, is reversed.

Proof: The transformation on the reality edges is the same as already discussed and the only difference is that here we are dealing with terminals instead of labels. In any case, no other reality edges are affected. Desire edges are not changed because they depend only on the target permutation. Finally, the reversal of terminals is a direct consequence of the action of a reversal on labels.

The following result tells us how the number of cycles is affected by a reversal. We need another important definition before stating and proving it. Let *e* and *f* be two reality edges belonging to the same cycle in RD(α). Think for a moment of both e and f as being oriented counterclockwise in the diagram. Each one of them includes an orientation of the common cycle. If the orientations induced by e and f coincide, we say that they are *convergent* or that they *converge;* otherwise we say they are *divergent*, or that they *diverge*.

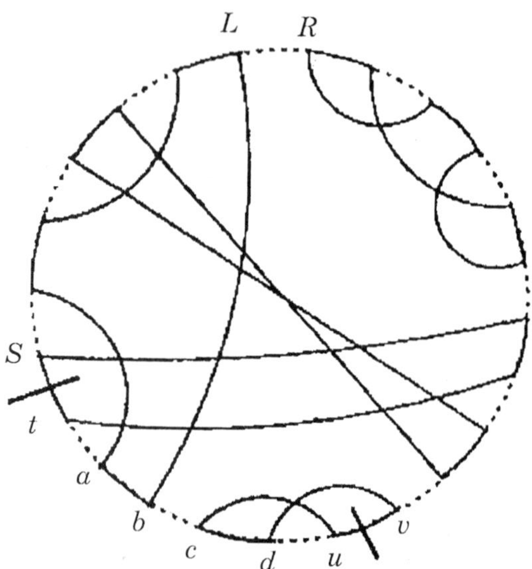

Fig. 12.11: **Reality and desire diagram before reversal indicated by two reality lines.**

Every time we need to think of reality edges as directed, we write them as an ordered pair. The first element is the tail and the second is the head of the edge. The orientation or reality edges is important to determine good cycles, as we will see below.

Theorem 12.2 : Let ρ be a reversal acting on two reality edges e and f of $RD(\alpha)$. Then

1. If e and f belong to different cycles, $c(\alpha\rho) = c(\alpha) - 1$.
2. If e and f belong to the same cycle and converge, then $c(\alpha\rho) = c(\alpha)$.
3. If e and f belong to the same cycle and diverge, then $c(\alpha\rho) = c(\alpha) + 1$.

Proof. Figure 12.13 will help to clarify the argument. The new edges in $RD(\alpha\rho)$ are indicated by dashed arrows there. Edges e and f are oriented as described above. Notice that we did not draw the cycles as part of the diagram, because here all that we need from $RD(\alpha)$ are the orientations on e and f.

As mentioned earlier, only the cycles containing e and f can be affected, as far as the number of cycles is concerned. The other cycles will not change as connected components. Let us then discuss each case separately.

If e and f belong to different cycles, these cycles are joined into a single one. Thus, the number of cycles decreases by one. If e and f converge, the new edges are such that the cycle remains connected. Thus the number of cycles remains the same. If e and f diverge, the cycle gets broken into two pieces, causing an increase by one in the number of cycles.

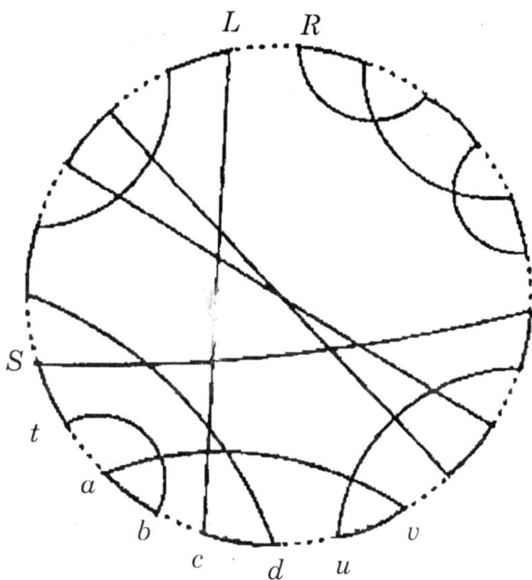

Fig. 12.12: Reality and desire diagram of permutation obtained from Fig. 12.11 after the indicated reversal.

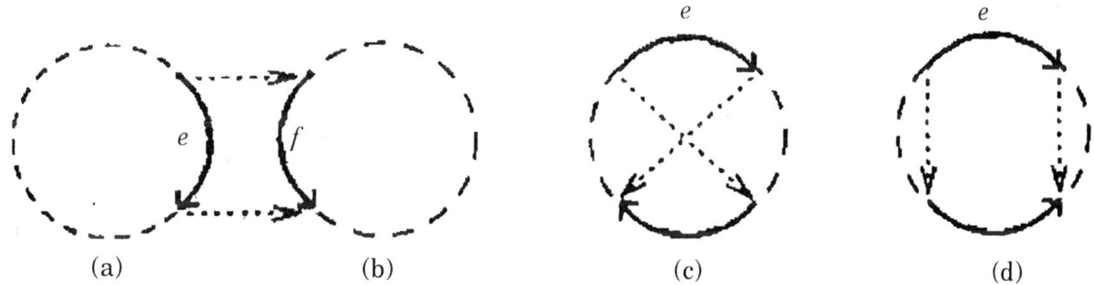

Fig. 12.13: The effect of a reversal in the cycles of a reality-desire graph: (a) edges from different cycles, (b) convergent edges, and (c) divergent edges.

Theorem 12.2 has an important consequence: The number of cycles changes by at most one with each reversal. This leads to another lower bound for the reversal distance.

Suppose that a series of reversals $\rho_1, \rho_2, \ldots, \rho_t$ takes α to β, that is,

$\alpha\rho_1\rho_2 \ldots \rho_t = \beta$.

Computing the number of cycles, we get

$$c(\alpha\rho_1\rho_2 \ldots \rho_t) = c(\beta) = n+1. \qquad (12.5)$$

On the other hand, by Theorem 12.2, we have

$$
\begin{aligned}
c(\alpha\rho_1) &- c(\alpha) &< 1,\\
c(\alpha\rho_1\rho_2) &- c(\alpha\rho_1) &< 1,\\
&\vdots\\
c(\alpha\rho_1 \ldots \rho_t) &- c(\alpha\rho_1 \ldots \rho_{t-1}) &\leq 1.
\end{aligned}
$$

Adding all this up and canceling terms, we get

$$n + 1 - c(\alpha) \leq t.$$

If $\rho_1, \rho_2, \ldots, \rho_t$ is an optimal sorting, then $t = d(\alpha)$ and

$$n + 1 - c(\alpha) < d(\alpha). \tag{12.6}$$

This lower bound is very good and for most oriented permutations, it comes very close to the actual distance. The reason that it does not always work is explained in the following sections.

Interleaving Graph

We can classify the cycles of $RD(\alpha)$ as *good* or *bad* in the following sense. A cycle is good if it has two divergent reality edges; otherwise it is called bad. Good cycles can be spotted in the reality and desire diagram because they have at least two desire edges that cross. We must be careful, however, because not all cycles with crossing desire edges are good. The presence of crossing edges is just an indication that the cycle might be good. When in doubt, we should check directly the definition; that is, we should look for divergent reality edges.

The characterization of crossing edges relies strongly on the arrangement of the diagram around a circle. If we were allowed to draw the graph any way we wished, we could make edges cross or not at will. But in our diagrams desire edges must be chords, and although we sometimes draw them a bit curved when they would be too close to the circumference, their crossing properties relative to other chords are conserved. This classification does not apply to cycles with one reality edge. Such cycles do not represent break-points and never need to be touched in sorting a permutation. Let us call *proper* cycles the cycles with a least four edges. It turns out that if we have only good cycles in a permutation α, then the lower bound is actually an equality, and we can sort α increasing the number of cycles at the rate of one cycle per reversal. Of course, we must be careful in choosing the reversals.

We will deal with this question later. Even when there are bad cycles, that is, cycles in which all reality edges induce the same orientation, it is sometimes possible to sort at the rate of one cycle per reversal. This happens because a reversal can twist a bad cycle, transforming it into a good one. Thus, while we are breaking one cycle, we may be twisting another one, so that we always have a pair of divergent edges to do the next step, until we reach the target permutation. However, twisting one cycle while breaking another is possible only if the two cycles are such that some desire edge from one of the cycles crosses some desire edge from the other one. We say that two cycles in this situation *interleave.* Notice that we are relying again on the fact that desire edges are chords on a circle.

Therefore, it is important to verify which cycles interleave with which other cycles in $RD(\alpha)$. We construct a new graph called the *interleaving graph* of α with respect to β taking as nodes the proper cycles of $RD(\alpha)$ and making two nodes adjacent if and only if the corresponding cycles interleave. This graph is denoted by $I_\beta(\alpha)$. Apart from the graph structure, we also label each cycle as good or bad. The connected components of $I_\beta(\alpha)$ are classified based on the cycles they contain, as follows. If a component consists entirely of bad cycles, it is called a bad component. Otherwise, it is a good component. Thus, a component with at least one good cycle is good. Examples of good and bad cycles and components are given next.

The diagram in Figure 12.14 shows a permutation with six cycles, each one containing two reality edges and two desire edges. Cycles C and F are good and all the rest are bad. Good cycles are represented in gray. The same figure also shows the interleaving graph for this permutation. There are two good components and one bad component in this case. The only bad component is the one containing the bad cycles A and E.

There are two interleaving bad cycles. There are no good cycles, and this means that no reversal will increase the number of cycles in the case. It follows that the reversal distance here is strictly greater than the lower bound $n + 1 - c(\alpha)$. Actually, this argument applies to any diagram that contains a bad component. Because reversals that act on a given cycle

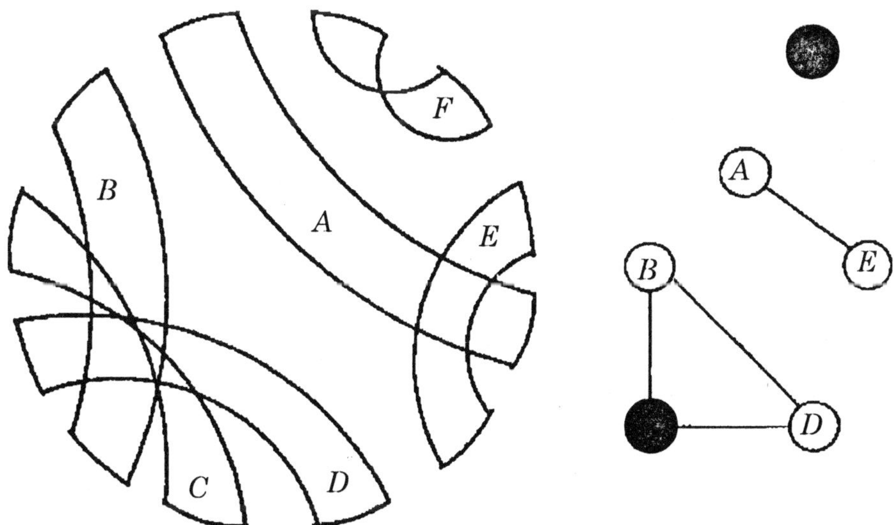

Fig. 12.14: Reality and desire diagram of a permutation. Cycles are labelled with the letters A through F. The interleaving graph of this permutation appears on the right, with good cycles in gray.

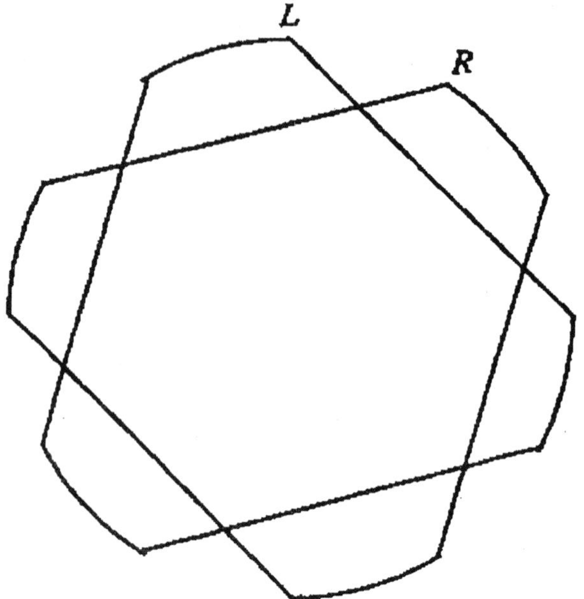

Fig 12.15: A diagram with no good cycles.

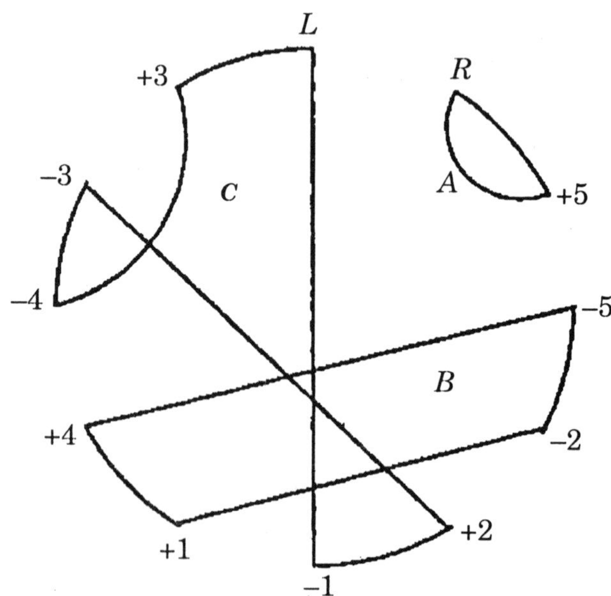

Fig. 12.16: A diagram with three cycles. Notice that only proper cycles take part in the interleaving graph.

A can only twist cycles that interleave with A, a bad component will remain bad as long as reversals act on cycles of other components. The only way to sort a bad component is by using a reversal that does not increase the number of cycles. Thus the lower bound $n + 1 - c(\alpha)$ will not be tight for permutations with bad components.

There we have two proper cycles, B and C, with B a bad cycle and C a good one.

We now use this last example to illustrate the sorting of a good component. Recall that we need to choose two divergent edges in the same cycle to define a reversal that increases the number of cycles. In Figure 12.16 the only good cycle has three reality edges:

$e = (L,+3)$, $f = (-3, -4)$, and $g = (-1, 2)$.

Edges f and g converge, so this pair is not a good choice. Edges e and g diverge, and the reversal defined by them produces two good components, each containing a single cycle. Edges e and f diverge as well, and produce a single good component a single cycle. Edges e and f diverge as well, and produce a single good component with two cycles. In both cases, we do not create bad components. Thus, there are two sorting reversals in this example.

We end this section with the following result, which summarizes our conclusions so far. The proof is simply a repetition of the arguments we used in the illustrative examples.

Theorem 12.3 : A reversal characterized by two divergent edges of the same cycle is a sorting reversal if and only if its application does not lead to the creation of bad components.

This theorem characterizes sorting reversals of a particular kind, but it does not tell us anything about their existence. In fact, it is true that as long as we have a good cycle, there will always be a sorting reversal of the kind that increases the number of cycles, but the proof of this will be omitted.

Bad Components

Having sorted all the good components with the methods we now turn our attention to the bad ones. Our goal in this section is to classify bad components in a hierarchy and, based on that, to provide an exact formula for the reversal distance of a permutation with respect to a target. Bad components are classified. We first distinguish hurdles from non-hurdles. Among hurdles, we distinguish super hurdles from single hurdles. We define all these concepts below. We say that a component B *separates* components A and C if all chords in $RD(\alpha)$ that link a terminal in A to a terminal in C cross a desire edge of B. Desire edges can be viewed as chords as well, so the usual definition of chord crossing applies here. Letters A to F indicate components, all of which are bad. In this example, B separates A and C. Other examples of separation include: E separates F and C, B separates F and D. Separation is an important concept, in that a reversal through reality edges in different components A and C will result in every component B that separates A and C being twisted.

A bad component becomes good when twisted, but a good component can remain good or become bad upon twisting. To avoid complications, we resort to twisting only when there

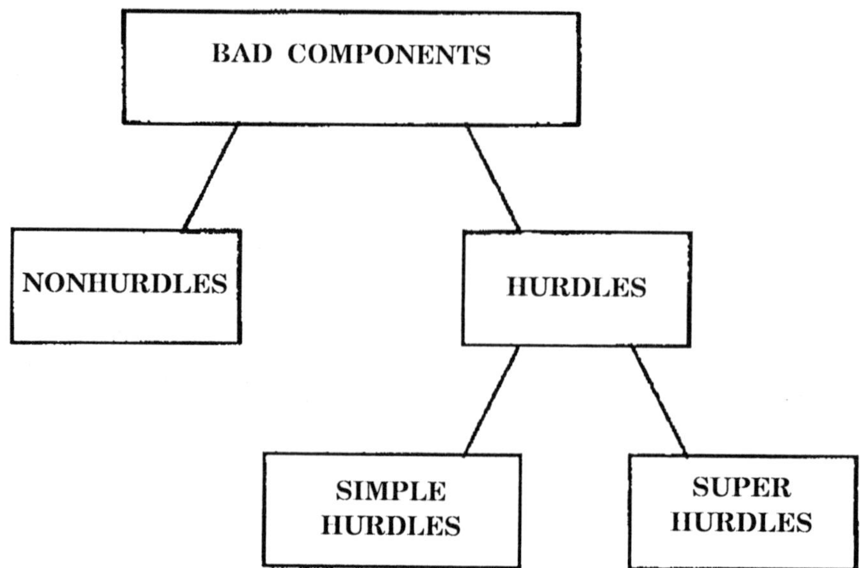

Fig. 12.17: Classification of bad components.

are no good components around. A *hurdle* is a bad component that does not separate any other two bad components. If a bad component separates others, then it is a *non-hurdle.* We denote by $h(\alpha)$ the number of hurdles of α. We say that a hurdle A protects a non-hurdle B when removal of A would cause B to become a hurdle. In other words, B is protected by A when every time B separates two bad components, A is one of them. In Figure 12.18, F protects E, and this is the only case of protection in this diagram. A hurdle A is called a *super hurdle* if it protects some non-hurdle B. Otherwise it is called a *simple hurdle.*

Notice that a hurdle can protect at most one non-hurdle. In Figure 12.18 the only super hurdle is F. All other hurdles are simple. A permutation α is called a *fortress* when its reality and desire diagram contains an odd number of hurdles and all of them are super hurdles. Fortresses are permutations that require one extra reversal to sort, due to their special structure, as we will see below. A fortress must have at least 24 labels, given that we need at least two labels to make a proper cycle, four labels to make a bad component, eight labels to make a super hurdle and its corresponding protected non-hurdle, and at last three such super hurdles and accompanying hurdle pairs to make a fortress.

We are now ready to write the exact formula for reversal distance of oriented permutation is. We have

$$d(\alpha) = n + 1 - c(\alpha) + h(\alpha) + f(\alpha), \tag{12.7}$$

where $c(\alpha)$ is the number of cycles (including proper and non-proper), $h(\alpha)$ is the number of hurdles, and $f(\alpha)$ is 1 or 0 according to α being a fortress or not, respectively.

The term $n + 1 - c(\alpha)$ accounts for good components and any components that are initially

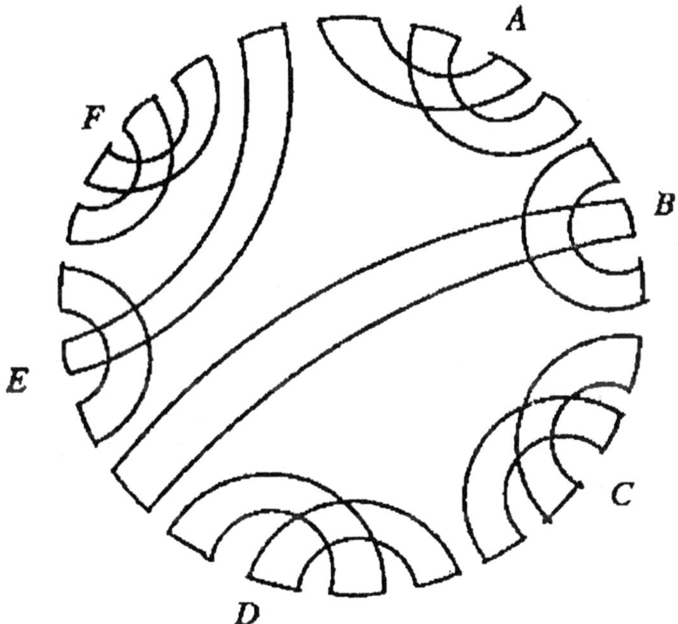

Fig 12.18: A diagram with no good cycles. Bad components that separate other bad components.

bad but become good in the process of sorting. The term $h(\alpha)$ is needed because bad components sometimes need an extra reversal to make them good. We do not need to include non-hurdles as these will be twisted. Finally, an extra reversal is needed for fortresses, because a non-hurdle becomes a hurdle at some point in the process.

Sorting Reversal

In this section we describe an algorithm for sorting an arbitrary oriented permutation, using the formula for reversal distance presented previously. The algorithm chooses at each step a sorting reversal, thereby guaranteeing that we reach the target permutation with the minimum number of reversals. We will later analyze in detail the complexity of the algorithm. Let us now verity that it is indeed a correct procedure. Before explaining the algorithm, we define special kinds of reversals that are used in the sorting procedure. As we saw, when there are no good cycles, we are forced to use either a several on two convergent edges or a reversal on edges of different cycles. In the first case, the number of cycles remains constant, whereas in the second case the number of cycles decreases by one. When choosing a reversal on convergent edges, it is better to do this in a hurdle than in a non-hurdle.

The reason is that this operation transforms the bad component into a good one, without modifying the number of cycles. If we get rid of a non-hurdle, we do not modify the number of hurdles or the fortress status of the current permutation, so the reversal distance will remain the same according to formula. It does not help to do it in a super hurdle either, because then the non-hurdle it protects will become a hurdle, keeping $h(\alpha)$ constant. Thus,

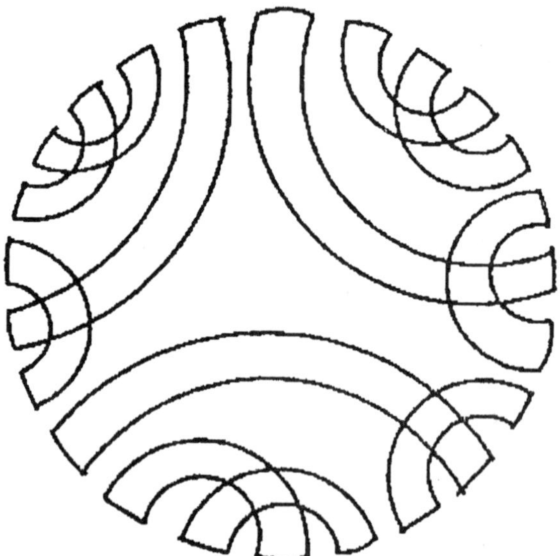

Fig. 12.19: A smallest possible fortress.

every time we have to use a reversal on convergent edges we will do so on a simple hurdle. It does not matter which cycle in the simple hurdle we use—they are all equivalent for this purpose.

We call a reversal characterized by reality edges of the same cycle in a hurdle a *hurdle cutting.* Hurdle cutting does not change $c(\alpha)$ and decreases $h(\alpha)$ when the hurdle is simple, but we have to be careful or $f(\alpha)$ many increase. To eliminate this possibility, we will use hurdle cutting only when $h(\alpha)$ is odd. Then the remaining hurdles are even in number and we can rest assured that the resulting permutation is not a fortress. Therefore, hurdle cutting is a sorting reversal when the hurdle is simple and $h(\alpha)$ is odd. The other possibility is to use a reversal on edges of different cycles. This will decrease the number of cycles, that is, it will increase the term $n + 1 - c(\alpha)$ in the reversal distance, but we can still make progress if we manage to reduce $h(\alpha)$ by two with this reversal. The only way to do it is by choosing edges in different hurdles. A reversal characterized by reality edges of different hurdles is called a *hurdle merging.* Its effect is that these two hurdles become good components, as well as any non-hurdle that separates them. There is the danger of a hurdle merging transforming a non-hurdle into a hurdle. This happens when the two hurdles being merged "protect" the non-hurdle in a similar way as a super hurdle protects a non-hurdle.

To solve this problem we need an additional definition. We call two hurdles A and B opposite when we find the same number of hurdles when walking around the circle counterclockwise from A to B as we find walking from B to A counterclockwise. Notice that opposite hurdles exist only if $h(\alpha)$ is even. Choosing opposite hurdles A and B we are safe, because this guarantees that we are not creating new hurdles when merging A and B, and

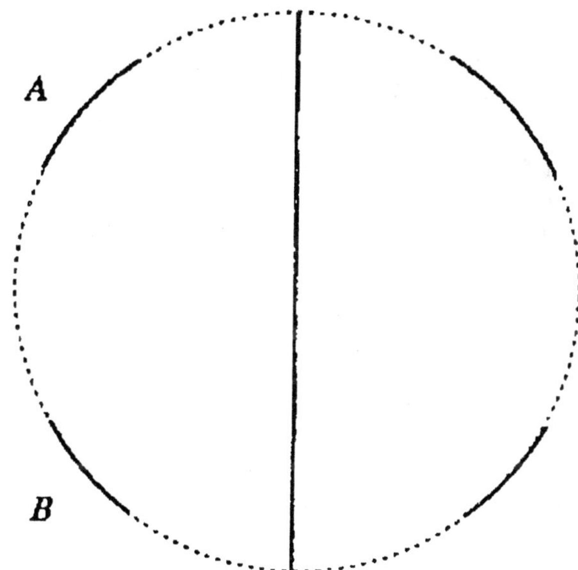

Fig.12.20: Merging hurdles A and B will transform non-hurdle C into a hurdle. A sorting reversal must merge either A or B with some hurdle separated from it by C.

hence $n + 1 - c(\alpha) + h(\alpha)$ will decrease by one. We still have to be careful with $f(\alpha)$. If we have even $h(\alpha)$, there is nothing to worry about, for hurdles merging on opposite hurdles will then decrease $h(\alpha)$ by exactly two, keeping it even. If $h(\alpha)$ is odd and there is a simple hurdle, there are no opposite hurdles, but we apply a sorting hurdle cutting and then we have even $h(\alpha)$. If $h(\alpha)$ is odd and there are no simple hurdles, we have a fortress, so $f(\alpha)$ is already equal to 1 and cannot increase.

We conclude that hurdles merging is a sorting reversal as long as we pick two opposite hurdles to merge. We summarize all this in the routine *Sorting Reversal.* This function finds a sorting reversal given an oriented permutation α.

The complete algorithm to sort by reversals consists merely of a loop in which we call the function and apply the returned reversal until we reach the target. Let us now give a rough estimate of the running time in terms of the permutation size n. We start by analyzing the construction of all structures defined in this section that are relevant for the algorithm. Constructing the diagram $RD\beta(\alpha)$ takes linear time, since all we have to do is to determine the reality and desire edges. Finding the cycles in this diagram also takes $O(n)$ time. For each cycle we must determine whether it is good or bad, and this takes $O(n)$ per cycle, for a total of $O(n^2)$. Next, we must determine the interleaving relationships between cycles. This can be done by examining each pair of desire edges and by storing the interleaving graph as an adjacency matrix in $O(n^2)$, time.

Counting the number of good and bad components, non-hurdles, simple hurdles, and super hurdles can all be done in linear time once we have the information on good and bad cycles. Therefore, an overall time bound to construct the needed structures would be $O(n^2)$. In algorithm *Sorting Reversal* the most time-consuming part is the one involving good components. We need to try several reversals and see whether the structures of the resulting permutation—after the reversal is applied—have some property. Since a reversal is identified by a pair of edges, we have to try at most $O(n^2)$ reversals. For each one, $O(n^2)$ time has to be spent checking the resulting permutation. Hence, the worst case running time estimate for *Sorting Reversal* is $O(n^4)$. We need to call this function $d\beta(\alpha)$ times in order to sort a permutation α with respect to β, so the final time bound is $O(n^5)$. Faster implementations have been proposed.

Unoriented Permutation

In this version of the problem the orientation of blocks in the genomes is unknown. All we known is their ordering. A similar theory can be developed for this case, with some results analogous to what we had in the oriented case, but with many differences as well. One very important difference is the fact that the unoriented case results in an NP-hard problem, whereas the oriented case, as we saw, can be solved in polynomial time. We start by giving some definitions, most of which are analogous to the ones given for the oriented case. An *unoriented permutation* α is a mapping from {1, 2, . . ., n} to a set L of n labels. Usually L is the same as the domain of α.

```
Algorithm Sorting Reversal
    input: distinct permutations α and β
    output: a sorting reversal for α with target β
    If there is a good component in RDβ(α) then
        pick two divergent edges e, f in this component,
            making sure the corresponding reversal does not
            create any bad components
        return the reversal characterized by e and f
    else
        if h(α) is even then
            return merging of two opposite hurdles
    else
        if h(α) is odd and there is a simple hurdle
            return a reversal cutting this hurdle
        else
        // fortress
            return merging of any two hurdles
```

Fig. 12.21: An algorithm that returns a sorting reversal for a with target β, provided α # β.

A *reversal* is defined as in the case of oriented permutations, except that it does not flip arrows, as there ar no arrows. It just reverses the order of a segment of consecutive labels. Notice that here it does not make sense to have a unitary reversal, because it does not change α at all. The goal is again to obtain a shortest possible series of reversals that transform α into β. If $\rho_1, \rho_2, \ldots, \rho_t$ is a shortest series of reversals such that

$$\alpha\rho_1\rho_2 \ldots \rho_t = \beta,$$

then *t* is the *reversal distance* of α with respect to β, denoted by $d_\beta(\alpha)$. Sorting reversals are defined accordingly. The same argument used in the oriented case holds to show that $d_\beta(\alpha) = d_\alpha(\beta)$. If the target is clear from the context, we drop explicit mention to it. The identity in this context is the permutation defined by $I(_i) = i$ for all $_i$ in 1 . . *n*. The following example should clarify our discussion.

Example 12.1: Figure 12.22 shows two chromosomes with homologies between blocks indicated by lines. We assigned labels 1 through 8 to the blocks in the lower chromosome, and then transferred the labels to the upper one giving equal labels to homologous blocks. Thus we obtain a starting permutation in the upper chromosome and our goal is to sort it into the lower one, the identity. A series of four reversals that sort this permutation is given in Fig. 12.23.

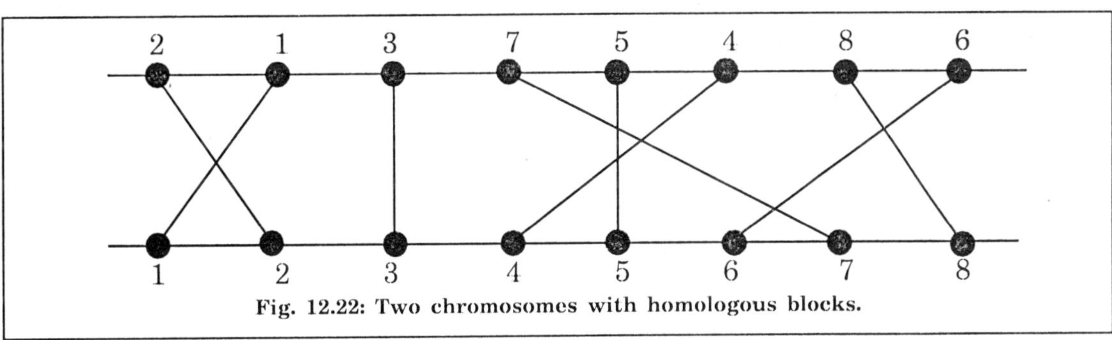

Fig. 12.22: Two chromosomes with homologous blocks.

Can we obtain a shorter series of reversals? Not in this case. To make a convincing argument that four reversals are necessary, we look at the breakpoints of this permutation, just as we did in the oriented case. For simplicity, we assume that the target is the identity permutation for the rest of this chapter.

A breakpoint of an unoriented permutation α is a pair of labels adjacent in α but not in the target, or, equivalently, adjacent labels that are not consecutive. Breakpoints are to be considered in the extended version of α, which includes extra left (*L*) and right (*R*) elements. The expression $b(\alpha)$ denotes the number of breakpoints of α. The permutation in question has seven breakpoints. The only pairs that are not breakpoints are 2, 1 and 5, 4.

A reversal can remove at most two breakpoints, just as with oriented permutations. Hence, the same bound we had previously for oriented permutations applies here, which is

$$d(\alpha) \geq \frac{b(\alpha)}{2}. \qquad 12.8$$

2	1	3	7	5	4	8	6
1	2	3	7	5	4	8	6
1	2	3	4	5	7	8	6
1	2	3	4	5	7	6	8
1	2	3	4	5	6	7	8

Fig. 12.23: Sorting permutation 21375486.

With this in mind, it now becomes obvious why one cannot sort the permutation α of Fig. 12.24 with less than four reversals. From Eq. (12.8) we see that $d(\alpha) \geq 3.5$, but, because $d(\alpha)$ is an integer, this implies $d(\alpha) \geq 4$. So, our solution with four reversals is optimal.

L • 2 1 • 3 • 7 • 5 4 • 8 • 6 • R

Fig. 12.24: Breakpoints of an unoriented permutation indicated by the symbol •.

Consecutive Labels

Consider the permutation α displayed below in its extended form, and with breakpoints marked by bullets:

L • 4 5 • 3 2 1 • R

Observe that between two consecutive breakpoints we have a sequence of consecutive labels, either increasing or decreasing. A bit of thought persuades us that this is actually a general phenomenon. Indeed, if we have two adjacent labels that do not make a breakpoint, they must be of the form

$\ldots x(x + 1) \ldots$

or

$\ldots x(x - 1) \ldots$

In the first case, if $x + 1$ is not part of a breakpoint, then the label to its right must be $x + 2$, because its other consecutive label, x, has been used already. Thus, the increasing sequence extends to the right as for as the next breakpoint or until the end of the permutation is reached. A similar argument can be applied to the left of x and also to both sides of the second case above.

A sequence of consecutive labels surrounded by breakpoints but with no internal breakpoints is called a *strip.* We have two kinds of strips, increasing and decreasing ones. A single label surrounded by breakpoints is itself a strip, and in this case it is both an increasing and a decreasing strip. Exceptions to this rule are the labels L and R, which are always considered part of an increasing strip, even if they are all by themselves in a strip. In fact, we identity these extreme elements into a single one for the purpose of defining strips. So, for instance, if 0 is involved in a longer increasing strip 0, 1, . . . and $n + 1$ is also part of a longer strip . . ., n, $n + 1$, we consider these two sequences as a single strip. They are linked by the common element $L = R$.

Example 12.2: The following example has five strips: two increasing ones ($L = R$, 1, 2 and 5, 6), one decreasing strip (8, 7), and two strips that are both increasing and decreasing (3 and 4).

L 1 2 • 8 7 • 3 • 5 6 • 4 • R

Looking at strips can help us remove breakpoints, as the following result shows.

Theorem 12.4 : If label k belongs to a decreasing strip and k–1 belongs to an increasing strip, then there is a reversal that removes at least one breakpoint.

Proof: Labels $k - 1$ and k cannot belong to the same strip, because the only strips at the same time increasing and decreasing are composed of a single element. Therefore, they belong to different strips. This in turn implies that each one is the last element in its strip, that is, each one is immediately followed by a breakpoint,

$$. . . (k - 1) \bullet . . . k \bullet . . .$$

$$. . . k \bullet . . . (k - 1) \bullet . . .$$

Fig. 12.25 : Possible relative positions of strips of k and $k - 1$.

This figure presents the two possible schemes. The strip of $k - 1$ may either follow or precede the strip of k in the permutation. In both cases, the reversal that cuts the two breakpoints immediately following $k - 1$ and k brings these two labels side by side, thus reducing the number of breakpoints by at least one.

We know that all permutations have at least one increasing strip, the one containing L (or R). In contrast, a permutation may not have any decreasing strip. If, however, a permutation does have a decreasing strip, then we can use the preceding result and conclude that there is a breakpoint-removing reversal. This happens because when there are both increasing and decreasing strips, they must “meet” somewhere, in the sense of their ordering in the identity permutation. Another way of seeing that is by taking the smallest label belonging to a decreasing strip, and calling it k. Certainly $k - 1$ belongs to an increasing strip, otherwise k would not be the minimum. A similar result holds with $k + 1$ substituted for $k - 1$ in the statement, as the following theorem shows.

Theorem 12.5 : If label k belongs to a decreasing strip and $k + 1$ belongs to an increasing strip, then there is a reversal that removes at least one breakpoint.

Proof: As in the previous theorem, k and $k + 1$ cannot be in the same strip. Thus, each one is the first element of its strip, and is immediately preceded by a breakpoint. The reversal cutting through these two breakpoints reduces the total number of breakpoints, since k and $k + 1$ will end up side by side.

These two results mean that, as long as we have decreasing strips, we can always reduce the number of breakpoints. Notice that this applies also to single-element strips. However, sometimes we do not have a decreasing strip. The following result provides a compensation when we have to create a permutation with no decreasing strips in the process of sorting.

Theorem 12.6 : Let α be a permutation with a decreasing strip. If all reversals that remove breakpoints from α leave no decreasing strips, then there is a reversal that removes two breakpoints from α.

Proof: Let k be the smallest label involved in a decreasing strip. As we saw in the proof of Theorem 12.4, there is a reversal ρ that unites k and $k - 1$. If $k - 1$ is to the right of k, ρ leaves a decreasing strip, contradicting our hypothesis. So, $k - 1$ must be to the left of k.

(a) $\ldots (k-1) \bullet \ldots k \bullet \ldots$

(b) $\ldots \bullet\ 1 \ldots \bullet\ (l+1) \ldots$

Fig. 12.26: Permutations without reversals that leave a decreasing strip.

Now let l be the largest label involved in a decreasing strip. There is a reversal σ that unites l and $l + 1$. If $l + 1$ is to the left of l, σ leaves a decreasing strip, so $l + 1$ must be to the right of l. Observed that k must be inside the interval reversed by σ, otherwise s would leave k's decreasing strip intact. Likewise, l must belong to the interval of ρ.

We claim that these two intervals are actually the same, that is, ρ = s. Indeed, if this is not true, there is a strip that belongs to the interval of only one of these reversals. Suppose for a moment that this strip belongs to the interval of ρ. If the strip is increasing, ρ would make it into a decreasing strip. If it is decreasing, then it will survive in $\alpha\sigma$, because σ can only augment it. So, neither case is possible. A similar argument can be drawn if there is a strip exclusive to σ. Thus, the claim holds. But in this case this reversal removes two breakpoints, because it joints $k - 1$ with k and l with $l + 1$.

Algorithm

We are now ready to present an algorithm that sorts a permutation using at most twice as many reversals as the minimum possible, $d(\alpha)$. This is therefore an approximation algorithm. It relies on the theorems we just saw. The idea behind it is simple. There is a main loop, where we look at the current permutation and select the best possible reversal to apply to it, following certain criteria. We then update the current permutation and report the reversal applied to it. The loop stops when the current permutation is the identity. The criteria we use to select reversals are based on the previous results. If the current permutation has a decreasing strip, we look for a reversal that reduces the number of breakpoints and leaves a decreasing strip. Good candidates for this are the reversals that appeared in the proofs of Theorems. If no such reversal exists, we know there is a reversal that encompasses all the decreasing strips and removes two break-points. Finally, if the current permutation has no decreasing strips, we select a reversal that cuts two breakpoints. We know that there are at least two breakpoints, otherwise the current permutation would be the identity. No permutation has exactly one breakpoint, so all permutations different from the identity have as least two breakpoints. Furthermore a reversal that cuts through two breakpoints is guaranteed not to increase the total number of breakpoints. We know that this algorithm terminates because the number of breakpoints is reduced at every iteration, except when we encounter permutations with no decreasing strips. But this situation cannot happen twice in a row. So we can certainly say that the number of breakpoints decreases at least one unit every two iterations. In fact, we have a larger average rate of breakpoints removed per iteration, as the following result shows.

Theorem 12.7 : The number of iterations in algorithm *Sorting Unoriented Permutation* is less than or equal to the number of breakpoints in the initial permutation.

Proof: Our goal is to prove that, on average, each iteration removes at least one breakpoint. The only difficulty is caused by the reversals chosen when there are no decreasing strips. But each time we reach this situation, the *previous* iteration just removed two breakpoints, sustaining the average of at least one breakpoint removed per iteration. This argument does not work for the first iteration, if we happen to get an initial permutation with no decreasing strip. But then the very last iteration, that produces the identity, removes two breakpoints and compensates for this special case.

```
Algorithm Sorting Unoriented Permutation
    input: permutation α
    output: series of reversals that sort α
    list ← empty
    while α # 1 do
        if α has a decreasing strip then
            k ← the smallest label in a decreasing strip
            ρ ← the reversal that cuts after k and after k – 1
            if αρ has no decreasing strip then
                l ← the largest label in a decreasing strip
                ρ ← the reversal that cuts before land before 1 + 1
        else
            ρ ← the reversal that cuts the first two breakpoints
    α ← αρ
    list ← list + ρ
return list
```

Fig. 12.27: An algorithm for sorting unoriented permutation with a guarantee: The number of reversals used is not greater than twice the minimum.

CONCLUDING REMARKS

The genetic material of organisms changes as they evolve. In this chapter we were concerned with large transformations at the genome level, as opposed to point mutations at the sequence level. We concentrated our study in one particularly common such transformation, the reversal. Given two genomes, we seek the minimum number of reversals that lead from one to the other. This is the reversal distance between them, and it represents the most parsimonious explanation of their differences. Depending on available information on gene orientation, we have two versions of this problem: one involving oriented permutations and another for the unoriented case.

A remarkable fact is that, while the reversal distance can be computed in polynomial time for the former, the problem is NP-hard for the latter. We presented an elegant theory that leads to a polynomial-time algorithm for the oriented case. This theory relies on simple concepts such as breakpoints and the reality and desire diagram, as well as on more complex ones such as hurdles and fortresses. For the unoriented case, we developed a number of tools that provide important information on the actual distance, then presented an approximation algorithm that generates a series of reversals at most twice as long as the optimal one.

Chapter 13

Metabolic Networks and Pathways

Life is a process far from thermodynamic equilibrium. Different enzymatic chains and networks have to operate continuously to prevent chemical or thermodynamic equilibrium form occurring, for that would mean the death of the organism. The analysis of metabolic pathways tries to understand and investigate the enzyme chains involved in this process. A qualitative focus considers which enzymes are involved, which pathways are possible in a complex network (a nontrivial question), and what happens if one enzyme is knocked out or a new one is added to the set.

Techniques developed in our laboratory to analyze metabolic pathways qualitatively will be the focus of this chapter. It should be noted, however, that there is also a detailed quantitative perspective from which to analyze metabolic pathways in their conversions and transitions—but it needs much more experimental data on the considered enzymes. It includes metabolite concentrations, substrate and product affinities, and kinetic data on enzymes such as their turnover number. Isotope labelling of metabolites and NMR are advanced tools to provide experimental data for such studies. These questions are only briefly discussed in this chapter. The interested reader is referred to the excellent textbooks and reviews on thee questions. Dynamics of metabolites may also be tackled in comprehensive cellular simulations.

Another aspect of all enzymatic reactions is that they can be divided into two types. There are biochemical reactions, that can proceed from substrate to product and vice versa: these are reversible reactions. There are also reactions that proceed in only one direction: these are irreversible reactions. In most cases the latter is achieved as the product is either moved from the cell into outer compartments or by rapid conversion of the product further down the enzymatic chain.

There are over 2000 known biochemical reactions. If there is pathway of enzymatic reactions, the flow through the system is called the metabolic flux (e.g., the rate of flow in micromoles per minute) of substrates converted into products. Another very general division of enzymes is between energy-producing (exergonic reactions) and energy-consuming (endergonic reactions). Similarly, on the metabolic level, metabolic pathways can conveniently be divided into catabolic (metabolite degrading; any type of metabolic "digestion') and anabolic (producing more complex biological metabolites). Anabolic pathways usually contain many endergonic reactions because they require a great deal of energy to create the more complex metabolites. They are generally more extensive in their network and enzymatic reactions, and occur more often in all autotrophic organisms (such as plants, where the required energy is supplied by photosynthesis.

Relatively few metabolites serve as starting materials for a variety for a variety of products. In contrast, all animals (including man), as well as fungi, parasites, and saprophytic organisms, rely for their energy supply on catabolic pathways, which usually contain more exergonic enzymatic reactions. Few common intermediates are produced from a large number of substrates. An important first end product is often acetyl-CoA, and ultimately, via oxidative pathways, oxidative phosphorylation of all available reduction equivalents (NADH, FAD, etc.) to generate a maximum of ATP. Similarly, NADPH and ATP serve as major high-energy components in transfer reactions in between. A metabolic pathways is thus a series of connected enzymatic reactions that produce a specific product. Ample examples are provided by the KEGG metabolic database. Furthermore, pathways may be branched or interconnected . Delineating a pathway from the metabolic network has until now mainly been done using traditional or chemical knowledge. Algorithms are very helpful in gaining rationality and precision. The abundance of enzymes varies with the kind of organism, cell type, nutrient status, and developmental stage. Another aspect, which can be party revealed by meticulous genome comparison and annotation, is the detection of isoenzymes: the same enzyme is present in different varieties, depending on tissue or developmental state: in prokaryotes, even in the same cell. Different regulatory domains often confer different regulation of isoenzymes, and thus tissue-specific or condition-specific metabolization for the same type of reaction can be achieved.

An example is lactate dehydrogenase, in which heart-specific and muscle-specific varieties exist, allowing for optimal adaptation of the metabolism in the two different types of tissues. Problems we will discuss are pathway identification and completeness, identification of alternative pathways and all possible fluxes through a given metabolic network, and better understanding of the importance and the specific biological significance of pathway variations. Pathways are the next level of complexity after the protein reading frame. On the other hand, they are only an abstraction, because in reality there are always metabolic networks. In that respect, it is important to have tools for their analysis as well as their synthesis.

METABOLIC NETWORK

Cellular reactions are highly connected and entangled. In order to get a good overview of the complexity of metabolism, one can use standard literature and databases available on the internet (e.g., KEGG, EcoCyc. However, metabolism in much more flexible than the well-known set of pathways available from biochemistry textbooks. Using known information about metabolism, computational calculations can help to find new possible routes in the biochemical network that can then be proven by experimental approaches such as isotope labelling of metabolites, use of metabolic inhibitors, and knocking out of enzymes.

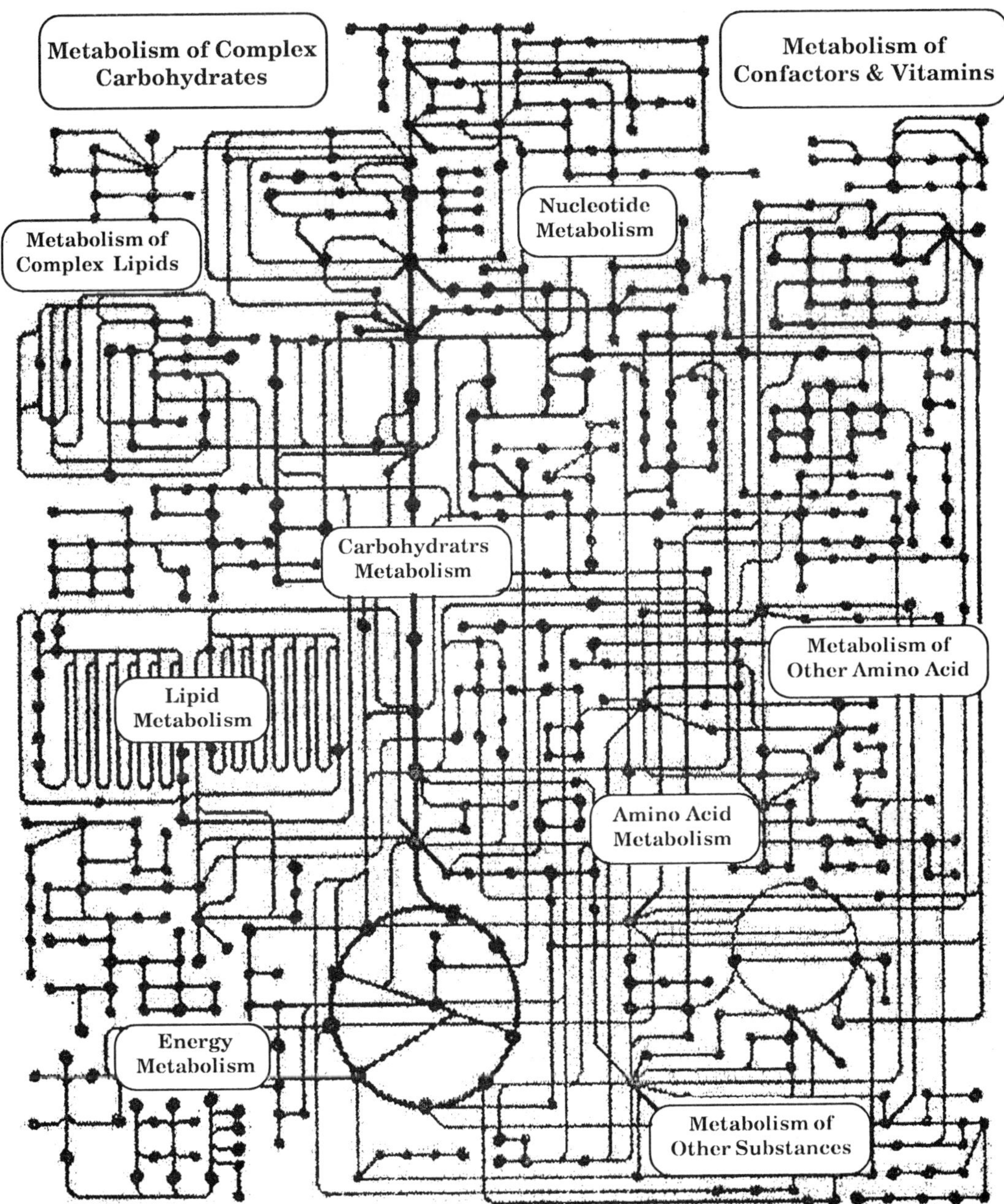

Fig. 13.1: The complexity and inter-connectedness of metabolic pathways is sketched in this simplified pathway map, when considers only textbook pathways.

Through chemical reasoning or tradition, standard pathways such as glycolysis are known for well-known metabolites. However, given the network nature of enzymatic interactions, are these the only possible routes through an enzyme network? In a similar vein, new genome expression data try to predict co-regulation of enzymes, and in this way provide a completely different approach to pathway definition. Can such coregulated enzymes be called a pathway? To decide this on a more rational basis, a more algorithmic definition of pathway is required.

Three different approaches are generally used to map the pathways in a biochemical network from scratch. They do not require information about enzyme kinetics, affinity, or metabolite concentrations—but this means each of them is only stating whether a certain metabolic route (a set of enzymes working together) is accessible to the set of enzymes present in the cell. It does not give information on the extent to which a certain metabolic path is used in the cell under a given metabolic condition. A first way to identify a pathway is by synthesis.

Transformation routes from a substrate to a product are found by successive addition of reactions, very similar to the way in which the major, well-known pathways were found by chemists. This can be done systematically, and details of this types of analysis are available in Seressiotis and Vailey (1986).

Elementary Flux Mode

A powerful tool is the use of elementary flux modes. This will be examined now in more details. An elementary mode is the minimal set of enzymes that can operate at steady state, with the enzymes weighted by the relative flux they carry. "Minimal" is to be understood in the sense that if only the enzymes belonging to this set are operating, complete inhibition of a further enzyme of this set would lead to cessation of any steady-state flux in the system. With the help of the elementary modes, it can be determined whether a stoichiometrically balanced path exists between a particular set of substrates and products – no matter how complex and entangled a metabolic network may be.

Actually, it is not sufficient to construct linear pathways just by following different metabolites, because by-products that cannot be excreted by the cell would then have to be balanced by additional reactions that produce further by-products, and so on. After this the full set of non-decomposable steady state flows that the network can show, including cyclic flows is enumerated. An actual steady-state flux pattern such as measured by an experiment, will always be a non-negative linear combination of these modes. Thus the living cell is described as a mixture of pure states of these elementary modes that form a unique set. The stoichiometry of the network is critical for identifying the elementary flux modes, the noncompensable modes that allow a steady-state flux from substrates to products.

At first the enzymes involved are abbreviated as is convenient, and from biochemical knowledge it is determined in which direction the reactions catalyzed would proceed. For example, in modeling a central part of the metabolism involving glycolysis and the pentose phosphate cycle, one could write.

—ENZREV

PGI ALD TPI . . . (for phosphoglucoseisomerase, aldolase, and triosephosphate isomerase, etc.)

—ENZIRREV

HK PFK . . . (hexokinase, phosphofructokinase, etc., these are irreversible)

Next the external and internal metabolites are defined. For an external metabolite it is assumed that there is sufficient buffering capacity either in cell or in the environment so that the steady-state condition is always fulfilled for this metabolite. In contrast, for any internal metabolite all reactions have to be properly represented by stoichiometric equations in the model because it has to be properly supplied in stable flux from the enzymes modeled in the system. For our example this would be

—METINT

E4P S7P . . . (erythrose 4-phosphate, sedoheptulose 7-phosphate, and so on are internal; the example metabolites are internally created by the pentose phosphate cycle)

—METEXT

CO_2 NADP NADPH . . . (and so on; for example cofactors are involved in so many other cellular reactions that they may be considered to be externally created and buffered)

Finally, all enzymatic reactions involved in the transformation are written down:

—CAT

HK: GLC + ATP = G6P + ADP

PGI: G6P = F6P

PFK: F6P + ATP = FDP + ADP

(the enzymatic reactions of hexokinase, phosphoglucose isomerase, phosphofructokinase, and so forth)

The algorithm calculating the elementary modes from this information is now briefly sketched. The algorithm has been implemented as computer programs in small talk. The programs start from a list of reaction equations and a declaration of reversible and irreversible reactions, and of internal and external metabolites. As in several other metabolic simulators, this list is automatically translated into a stoichiometric matrix.

This matrix is then transposed and augmented with the identity matrix, to give a matrix called initial tableau. From this, further tableaux are consecutively computed by pairwise linear combination of rows so that the columns of the transposed stoichiometric matrix successively become null vectors. This procedure corresponds to ensuring that the steady state constraint is satisfied for each metabolite in turn. Different metabolic pathways can be analyzed in this way. This assumes that there has already been a detailed stocktaking of the available enzymes. This will be explained by two methods, pathway alignment and pathway completion by genome annotation.

Apart from just enumerating paths, applications of elementary mode analysis include the following:

(i) Testing whether a set of enzymes allows production of a desired product.

(ii) Detecting non-redundant pathways and enzyme activities (important for drug design)

(iii) Detecting pathways with maximal molar yields (important for biotechnology and metabolic engineering) as well as alternatives with nearly maximal yields.

(iv) Genome comparisons – it rapidly summarizes whether pathways are complete, and detects gaps and inconsistencies in the annotation.

(v) Medical or pathophysiological assessment of the impact of enzyme deficiencies.

However, for several pharmacological applications, elementary mode analysis is quite interesting.

(vi) Identifying toxic metabolites .

(vii) Following pharmacologically active, intermediary metabolites and products accruing or created by the enzyme machinery of the body after, for example, painkilling drugs or sedatives are metabolized.

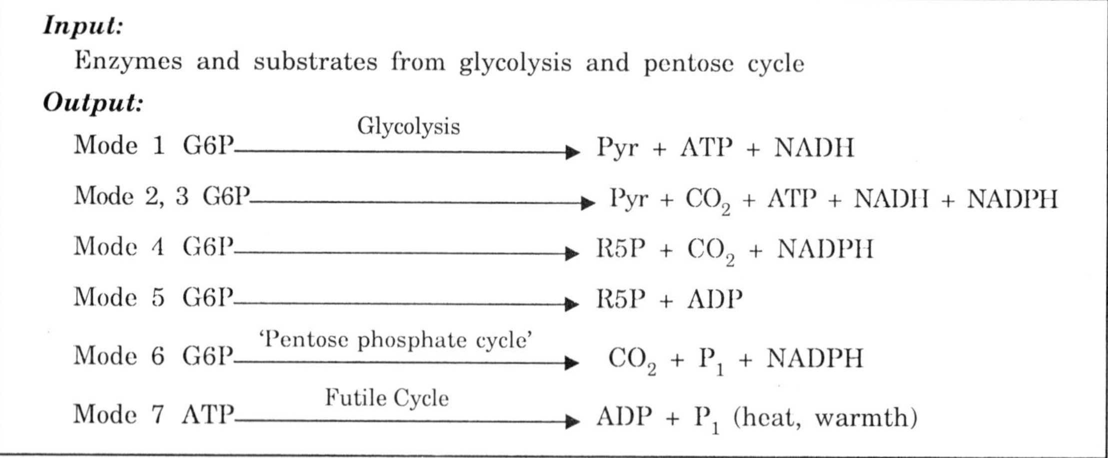

Fig. 13.2 **Sketch of the elementary modes calculated for the pentose phosphate cycle and glycolysis. After program input (for details, see text), the algorithm METATOOL calculates seven elementary modes (listed below the input; products are given only qualitatively; modes 2 and 3 differ in their enzyme connectivity and their exact stoichiometry). This illustrates that even such a simple network contains more modes of operation than biochemical textbook knowledge would suggest.**

In metabolic engineering, the method also allows *in silico* tests to determine whether artificially introduced pathways are likely to function stoichiometrically. This could concern redox balance. Furthermore, an improved analysis of he effects of adding other substrates the network is interesting. In some medically quite interesting cases this may lead to either potentiation or reduction of the effect of a drug. Textbook descriptions of standard pathways can now be made much more precise or complete by results from elementary mode analysis.

Thus, analyzing the system composed of glycolysis and the pentose phosphate cycle in some more detail, we could show that a total of seven modes can be obtained. Besides the textbook pathways (modes 3 to 6), some additional modes of conversions can be found: Mode 1 represents the usual glycolytic pathway. Modes 2 and 3 degrade G6p to pyruvate and CO_2, producing ATP, NADPH, and NADH. Mode 4 converts G6P into ribose-5-phosphate (R5P) and CO_2. It is of importance when the metabolic needs for reducing power for biosynthetic purposes and for R5P in nucleotide six pentoses. This is important when the need for R5P exceeds that for NADPH. Note that, depending on which type of utilization of R5P is considered, additional Atp consumption may occur in modes 4 and 5.

The term "*pentose phosphate cycle*" applies to mode 6 best, because carbons are cycled several times before ending up in CO_2. Like modes 2 and 3, it is relevant when much more NADPH than R5P is required, but does not lead to NADH and ATP production. Mode 7 is a futile cycle formed by phosphofructokinase and fructose biophosphatase. The futile cycle is biochemically relevant because it generates additional heat in the organism-if required by environmental conditions. Further complexities include various modes that arise if genome-specific variations are analyzed or the reversibility of reactions is changed.

Furthermore, from the start a smaller or larger network can be considered. Further details, descriptions, and conclusions on all these questions are found in Schusteret *et al.* (2000). An alternative to elementary mode analysis is to think of all possible metabolic fluxes as a type of space, the metabolic flux space. As in analytical geometry, one may then ask how this space is set up by vectors. These are the so-called basis vectors of flux space. This has a problem, however: there are many possible solutions for the set of vectors defining the flux space. Nevertheless, very nice studies have been done using such flux balance studies – for example, investigating the comparative robustness of the *E. coli* metabolic network to supply cell growth over a wide range of different flux conditions or the effect of *E-coli* gene mutants, as well as assessing the metabolic capabilities of *haemophilus influenza* Rd.

We will now compare pathways in different genomes. This yields important information on their evolution, on pharmacological targets, and on biotechnological applications. We will discuss a pathway alignment approach in which three methods are combined: (1) analysis and comparison of biochemical data, (2) pathway analysis based on the concept of elementary modes, and (3) comparative genome analysis. For the last, the rapid increase of completely sequenced genomes is very helpful.

Comparing 17 completely sequenced genomes for their conservation of the glycolytic pathway could reveal a surprising plasticity of this very central pathway. This concerns both presence and absence of key enzymes as well as specific patterns of isoenzymes in different species. More deviations than previously thought from the textbook standard become apparent. Pathway alignment opens up new routes for potential pharmacological targets and reveals interesting bypasses such as the Entner-Doudoroff pathway. In a similar vein, Archaean-bacterial-, and parasite-specific adaptations can be rapidly identified and described. Sequence alignment is a well established tool for investigating and comparing nucleic acid and protein sequences from different species and for identifying characteristic gaps, insertions, and dissimilarities.

The availability of full genomic sequences and the increasing amount of biochemical data open up higher-order possibilities for comparative different species is an important step toward a more global physiological comparison.

Recognition of such differences is interesting for biotechnology (identification of alternative enzymes) and pharmacology (difference in drug targets). Biochemical data are extended and compared with elementary mode analysis of substrate fluxes, comparative genome analysis, and pathway alignment. Other valuable methods to consider and include for large-scale comparisons of genomic data and pathways include clusters of orthologous sequences (COGs; and the application of related enzyme clusters.

Bioinformatics Tools

The concept of this approach is to compare a range of organisms for one specific pathway, indicate the presence or absence of the involved enzymes by two different letters (or zeroes and ones), and then compare the enzyme sets I the same way as sequences are normally compared, marking well and less well conserved regions. Pathway alignment focuses on the biochemical capacity of the organisms compared. The utilization of complementary data and tools extends the biochemical approach and improves the predictive power of such efforts. A solid knowledge basis from literature and experimental data is established. Extensive use of sequence and genome analysis allows the characterization of the full repertoire of enzymes present in the analyzed organisms in regard to the analyzed pathway.

The availability of full genomic sequences is exploited, and pathway fluxes and bypasses regarding the identified enzymes are tested for consistency by elementary mode analysis. Also, the determination of futile cycles in the different organism-specific enzyme sets is an indicator for allosteric regulation of key enzymes involved. The collecting of biochemical and genomic data required to establish presence or lack of enzymes prior to pathway alignment is greatly facilitated by the availability of databases. These systems are continuously developed further (e.g., the orthologue group tables for KEGG. Differential metabolic display, another helpful pathway visualization and exploration tool, is based on petri nets to generate all pathways satisfying certain constraints.

However, since databases are based on automated methods, including preformed pathway charts, these automated predictions are further refined for an accurate analysis. Available completely sequenced genomes should be extensively cross-compared to better identify all encoded enzymes in these organisms, because in different genomes, different non-orthologous sequences may encode the same enzyme. Each of these families can be recognized only if at least one member has been biochemically characterized; a fact that stresses again the need for both genomic and biochemical data bases and data. Genomic information is particularly valuable for organisms that are different to analyze biochemically.

Fro genome cross comparisons we recommend the extensive use of differential genome analysis, using assigned reading frames from one genome to compare its protein content against that of another genome and using application-specific chips for rapid sequence-to-sequence comparisons. Gene duplications, replacement by unrelated sequences (non-orthologous displacement; and gene neighborhoods also have to be taken into account.

Phylogenetic analysis has to be applied in order to further analyze gene duplication events and clarify substrate specificities of the encoded enzymes. Biochemical functions, sequence-based enzyme comparisons, and detailed enzyme specificities are made more accurate by a careful comparison of domain architecture and detection of duplication and gene displacement events, operon organization, and gene families. This is more exact than database overviews with a splendid amount of genome data that rely strongly on an automated assignment of enzyme functions by reciprocal gene similarities. As a medical application, identified enzymes from parasite-specific pathway variants may potentially be blocked pharmacologically.

Analysis of Glycolysis

Applying pathway alignment for the analysis of glycolysis, the metabolic flow and corresponding enzymes from glucose 6-phosphate onward present a relatively well conserved region of the glycolytic pathway. However, in a number of organisms several glycolytic enzymes are missing, according to comparative sequence analysis and/ or biochemical data. For example, phosphofructokinase and aldolase activity are absent in *Mycoplasma hominis*. Similarly, phosphofructokinase and pyruvate kinase seem to be lacking in *helicobacter phlori*. Different alternatives to processing glucose to pyruvate become apparent.

In glycolysis two molecules to triose are derived from one hexose, and the energy yield is 2 moles of ATP per more of glucose. The complete glycolytic pathway is present in *E. coli*. This is also the case in most eukaryotic (including human) cells. However, a first variation concerns the route of transfer of glucose into the cell. In contrast to higher organisms, the major *E. coli* pathway for glucose transport uses the PTS glucose transporter, which utilizes phosphoenolpyruvate as an energy-rich compound to directly phosphorylate glucose to glucose 6-phosphate. The connections of pathways to other parts of metabolism, such as the connection of glycolysis to the pentose phosphate pathway, allow for further plasticity (e.g., the Entner-Doudorooff pathway is used instead of glycolysis in some bacteria.

The ATP yield is only 1 mole per model of glucose. In *Methanococcus jannaschii,* all enzymes of the non-oxidative PPP were identified. In contrast, the two dehydrogenases required for the oxidative PPP could be identified neither in the genome nor by biochemical assays. Glycolysis appears to be operative although hexokinase, phosphofructokinase, aldolase, and phosphoglycerate mutase could not be identified in the genome. Assuming that these enzymes or other enzymes are present (for instance, there are some indications of ADP-dependent sugar kinases in archaea.

The elementary modes for this system can then be calculated. One of these modes in fact allows the transformation of glucose 6-phosphate into ribose 5-phosphate, which is required for nucleotide biosynthesis. Thus, the oxidative PPP is not needed for converting hexoses into pentoses under steady-state conditions. This is in agreement with results of Pandolfi *et al.* (1995) on glucose 6-phosphate dehydrogenase deficiencies. A further tool for this type of analysis is enzyme hierarchies to find common patterns between pathways based on the EC classification – which should be critically further classified by biochemical knowledge and comparisons. Pathway alignment reveals species-specific peculiarities and can become particularly powerful if coupled to metabolic engineering. Examples, would be engineering *Corynebacterium glutamicum* mutants to demonstrate two independent pathways not only for lysine but also for DL-diaminopimelate as an important building block of the murein cell wall of this bacterium, or to better understand acetate production in *E. coli* by comparing it against other species and reducing its acetate synthesis by expressing a *B. subtilis* acetolactate synthase. Planning criteria for such projects that take results of metabolic flux analysis and a modular perturbation method into account are explained by Kholodenko *et al.* (1998); experimental suggestions for isotope labelling are given by Roscher *et al.* (2000).

GENOME ANALYSIS

Genome analysis is a complex and continuous process because the interpretation of the original genome sequence available after a major sequencing project and the experimental data available change during subsequent years. Therefore, before discussing additional techniques, we will illustrate pathway identification in the process of genome annotation and, with more information becoming available, subsequent reannotation. For analysis the sequence has first to be available in electronic format. For a new sequencing project this is a rather cumbersome process and requires a great deal of technology. However, for studying and other purposes, many full genomic sequences and continuing public sequencing projects are deposited in public Web sites. In the next step, determining the position of potential reading frames, one has to keep in mind the need to distinguish between protein-coding and RNA-coding reading frames. This step is far from trivial. Important enzyme reading frames necessary to identify a complete pathway may, for example, escape detection because an extension of a reading frame escaped detection or an important enzyme activity is encoded on the opposite strand-not to mention notorious complications, particularly in higher eukaryotes, from alternative splicing or even editing of the mRNA. If reading frames have been established and translated (note whether the proper genetic translation table has been used), a function has to be identified for the reading frame.

Bioinformatics allows rapid comparison against existing data to achieve this. Thus protein reading frames can be assigned a putative function by sensitive sequence similarity comparisons. For detailed protocols and methodology the reader is referred to Bork *et al.* (1998) and Bork and Gibson (1996). Owever, to illustrate how this process allows completion of pathways, including identification of previously unnoted enzymatic activities, we will review some examples from a recent genome reannotation effort we conducted for *Mycoplasma pneumoniae.*

Reannotation of Molecular functions

The reannotation of molecular functions can potentially provide some answers regarding higher levels of cellular interactions such as metabolism (but also, for example, regarding pathogenicity factors). *Mycoplasma pneumoniae* is a pathogen with a compact genome that causes atypical pneumonia in man. We analyzed the complete genome of *M. pneumoniae* is a pathogen with a compact genome that causes atypical pneumonia in man. We analyzed the complete genome of *M. pneumoniae* five years after the original sequence was published. The reannotation could rely on more data, in particular from molecular and biochemical analysis of *M. pneumoniae* during the intervening five years as well as better software for sequence analysis developed during that time. A well known example is PSI-BLAST, an iterative sequence search algorithm including sequence alignment information from similar sequences found during the search. In this way all individual reading frames of *M. pneunmoniae* were reexamined.

We will review here some examples regarding better pathway identification or completion during the effort. For example, the reading frame $MPN547_{(MP295)}$ [the new MP genome identifier is given first, followed by the (old MP genome identifier)] was originally annotated as a homologue of MG369, a conserved hypothetical protein of unknown function. This function was also indicated in the update of the *M. genitalium* genome sequence in December 1999. However, detailed sequence analysis in the meantime revealed more about the function. For example, there was a sequence similarity apparent in PSI-BLAST searches.

The expected odds of observing a similarity of this extent by chance was estimated to be 1 in 10 million (nevertheless, such expected values, E-values for short, are nontrivial to estimate and should be taken only as an indication for stronger confidence and for finding a biological confirmatory context for the much higher than chance similarity detected between the two sequences by the algorithm). There was similarity of the N-terminal 300 amino acids to experimentally characterized dihydroxyacetone kinases from different bacteria and fungi. Another way to show this relies on clusters of orthologous genes, COGs for short.

The *M. pneumoniae* sequence can be shown to be a member of such a cluster of orthologous genes in which biochemically characterized members are dihydroxyacetone kinases. In *M. pneumoniae,* the dihydroxyacetone kinase domain could yield ATPby transforming dihydroxyacetone phosphate and ADP into dihydroxyacetone and ATP. This finding should be interpreted in the context of other reading frames. Thus the predicted activity can be metabolically connected to the remaining phospholipid metabolism predicted for *Mycoplasma pneumoniai.* It could provide the necessary supply of dihydroxyacetone phosphate via MPN051 $_{(MP103)}$ (glycerol 3- phosphate dehydrogenase reading frame, confirmed in reannotation). In addition, the remaining sequence of the MPN547 $_{(mp295)}$ reading frame (total length 558 amino acids) may in addition regulate or add to this predicted enzyme activity.

To give another example, a ribulose uptake pathway is apparent from the reannotation. Small oerons had been known previously for fructose (MPN078 $_{(MP077)}$, to MPN079 $_{(MP076)}$) and mannitol (MPN651 $_{(MP191)}$ to MPN653 $_{(MP189)}$). Ribulose was now found to be transported (MPN496 $_{(MP346),}$ MPN494 $_{(MP347)}$) and channeled via D-arabinose 6- hexulose 3-phosphate synthase (MPN493 $_{(MP348)}$) and D-arabinose 6- hexulose 3-phosphate isomerase (MPN492 $_{(MP349)}$) into fructose 6-phosphate and glycolysis. Of these proteins, MPN496 and MPN493 has not been functionally annotated before. MPN494 had been annotated as a hypothetical phosphotransferase.

These new functional assignments can also be made apparent by integrating data from SWISS-PROT annotations, (the well-known large database of protein reading frames) with further direct experimental data published for homologous proteins. In addition, we could now add the description of and direct experimental data (including mass-spectroscopy and sequence determination) for a small pentitol BC subunit of the ribulose transporter (MPN495 $_{(MP346.1)}$) not included preciously. An example of a more complex and non-metabolic pathway identified by genome reannotation in this study was the protein secretion system in *M. pneumoniae.*

Himmelreich *et al.* (1996) noted that they had identified the trigger factor, DnaK, SRP an FtsY as well as SecA, whereas from the channel-forming proteins only Secy coud be assigned, leaving the secretion pathway in complete. Now using extensive sequence analysis methods, and integrating recent literature, bioinformatics, and biochemical data, the reannotation identified the reading frames similar to SecD, SecE, and SecG, yielding a new, more complete picture of this secretory pathway in *M. pneumoniae.* Since several pathogenicity factors are secreted, the respective protein channels are potential drug targets. Similar analyses have been done on a number of different genomes, for example, a reannotation of the *Thermotoga maritima* genome.

Another aspect is the inclusion of structure and structural genomics, allowing analysis of structure-function relationships. This can be done either in a more summary way for whole genomes (e.g., comparing the spirochetes *T. pallidum* and *B. burgdorferi.* These lack, among other things, several parts of lipid biosynthesis and abundantly use, as is common in many other bacteria, TIM barrels and P-loop NTP hydrolases. In addition, detailed examination of the enxymatic function, structure, and evolution of the different TIM barrel families, comparing all available data, shows widespread, recruitment of enzymes between the central metabolic pathways.

PERSPECTIVE

We have given some examples of bioinformatic analysis of metabolic pathways, focusing on qualitative analyses; the detailed examples presented were mostly from our own work. The additional references given should make it clear that similar efforts have occurred in many laboratories around the world, and this chapter should serve as an incentive for further pathway analysis. Pathway analysis has become more and more important because sequencing, and in particular the number of newly sequenced genomes, is rapidly increasing.

Genome content is biologically meaningless present an important part of this information. The description of the enzymatic capabilities based on functional genomics (e.g., via a tool such as pathway alignment) make it possible to classify the great variety of organisms according to their different biochemical makeup. Medical aspects can now be tackled in unprecedented detail (e.g., analyzing metabolic changes in the liver, such as strong induction of the citric acid cycle and gluconeogenesis after major burns, which seems to be deranged mainly by strong induction of the liver antioxidant defense pathways. Missing enzyme activities and functional misassignment can happen all too easily.

If possible, several tools (e.g., not only PSI-BLAST but also domain analysis, COGs, etc.) should be combined. It has been explained how wevera indications for protein function from genomics, in particular regarding interactions, are combined for better predictions, including involved pathways. Similarly, different lines of evidence should be combined (e.g., not only the genome sequence but also biological data). It should also be taken into account that any interpretation of biological data may change over time as more knowledge becomes available. Experimental data for metabolic pathway analysis are almost always incomplete.

An interesting study by de Atauri *et al.* (2000) predicts the effect of uncertain boundary values in modeling a metabolic pathway. Such limitations and errors increase with integration to networks, cellular processes, or even the complete cell. In consequence, most models of this sort have to be interpreted with even more care. The large-scale structure of metabolic networks has been investigated by several authors. A fascinating observation is that metabolic networks are connected as scale free networks, meaning an exponential decrease in the number of more highly connected central metabolites. More complex organisms are more strongly connected than less complex ones. This improves adaptability to changing environmental conditions. The "diameter" of reactions (i.e., the number of enzymatic reaction of the network to a randomly chosen other one is kept close to the theoretically smallest possible number.

In this way environmental perturbations rapidly spread over the network and are optimally buffered by the different enzyme activities. The effect is thus a "small world" way of connecting. Scale-free networks also apply to internet hyperlinks or the U.S. Power grid and to the "small world" everyday experience that human relations are surprisingly well connected via common friends or acquaintances. The analysis of metabolic pathways is an overture to higher levels of simulation not only the fascinating regulation of the metabolic pathways (level + 1, so to speak), but ultimately the whole cell.

A further valuable efforts to bridge this gap while learning (or at least speculating) about prebiotic evolution is the analysis by Segre *et al.* (2000), modeling kinetically enhanced recruitment of simple amphophilic molecules leading to evolving and splitting catalytic non-cobalent assemblies. The road to exploring the full metabolic network of the cell is open, and there are a number of very useful tools to analyze metabolic pathways. Nevertheless, there is still a very long way to travel before we can grasp the bewildering complexities of living cells.

NETWORK CONTROL

Molecular biology, biotechnology, and bioinformatics have started to focus on the problem of gene-regulated metabolic network control. This problem cannot be circumvented, because no open reading frame can be expressed without he appropriate regulatory sequences. Moreover, some genes code for proteins that turn other genes on and off. Groups of these genes constitute networks with complex behaviors. Thee networks control other genes whose protein products catalyze specific biochemical reactions. Hence the small molecules that are substrates or products of these reactions can in turn activate or deactivate proteins that control transcription or translation.

Therefore, gene regulation can be said to indirectly control biochemical reactions in cellular metabolism, and cellular metabolism to exert control of gene expression. For these reasons, the interdependent biochemical processes of metabolism and gene expression can, and should, be interpreted and analyzed in terms of complex dynamical networks. Therefore modeling and simulation are necessary. To solve this problem, we have to bring together information from gene regulation and metabolic pathways. These data have been, and will be, stored systematically in specific databases that are accessible via the internet.

Recently many firms have been established that provide essential data for the solution of scientific and industrial problems and, even more important, the corresponding infrastructure. As a result, there are more than 200 databases available via the Internet for all known sequenced genes (e.g., EMBL), proteins (e.g., SWISS-PROT, PIR, BRENDA), transcription factors (e.g., TRANSFAC), biochemical reactions (KEGG), and signal induction reactions (e.g., TRANSPATH, GeneNet). Besides databases, simulators for metabolic networks, which employ most of the currently popular modeling methods, are also available via the internet. In addition to the classical methods of differential equations, discrete methods have become quite important.

The integration of relevant molecular database systems and the analysis tools will be the backbone of powerful information systems for biotechnology. This chapter introduces the basics of database and information systems for molecular biology. Based on this introduction, we will discuss the important topics of database integration for the automatic fusion of molecular knowledge. Furthermore, computer science is developing and implementing tools for the analysis of molecular data. The analysis of molecular data and of metabolic networks are of equal importance. The chapter presents the rule-based model of metabolic processes, which is the core of our simulation shell, and also shows the prototype of the MARG-Bench system, which integrates basic molecular database systems and the simulation tool.

Databases

For the physical storage and maintenance of data by means of computers, specific software is necessary. In the simplest case, the functions of the operating system are used and the data are stored in simple files. Therefore the biological software tools themselves specify the internal structure and sequentially write the data into related files. The tools are as different as the file formats used. These range from unreadable binary format to well standardized XML (Extensible Markup Language). In this respect the software engineer of each tool is responsible for the data storage and must implement methods for data access, updates and backup. This kind of data storage and access is often mistaken for a data base management system (DBMS).

Some important disadvantages of such cumulatively acquired data collections are lack of performance, data redundancies, and lack of standardized query languages, synchronous data updates, and explicit scheme information that can be called up. These disadvantages are often unimportant for individual uses, but in the use of multiple tool usage and massive global and public use, they present a problem. Nevertheless, the necessity of their use is generally accepted, because all collected data have to be analyzed, using several tools and techniques. It is important to enable efficient data access that is independent from the implementation of the storage mechanism. Consequently, for a common and wide-ranging

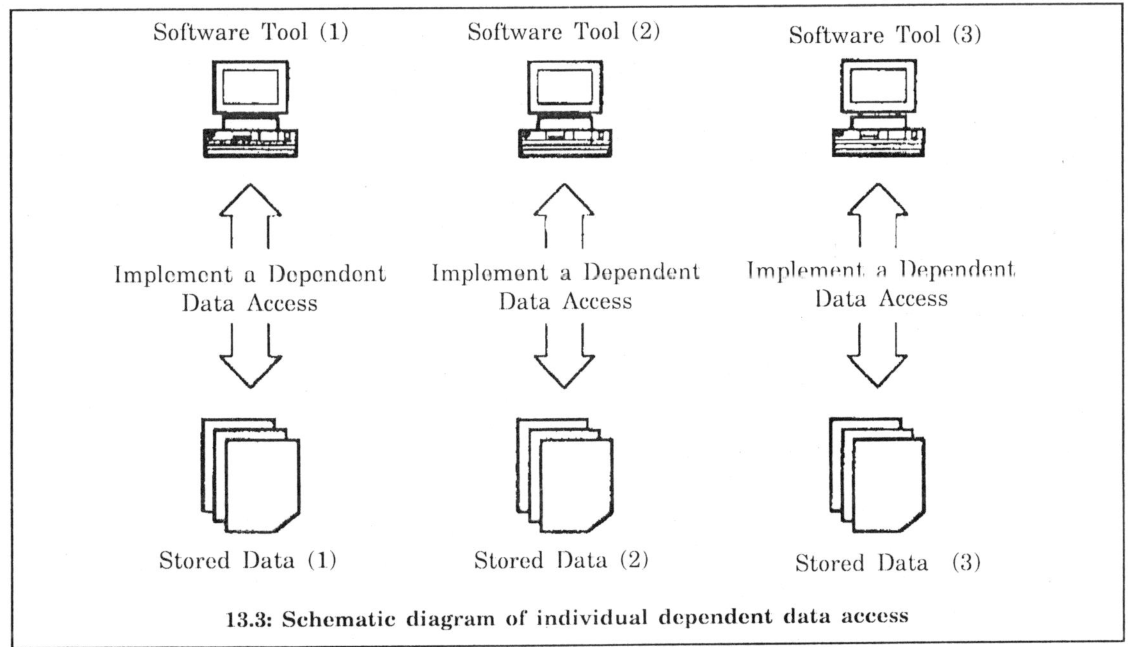

13.3: Schematic diagram of individual dependent data access

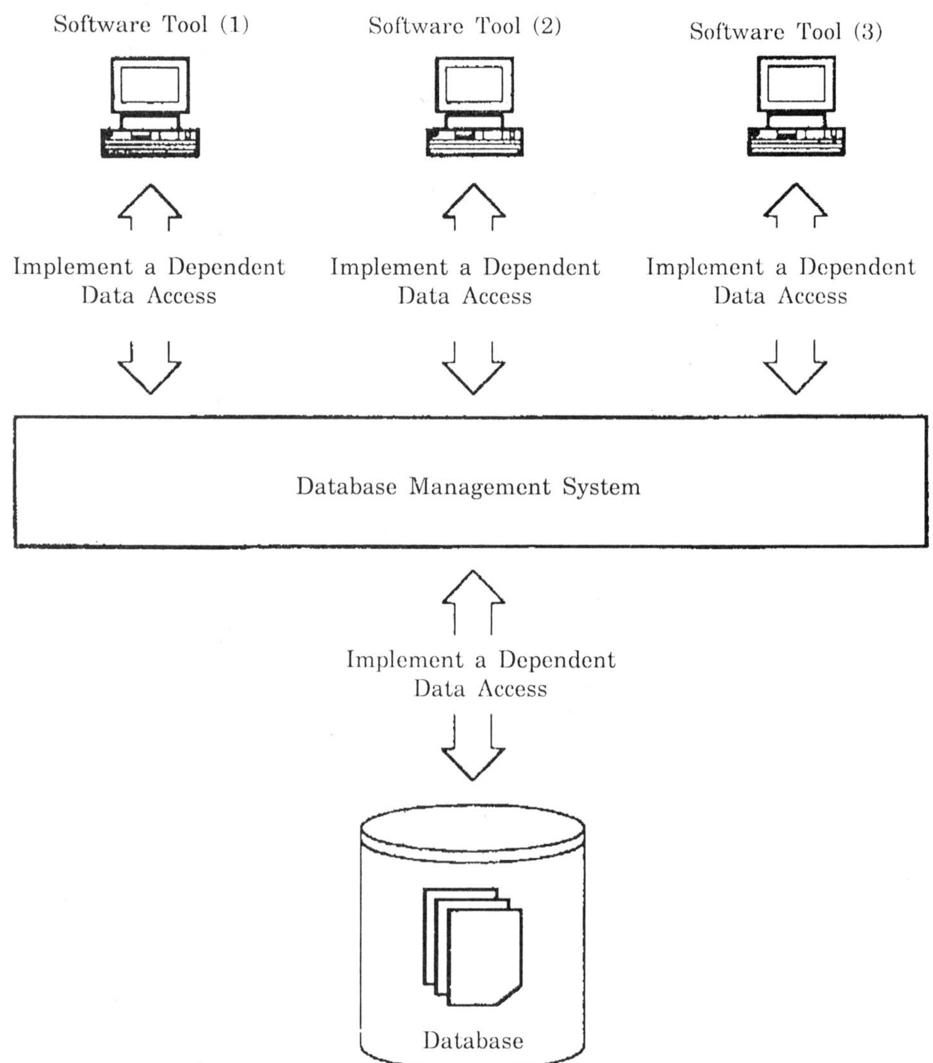

Fig. 13.4: Schematic diagram of data access using a database managemnet system

use of data, a homogeneous kind of data storage and access is essential. Therefore the main tasks of a DBMS are the following:

(i) Uniform management of all data

(ii) Provision of data operations such as storage, update, and search

(iii) Covering a unique data description of all stored data (scheme)

(iv) Consistency check

(v) Authorization control

(vi) Transaction management

(vii) Backup mechanism

Today many DBMS are available that differ in price, data model (hierarchical, relational, object-oriented, object-relational etc.), features (supported operators, transactions, indexes etc.), and query mechanism and languages (SQL, OQL, QBE, etc.). Popular examples of such DBMS are Oracle, INFORMIX, DB2, Microsoft SQL server, POET, and Object Store.

The main advantages of such DBMS are very efficient searching and access opportunities, common usable data, few data storage redundancies, and easy data access by software tools. Thus, DBMS are increasingly used for biological data. In most cases stored biological data are not independent. Rather, they are processed by related biological tools. That means that tools and related data represent a coupled system, which is called an information system (IS). An IS is a complex, coupled software system for information processing. In other words, we can define an IS as a logical union of he data itself, tools and the import of external data. Consequently a DBMS may be identified as the central substructure of persistent data storage of such a system.

Molecular Databases

In molecular biology a large number of research projects are currently producing an exponentially increasing amount of data. A popular example is the Human Genome Project. In this context alone about 3×10^9 base pairs and mapping data must be stored. The most popular public sequence database, EMBL, includes 4.7 million entries of primary nucleotide sequences and related data.

If these databases are set in relation t the count of molecular databases, an impression of the data volume may be given. These data are often published in publicly accessible sources. Often the categories are very different. One investigation has divided sequence and related data into 16 categories. In correspondence with this, about 200 WWW-based data sources are listed. This is meant to be only a fraction of the overall number of databases available. In most cases these data are stored for processing, analyzing, and other functions.

A common approach is to collect all available data, then decide which are dispensable. However, storage capacity is not a problem today. Nowadays the majority of the molecular database provides store their research results in "quick-and-imperfect" systems. In most cases simple flat files are used, which are managed by the directory and file system structures of the operating system. Data access is achieved with software that can handle the individual file format. As an interface for external access, use of the WWW is popular. In those cases HTML forms coupled with CGI scripts are used, and the data obtained by this method are presented in HTML pages, using tables or other formats.

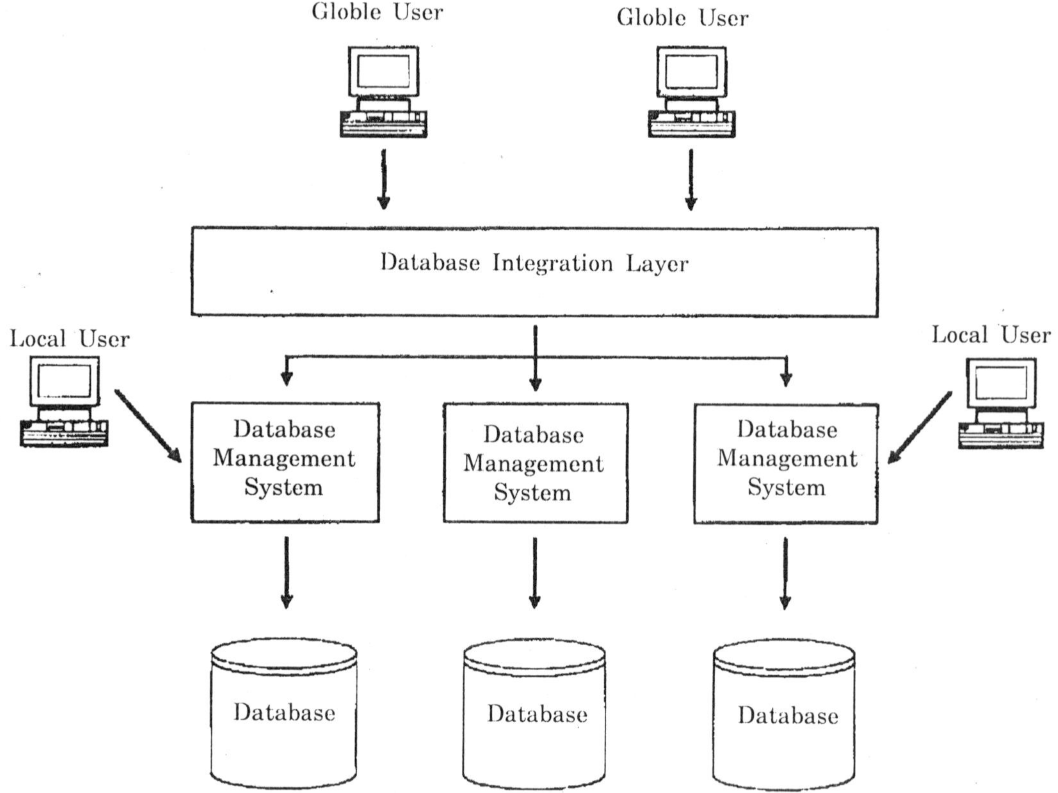

Fig. 13.5: Integrative data access using database intrgation software

A popular example is the KEGG system. One of the reasons why the content of these databases constantly increases is clarified in a special issue of *Nucleic Acids Research* that is published annually in January and gives an overview of databases as well as presenting selected systems in more detail. In order to provide a rough picture of the extent currently available databases, a selection of important databases is listed in the appendix.

Information Fusion

In order to use these special databases, the user has to connect to each database separately. However, the integration of databases can help in detection of new information. An example is the interaction between aligned gene sequences of an organism to supplement an unknown part of a metabolism or suggest and predict alternative pathways. At first sight, two basic problems may arise when handling such a distributed data retrieval:

(i) How does the user access each relevant database?

(ii) How can the several query results be merged into a joined data set?

These problems lead directly to the field of database integration. The integration of databases has several parts. Data merging is the overcoming of the distributed data storage. On the one hand, the data overlap between the several databases and on the other hand, the data are totally different. To produce a global data set, it becomes necessary to implement a unique global data store. There re two different ways to achieve this: copying all needed data into one large database (materialized and leaving the data where they are and merging them virtually (non-materialized). Derivation of an integrated scheme starts from the fact that every local database provides its own data scheme. But in the case of global access, a unique scheme over all databases is needed.

The method used to solve this problem specifies the degree of scheme integration. Consequently, it is possible got define one global scheme over all local schemes, meaning that all locally modeled data are integrated into a global scheme (bottom up). The other kind of scheme integration is to specify which data are needed by individual application scenarios, thus making it possible to model several partly integrated global schemes (top down). For adequate access to the integrated data, a unique data access and querying method is needed.

For this, the mapping of local data models and access methods to global ones has to be carried out. In the case of read-only data access, this no problem, but for writing operations, problems occur that include transactions of global management methods for simultaneous writing on the same object, data consistency mechanisms, and global integrity policies. Besides the data retrieval, the quality of the integrated data is important. Therefore, it must be determined how reliable the data actually are.

It is generally accepted that databases include faulty or low-quality entries such as incorrect sequences, missed annotation, and wrongly assigned enzyme numbers. Consequently, these quality problems are propagated during the database integration to the global data set. Hence, a mechanism for quality control of the databases and their entries must be established.

A software layer has to be provided that offers methods for data integration. Approaches for software layers deal with such problems an inadequate support of standardized query languages, non-appropriate programming interfaces, and insufficient, consideration of individual user requirements. These problems catalyzed the implementation of an alternative approach, called MARG-Bench. Moreover, new concepts of "information fusion" have been developed that are focused on "information discovery" and on the bases of database integration. The notion of transformation includes integration, filtering, analysis, and preparation of data aimed to discover and represent the hidden knowledge.

The main difficulties are the unique access to proprietary flat files, Internet interfaces, and the large amount of data.

One of the most developed technologies of web integration of molecular databases used SRS (Sequence Retrieval System). It is based on local copies of each component database, which have to be provided in a text-based format. The results of the query are sets of Internet links, through which the user can navigate. As of 2000, more than 100 databases on molecular biology are integrated under SRS. However, within the limitations of this approach, data fusion is still the task of the user.

Real data fusion (i.e., data for one real-life object, such as an enzyme) from two different databases (e.g., KEGG and BRENDA) is not found to be represented twice by different web pages. Therefore, research groups try to integrate molecular databases on a higher level than the SRS approach. Results of current database research, including federated database systems, data warehousing architectures, and data mining techniques, are applied. Many problems in bioinformatics require the following:

1. Access to data sources that are large in volume, highly heterogeneous and complex, constantly evolving, and geographically dispersed.
2. Solutions that involve multiple, carefully sequenced steps
3. Smooth passing of information between the steps
4. An increasing amount of computation
5. An increasing amount of visualization

BioKleisli is an advanced technology designed to handle the first three requirements directly. In particular, it provides, the high-level query language, CPL, that can be sued to express complicated transformation across multiple data sources clearly and simply. In addition, while BioKleisli does not handle the last two requirements directly, it is capable of distributing computation to appropriated servers and initiating visualization programs.

Integration Approach

In consideration of the advantages of integrative data access and the existence of inflexible approaches, a new system for database integration was developed. According to the latest findings, the system offers a scalable and flexible approach. This is achieved by the concept of wide ranging database access by a wrapper (adapter) technology. The data merging is non-materialized, using set operations. Specific data integration-related schemes for the user can be defined, and the standardized access to the integrated data can be performed by an SQL-like language. For comfortable use, a JDBC driver is available.

Modelling of Metabolic Networks

Using the rule-based modeling of metabolic processes, the simulation environment MetabSim for he analysis and visualization of gene-controlled metabolic processes was implemented. The advantages of this concept is the integration of relevant molecular database systems that are available via the Internet.

Related Work

The availability of the rapidly increasing volume of molecular data on genes, proteins, and metabolic pathways improves the capability to study cell behavior. To understand the molecular logic of cells, it must be possible to analyze metabolic processes and gene networks in qualitative and quantitative terms. Therefore, modeling and simulation are important methods.

Mathematical models may be subdivided into two categories: analytical and discrete. Analytical models perform the processes of elements, acting as functional relations (algebraic, integral-differential, finite-differential, etc.) or logical conditions. An analytical model may be studied by qualitative, analytical, or numerical methods. Analytical models are generally based on integral and differential systems of equations. The paper published by Waser *et al.* 1983) presents a computer simulation of phosphofructokinase. This enzyme is a part of the glycolysis pathway. Waser and coworkers model all kinetic features of the metabolic reaction, using computer simulation. This computer program is based on the rules of chemical reaction, which are described by differential equations.

Franco and Canelas (1984) simulate purine metabolism by differential equations; each reaction is described by the relevant substances and the catalytic enzymes, using the Michaelis constant of each enzyme. Discrete models are based on state transition diagrams. Simple models of this class are based on simple production units, which can be combined. Overbeek (1992) presented an amino acid production system. A black box with an input set and an output set displays specific production. The graphical model of Kohn and Letzkus (1982), which allows the discussion of metabolic regulation processes, is representative for the class of graph theoretical approaches.

Kohn and Letzkus expand the graph theory by a specific function that allows the modeling of dynamic processes. In this case, the approach of Petri nets is a new method. Reddy *et al.* (1993) presented the first application of Petri nets in molecular biology. This formalism is able to model metabolic pathways. The highest abstraction level of this model class is represented by expert systems an object-oriented systems. Expert systems and object-oriented systems are developed by higher programming languages (Lisp, C++) and allow the modeling of metabolic processes by facts/classes (protein and enzymes) and rules/classes (chemical reactions). The grammatical formalization is able to model complex metabolic networks. Within this class of models one may consider the cell model (E-CELL), developed by Tomita *et al.* E-CELL is the generic computer software environment for modeling and simulation of whole cell systems.

It is an object-oriented environment for the simulation of molecular process in user-definable models, equipped with interfaces that allow observation and intervention, and written in C++. Using E-CELL, one could construct a hypothetical cell with a definite number of genes sufficient for transcription, translation, energy production, and phospholipid synthesis.

Rule-based Modelling

The model is an extension of the Semi-Thue system. Using a universal rule, this formalization allows the representation of genetic, biosynthetic, and cell communication processes.

Furthermore, it is necessary to expand this discrete model by adding concentration rates for each metabolite. Metabolites are substances or substance concentrations that can be modified by biochemical reactions. Enzymes are specific proteins that catalyze biochemical reactions. Inducers and repressors are metabolites that can accelerate or slow down/prevent biochemical reactions. The biochemical space (cell state) of a cell is a mixture of these components. The set of all cell states will be denoted by Z. by these definitions the abstract metabolism is defined by the actual cell state and the biochemical reaction rules. The metabolic rules is the basic unit of the metabolic system.

Let Z be a finite set of cell states. A 5-tuple (B, A, E, I, p) with $p \in [0, 1]_Q$ and $B, A, E, I \in Z$ is called a metabolic rule. p is rule probability; B (Before) is a set of preconditions; A (After) is a set of post-conditions; E (Enzyme) is a set of catalyzed conditions; and I (inhibitor) a set of inhibitor conditions.

Example 13.1. The reversible biosynthesis product glucose-6-phosphate ↔ fructose-6-phosphate will be catalyzed by the enzyme glucosephospat-isomerase. This process can be described by two rules:

- ({glucose-6-phosphate}, {fructose-6-phosphate}, {glucosephosphate-isomerase}, {}, p)
- ({fructose-6-phosphate}, {glucose-6-phosphate}, {glucosephosphate-isomerase}, {}, p).

On the basis of the metabolic rule, the basic model can be defined.

$G = (Z, R)$ is a metabolic system. Z is a finite set of cell states, $S \in Z$ is the start state, and R is the metabolic rule set.

In the following paragraph, the semantics of the metabolic system are defined. The integration of the analyzed metabolic features is the basis of this formalization and is the reason for specifying a stochastic parallel derivation mechanism that will describe the change of actual cell states, depending on the specified rule set. Therefore, the set of all activated rules must be fixed. This is the first step of the derivation process. A rule is "*activated*" if its preconditions are elements of the actual state $z \in Z$. Moreover, effects of inducer and inhibitor elements must be considered. If such metabolites are elements of the actual state z, then the probability of this rule will be modified by inhibitor and inducer effects. The special function CALCULATE (z, r) will determine the absolute probability value as rule r depending on state z.

A random generator (RANDOM), using the absolute probability value of the input, works as a Boolean function and will produce either positive or negative results (true or false).

Once the Boolean value is true (false), rule $r \in R$ is described as activated (deactivated) and goes into action.

Let $G = (z, R)$ be a metabolic system, $r = (B, A, E, I, p) \in R$ a rule, and $z \in Z$ a cellular state. r will be activated by z (r_z), if $Ax \subset B$, $x \in z$. $A(z) = \{r \in R : r$ is activated by z$\}$ is the set of rules activated by z.

Example 13.2: Let $G = (Z, R, z_0)$ be a metabolic system and $z_0 = \{S, D\}$ and $R = \{r_1, r_2\}$ with

$r_1 = (\{S\}, \{H, S\}, \{D\}, \{L\}, 0.8)$

$r_2 = (\{D\}, \{X\}, \{E\}, \{D\}, 0.6)$

For the configuration z_0 $A(z_0) = \{r_1\}$ the rule r_2 is not activated because the repressor is available.

Any activated rule $r \in R$ can go into action. The action of r will modify the actual cellular state of the metabolic system. Elements of the actual cellular state, which are elements of the Before set of rule r, will be eliminated in z and all elements of the After component will be added to z. Therefore, the action of rule r can produce a new state, $z' \in Z$.

Let $G = (Z, R)$ be a metabolic system, $z \in Z$ the actual cellular state, and $r_z = (B, A, E, I, p) \in R$. The action of r_z is defined thus: If RANDOM (CALCULATE (z, r)) = true, then $z' = (z - B) \cup A$. The action of r_z will be written as $z \to_r z'$.

According to the metabolic system defined in Example 2, the action of r_1 will produce the state z' = {H,S,D}.

The one-step derivation of a metabolic system is defined by the (quasi) simultaneous action of all activated rules. Therefore, the set of all activated rules is considered and two new sets are determined: the Before set and the After set. The Before set includes all B elements of the activated rules. The definition of the After set is analogous. Using these sets, the one-step derivation can be interpreted as an addition and subtraction of elements.

Example 13.3: Let $G = (Z, R, z_0)$ be a metabolic system and $z_0 = \{B, C, E\}$ and $R = \{r_1, r_2, r_3\}$, with

$r_1 = (\{B\}, \{S, B\}, \{C\}, \{\ \}, 0.9)$

$r_2 = (\{C\}, \{F, C\}, \{\ \}, 0.3)$

$r_3 = (\{C, F\}, \{B, C\}, \{E\}, \{\ \}, 0.3)$.

$A(z_0) = \{r_1, r_2\}$ is the set of activated rules. For z_0 we can identify one step derivations that will produce the following states:

$\{B, C, E, S\}$ (action of r_1)

$\{B, C, E, F\}$ (action of r_2)

{B, C, E, S, F} (action of r_1 and r_2)

{B, C, E} (empty action).

Let $G = (Z, R)$ be a metabolic system, $z \in Z$ the actual cellular state, $A(z)$ the set of all activated rules under z, and $B_z = \{B : \exists r \in A(z) \wedge B \in r_z\}$ and $A_z = \{A : \exists r \in A(z) \wedge A \in r_z\}$. The simultaneous action of $A(z)$ is called one-step derivation if $z' = (z - B_z) \cup A_z$. It is written $z = z'$.

Each action can be interpreted as an independent event. Therefore, the probability of each one-step derivation can be calculated in terms of the absolute probability values of all activated and deactivated rules. In the simulation system, this will be done by multiplying these values.

However, based on the one-step derivation, the derivation can be defined inductively, and a probability for any derivation can be calculated.

Let $G = (Z, R)$ be a metabolic system. $x \in Z+$ is a derivation in G iff $|x| = 1$ or $|x| > 1$; $Ey' \in Z^*z'$, z'' e Z : $x = z'z''$ y and $z''y$ is a derivation; and $z' \Rightarrow z''$.

Example 13.4: Let be G the metabolic system defined in example 3. The following sequence of cellular states describes a derivation:

{B, C, E} → T {B, C, E, S} → T' {B, C, E, S, F} → T'' → {B, C, E, S, F} ...

where T = {r_1}, T' = {r_1, r_2}, and T'' = {r_2, r_3}.

In the case of analytical modeling, it is necessary to expand the model, using abstract concentration rates. To realize this requirement for each component (metabolite), specific interger values must be assigned. These values can be interpreted as abstract concentration rates. Since these impacts can be included in the metabolic system, using the formalization of multi-sets, the definition of the metabolic system must be modified. Regarding the rule activation, the concentration rate of any Before component must be satisfied in regard to corresponding metabolites of the actual state.

The concentration rate of these metabolites must be higher than or equal to the corresponding Before component of this metabolic rule. Moreover, the function CALCULATE must be modified. In this case, the influence of all concentration rates of inductor and repressor metabolites will determine the absolute rule probability. All activated rules can go into action simultaneously. With regard to corresponding metabolites of the actual state (integer values), the addition and subtraction of the concentration rates of all Before and After components is necessary. In this chapter, only the fundamental part of this formalization is presented.

Let $z \in Z$ be a state. The multi-set $k: z \rightarrow \mathbb{Z}_0$ is metabolic concentration. K denotes the set of all metabolic concentrations.

Example 13.5: Let z = {glucose, lactose, RNA-polymerase}. |34 glucose, lactose, 15 RNA-polymerase| defines a metabolic concentration of 34 molecules glucose, one molecule lactose, and 15 molecules RA-polymerase.

Based on the formalization of multi-sets, the analytical metabolic system, which enables the discussion of kinetic effects, can be defined. G = (Z, R, k) with A $\in$ Z the start state, k $\in$ K a multi-set (K : A $\rightarrow$ N), and R a finite set of metabolic rules, where r = (B, A, E, I, p) $\in$ R with $p \in [0, 1]_Q$ and B, A, E, I are metabolic concentrations, is the metabolic system.

The definition of activation is necessary. Regarding multi-sets, the activation of any rule r $\in$ R depends on the specified concentration rates of each rule component, the concentration rates of the actual state, and the absolute rule probability.

Let G = (Z, R, k) be an analytical metabolic system, r $\in$ R a rule, and z $\in$ Z a cellular state. *r* is activated by z (r_z), iff $\forall x \in B$, x $\in$ z, and k(x) < k(z_x). A(z) = {r $\in$ R : r is activated by z} is the set of rules activated by z.

Based on this definition, the one-step derivation can be modified. All activated metabolic rules can go into action simultaneously. Regarding the set of activated rules, all After concentration rates must be added to the actual state and all Before concentration rates must be subtracted from the actual state.

G = (Z, R, k) is an analytical metabolic system; *z*, the actual cellular state; A(ẑ), the set of all activated rules under *z*; and

$$B_Z = \{B : \exists r A(z)\ B \in r_Z\} \quad and \quad |B_Z| := \sum_{b \in B} k(b)$$

$$A_Z = \{A : \exists r A(z)\ A \in r_Z\} \quad and \quad |A_Z| := \sum_{b \in A} k(a)$$

The simultaneous action of *A*(*z*) is called one-step derivation iff $z' = \{x : x \in A_z$ or $\forall x \in z$, y $\in$ $B_z x$ = y and k(x) – k(y) > 0}. The one-step derivation is written z $\rightarrow$ z'.

Using the one-step derivation operator, one can define the derivation of an analytical metabolic system inductively.

Example 13.6: Let G be an analytical metabolic system with k = [6A, 8B, 3E] and the rule set R with

r_1 = ([2A], [3B], [E], [X], 0.8)

r_2 = ([2B], [3D], [E], [X], 0.5).

A(k) = {r_1, r_2} and the set {[6A, 8B, 3E], [4A, 11B, 3E], [6A, 6B, 3D, 3E], [4A, 9B, 3D, 3E]} describes the different one-step derivations.

Application

The implemented universal metabolic rule allows the formalization of biosynthesis, gene expression, gene regulation, and cell communication processes. Regarding biosynthetic processes, the before, after, inducer, and repressor components are used. For example, Enzyme E_1 will catalyze the biochemical process S_1 into S_2. This can be expressed by B = S_1, = S_2, and E = E_1. Moreover, one can add any concentration rates. For example, B = $15S_1$, A = $12S_2$, E = $2E_1$. The probability value will model the flux of this biochemical rule, which can be influenced by specific inducer and repressor metabolites, depending on their concentration rates.

In the case of simple cell communication processes, only the before or after component will be used. From this, we obtain the following interpretation: Substance A enters the cell by endocytotic processes. Therefore, a rule must be defined in which only the after component will be assigned specific substances. Moreover, such processes can be influenced by specific events that can be formulated regarding inducer and repressor components.

Normally, metabolites will disintegrate after a specific time interval. This can be expressed by a rule that represents only the specific before component. Moreover, concentration rates and specific influence components can be defined.

In the case of gene regulation, the activity of operons can easily be modeled. If we choose an operon that represents two structure genes (S_1, S_2), two operator genes (O_1, O_2), and one enhancer sequence, then this can be expressed by

B = [RNA-polymerase, ribosome, amino acid, tRNA]

A = [S_1, S_2], E = [IO_1, IO_2] and I = [O_1, O_2].

However, this model allows the simultation of complex metabolic networks, and the grammatical formalization allows the definition and implementation of different languages. These languages represent specific biological aspects. For example, it is possible to produce the set of all possible pathways, to produce metabolic pathways depending on specific conditions (for example, the probability value), to search for the appearance of specific substances (for example, toxic substances), and so on.

Theoretical Aspects

The set of the attainable configurations is an infinite set, and the set of all derivations is enumerable. Moreover, the set of all configurations is undecidable (Hofestadt, 1996). Use of concentrations (multi-sets) is the main reason for the undecidability. However, this result implies that no interesting question can be solved in the field of biotechnology research.

In practice, biochemical systems are restricted. In the model, the depth and width of the derivation process can be restricted. Therefore, important questions are decidable, and the complexity of the derivation algorithm must be discussed. If the derivation depth is restricted, the language $L(G, i)$ is decidable. $L(G, i)$ is the set of all configurations that can be produced from the start configuration by the application of up to i derivation steps. Hence, for a metabolic system with the generation depth i, it is decidable if k is a member of $L(G, i)$. Based on the exponential complexity of the derivation process, this question cannot be solved in practice if i is high. Thus the calculation of a derivation tree is not possible in practice because the one-step derivation represents in exponential time complexity. However, the simulation tool is used, the derivation depth and/or width must be restricted.

MetabSim

The simulation system MetabSim is the current implementation of the rule-based model described above. It consists of two main parts. The first part, the object-oriented data structure, describes the metabolic grammar. The main structure here is the metabolic rule. A rule contains the stoichiometry of substrates and products, enhancers, inhibitors, and factors, and the elasticity coefficients of one complex reaction. The user can instantiate the rule type to build up rules according to reactions of a metabolic network. The second data type is the metabolic state. The user can define a metabolic state to act as a root configuration. All later calculations will be based on this state.

The whole date structure is mapped into a database so that all rules and states are stored therein. The second part of *metabSim* is the derivation logic. Because the system has been designed to be modular, several derivation modules can be implemented and applied independently. Figure 13.4 shows the information flow in the rule-based simulation system. After the rule set and the root configuration (default cell states) are defined, the derivation logic can be applied to the data. In the first step, the Rule Selection module accesses the correct state and calculates the rules that can be applied because their premises are true

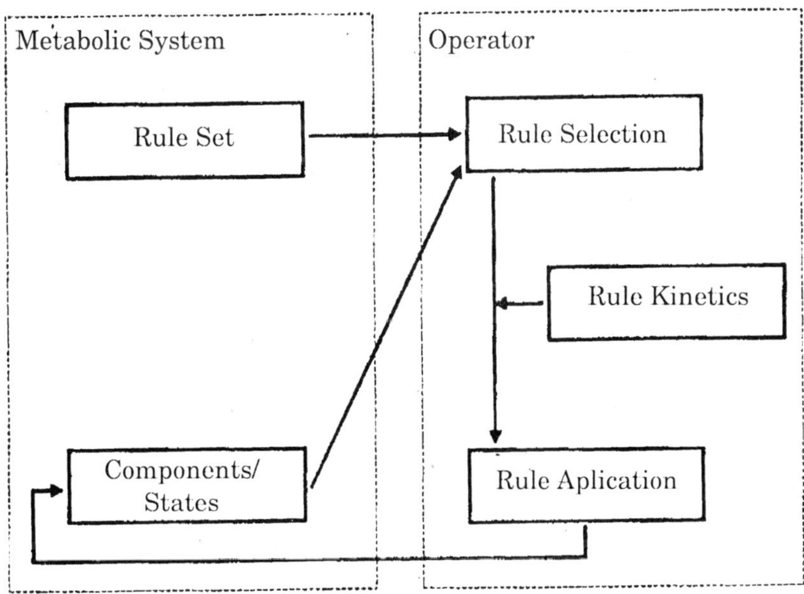

Fig. 13.6: Mechanism of the derivation process in MetabSim operators.

in relation to the current state.

The Rule Application module calculates the successive configuration(s). Optionally, the reaction time is applied by a Rule Kinetics module. The new configurations (states) are the input for the next derivation step. In biochemistry, for a set of single metabolic systems, the systems behavior is described by the use of differential equations. But the problem is that this approach does not discuss all interactions in large networks. It was decided to apply a simpler and more abstract method as a compromise between the mathematical modeling and the data outcome of molecular databases. The BRENDA database stores kinetic parameters of enzymes.

From this system one can get the Michaelis-Menten value, the turnover rate, and the type of regulation of the enzyme. In MetabSim several kinetic behaviors are implemented the constant flux and the linear, hyperbolic, and sigmoid dependencies from the substrate concentration. Their calculation requires the Michaelis-Menten constant (Km), the maximum reaction rate (V_{MAX}), and the enzyme type (allosteric, hyperbolic, etc.). The types read from the databases are mapped into the appropriate kinetic behavior and the calculation is done using the given parameters.

As an example, a rule network for the glycolysis pathway has been built up to simulate the consumption and the production of ATP. The following rules illustrate the stoichiometry of the metabolic system.

r_0 = ({D-glucose, ATP}, {glucose-6-phosphate, ADP}, {hexokinase, hyp})

r_1 = ({glucose-6-phosphate}, {(fructose-6-phosphate}, {phosphohexoseisomerage, hyp})

r_2 = ({fructose-6-phosphate, ATP}, {fructose-1, 6-bisphosphate, ADP}, {phosphofructokinase, sgm})

r_3 = ({fructose-1, 6-bisphosphate}, {glyceraldehyde-3-phosphate, dihydroxyacetone phosphate}, {aldolase, hyp})

r_4 = ({dihydroxyacetone phosphate}, {glyceraldehyde-3-phosphate}, {triosephosphate isomerase, hyp})

r_5 = ({glyceraldehyde-3-phosphate, Pi}, {1, 3-bisphosphoglycerate}, {phosphoglyceraldehyde dehydrogenase, hyp})

r_6 = ({1,3-biphosphoglycerate, ADP}, {3-phosphoglycerate, ATP}, {phosphoglycerate kinase, hyp})

r_7 = ({3-phosphoglycerate}, {2-phosphoglycerate}, {phosphoglycerate mutase, hyp})

r_8 = ({2-phosphoglycerate}, {phosphoenolpyruvate}, {enolase, hyp})

r_9 = ({phosphoenolpyruvate, ADP}, {pyruvate, ATP}, {pyruvatekinase, hyp})

In this rule set, metabolites and enzymes are included. The metabolite D-glucose is consumed by the rules r_4 and r_5. One ATP is consumed in the first part of the glycolysis, and 2 ATP are produced in the second part. This concludes with the double application of rules r_7 and r_9. In Fig. 13.8 the system is drawn as a graph by the Metabuis environment. The application of the operator leads to the substance concentration development.

MARGBench

The purpose of our project, which is supported by the German Research Council, is to present a virtual laboratory for the analysis of molecular processes. Therefore, we provide a full scalable system for a user-specific integration of different heterogeneous database systems and different interfaces for accessing the integrated data with special analysis tools (e.g., the simulation environment MetabSim, which was described earlier in this chapter).

Architecture of MARGBench

The system is divided into two main parts, the integration layer and the application layer. The integration layer consists of three modules: data acquisition, data storage and graphical data management.

In general, the BioDataServer realizes a logical nonmaterialized database integration based on the concept of federated databases. A workable Internet access to the molecular biological databases is the main prerequisite for a database integration. Within this context several problems must be solved:

- Different interfaces (CGI, JDBC, etc.)
- Different query languages (SQL, OQL, nonstandardized,etc.)
- Different data structures (HTML, flat files, database objects, etc.)
- Different data models (ERM, OO, etc.)

For such a homogeneous access to the several data sources, a functional interface was defined that is implemented by special software modules, called adapters. Using semiautomatic adapter generation, it will be possible to dynamically connect relevant data sources. In the case of an HTML data source, the adapter accesses the specific URL and parses the resulting HTML page. For this generation a description file is necessary. Such a description file contains structure and syntax information about the data source, which should be integrated into our system.

This information enables a mapping between the data fields of the docked data sources and the attributes of an integrated user scheme. Regarding the user requirements to the integrated data, it is possible to specify integrated data schemes. These schemes can be defined or manipulated, using a special language. They describe the accessible attributes of integrated data sources in transparent form. In order to obtain a complete and wide spectrum of data, integration of as many databases as possible is recommended. Using this special integrated user scheme, the BioDataServer combines the outcomes of adapter queries into integrated global results, known as information fusion.

The scheme is relational and defines the source for each attributed and its internal dependencies. This is the basis for performing logical data integration and provides a relational view of the fused data. Thus it is possible to retrieve integrated data by a subset of SQL queries. As can be seen, an automatic mechanism is necessary to merge the partitioned data values from the various databases. This is one take of the BioDataServer, and can be solved using mathematical set operations. The prerequisite to access the database integration server by computer programs is the definition of an interface. Because the server should be accessible via the Internet, a communication protocol and a query language must be specified.

Many database systems support a subset of SQL as querry language, which in turn is based on the relation model and is well standardized. This was the reason for supporting SQL elements by the BioDataServer. Different techniques in the field of interfaces for remote database access have been established (e.g., JDBC and ODBC). ODBC is supported only by Microsoft platforms. Therefore, the BioDataServer currently offers a JDBC driver, which provides a standardized database access to JAVA applications. Consequently, any JAVA platform can access the BioDataServer by related JAVA programs.

The main advantages of this BioDataServer are the transparent physical database access, the dynamic building of new nonmaterialized, logical integrated databases, a standardized access interface, a client-server capability, and platform independence. With the implemented JDBC driver the integration service of the BioDataServer is also independently useful for other external JAVA applications. The next level in the integration layer is the BioDataCache, which handles the local storage of the fused data in a user-specific integration database. Thus it is possible to build individual integrated databases that reflect the individual user's application requirements, and to perform data analysis, cleaning, improvements, and enrichment.

The user is able to interactivity specify and create the integration database in interface definition language (IDL) syntax. If the IDL is ready, the service modules will be generated automatically. The individual data import is based on specific, integrated user schemes of the BioDataServer, which must be defined previously. The access to these integration databases is possible by the Common Object Request Broker Architecture (CORBA) and OQL. The BioDataCache provides the materialized layer for the data integration that is based on CORBA.

The access for importing integrated data from the BioDataServe is possible using the JBDC driver of the integration layer. Furthermore, the BioDataCache has an integrated user scheme for the selection of attributes that should be integrated. An empty database related to the integrated scheme is generated automatically. Once data from the integration service are read, they are stored in the underlying standard object-oriented database. By storing the fused information in the cache, an separate new data source will be created. This new user-specific integration database represents a metadatabase and is comparable to a data warehouse.

The offered CORBA interface, similar to the BioDataServer, enables other software tools to access the BioDataCache. The third part in the integration layer is represented by the BioDataBrowser. It is automatically generated during the generation of an integration database. This module allows the user to graphically manage and browse the fused data. Its functions are similar to a Windows file browser. Furthermore, a JAVA interface is offered to access this component within JAVA applications. The development of DBMS-supported applications forces the programming of database-related user-interactive components to establish database connections, query the data, transmit the results, store the data, and so forth. The BioDataBrowser provides this feature and can be included as a component in different JAVA applications.

Application of MARGBench

The application of MARGBench is done at different levels, because the system consists of plugable components. Every user is able to access the components that are essential for his or her specific integration problem. One possibility is the use of the online integration provided by the BioDataServer component. Here a uniform SQL interface with client/server and multiuser/multiclient feature is available. The software METATOOL is an independent system that has been written in the C programming language. Separate programs are necessary to generate the input file and to read the output files.

The input for the integration architecture is obtained by the BioDataServer. This client program has been implemented using the PHP programming language, and enables the program to be used via the Internet. The second way to use MARGBench is to access automatically generated integration databases. Once the integration database has been established, the user can load his or her integration database with integrated data. For coupling application programs, the CORBA interface and the BioDataBrowswer can be used. As a reference application, MARGBench is used with the simulation tool MetabSim.

It accesses a specially configured BioDataCache to obtain data and to build the rule network. First, a BioDataCache for storing information about metabolic networks is configured. After the data structures for the cache are established, the BioDataCache installation tool is used to compile and install an integration database (cache) called MetabNets. From the tool a CORBA interface is obtained for accessing the data structures previously defined. This interface can be used to submit database queries and to operate on cache objects by creating, modifying, and deleting operations. Data structures of MetabNets are, for instance, the classes metabolism, pathway, reaction, and enzyme. Once the cache scheme is defined, the user can start the integration procedure. During this process the BioDataServer is queried and objects of MetabNets are instantiated with the query results. The data structures in the MetabSim program are also defined in CORBA.

Now it is necessary to implement a mapping algorithm to produce rules out of the MetabNets objects. The program containing this algorithm accesses the BioDataBrowser and

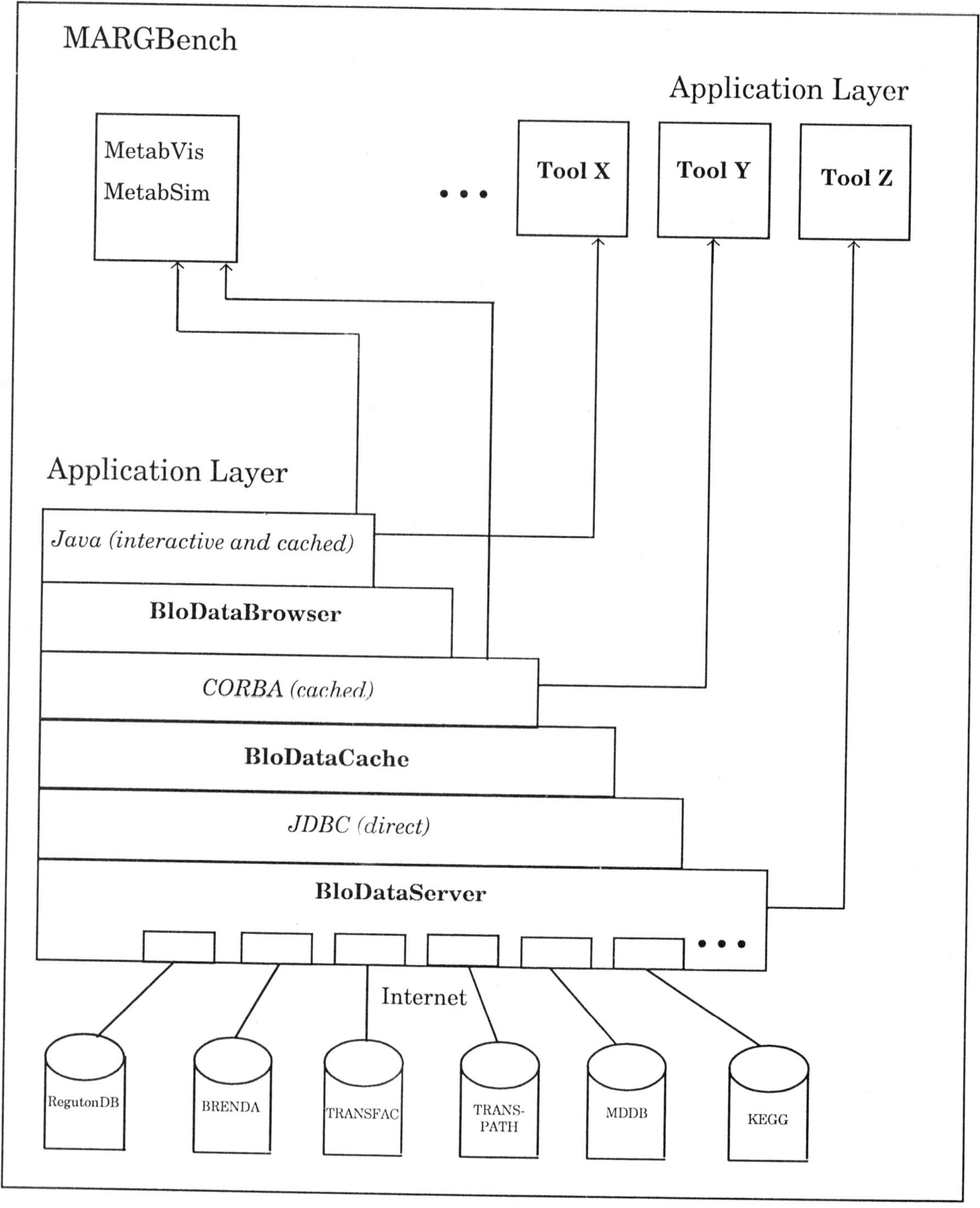

Fig. 13.8: Architecture of MARGBench

the BioDataCache. In this context, the browswer is used for the interactive selection of metabolic pathways and single reactions. An example of enzyme information viewed with the BioDataBrowswer is shown in Figure 13.9. An active interface for the data exchange between BioDataBrowser and application is available.

The application program (MetabSim) implements this data exchange interface. When the user selects one or more objects from the cache, the data exchange interface is called and the algorithm in MetabSim processes the transformation into MetabSim rules. In order to handle large networks, all entities of the MetabSim simulation model are stored in a standard object-oriented database system. The MetabVis program is the front end of the simulation. With this graphical user interface, one can initiate the import of MetabNets data, modify the rule, run the simulation, and display the results.

DISCUSSION

The scientific world is at the beginning of the century of biotechnology. The progress of this new technology depends on the application of methods and concepts of computer science, because the exponential growth of experimental data must be handled. In other words, molecular database systems and analysis tools must be implemented. Apweiler *et al.* show in chapter 2 that more than 20 molecular database systems and hundreds of analysis tools are available via the Internet. For the analysis and synthesis of molecular processes, the integration of database systems is important. Therefore, the main goal of bioinformatics is to develop and implement the information technological intrastructure for molecular biology. Various companies have already entered the market.

The most important are Human Genome Sciences (USA), Celera Genomics (USA), INCYTE Genomics (USA), Lion Bioscience (Germany), and Informax (USA). The backbone of such information systems is the integration of heterogeneous molecular database systems and analysis tools. This chapter has described the current methods of database integration and the architecture of a molecular information system for the analysis and synthesis of gene-controlled metabolic networks.

An integration shell was implemented that allows the semiautomatic implementation of a user-specific integration database which represents an information fusion process based on the integration of different molecular database systems. The implementation integrated, as a case study, seven different molecular database systems and the rule-based simulation tool MetabSim, which allows the simulation of gene-controlled metabolic networks. The rule-based method is easy to handle and also allows the abstract simulation of analytical effects. It does not have the vision of the virtual cell like Tomita, because the theoretical results show that the complexity of the simulation of the complete metabolic processes is exponential. That means that only parts of the metabolism processes is exponential. That means that only parts of the metabolism can be calculated.

However, simulation is, and will be, one important point in understanding the function of gene-regulated metabolic pathways. Moreover, the algorithmic analysis of metabolic networks must be achieved. The state of the art of the algorithmic analysis of metabolic networks. To understand the logic of life, algorithms for the calculation of alignments of metabolic pathways or for the prediction of metabolic pathways based on rueimentary knowledge must be developed. Finally, the analysis process needs information systems that integrate analysis tools and simulation environments based on the integration of molecular database systems.

Index

❑❑❑